Cold Plasma in Food and Agriculture

Cold Plasma in Food and Agriculture

Fundamentals and Applications

Edited by

NN Misra

Oliver Schlüter

PJ Cullen

AMSTERDAM • BOSTON • HEIDELBERG • LONDON
NEW YORK • OXFORD • PARIS • SAN DIEGO
SAN FRANCISCO • SINGAPORE • SYDNEY • TOKYO
Academic Press is an imprint of Elsevier

Academic Press is an imprint of Elsevier
125 London Wall, London EC2Y 5AS, United Kingdom
525 B Street, Suite 1800, San Diego, CA 92101-4495, United States
50 Hampshire Street, 5th Floor, Cambridge, MA 02139, United States
The Boulevard, Langford Lane, Kidlington, Oxford OX5 1GB, United Kingdom

Notices

Knowledge and best practice in this field are constantly changing. As new research and experience broaden our understanding, changes in research methods, professional practices, or medical treatment may become necessary.

Practitioners and researchers must always rely on their own experience and knowledge in evaluating and using any information, methods, compounds, or experiments described herein. In using such information or methods they should be mindful of their own safety and the safety of others, including parties for whom they have a professional responsibility.

To the fullest extent of the law, neither the Publisher nor the authors, contributors, or editors, assume any liability for any injury and/or damage to persons or property as a matter of products liability, negligence or otherwise, or from any use or operation of any methods, products, instructions, or ideas contained in the material herein.

Library of Congress Cataloging-in-Publication Data
A catalog record for this book is available from the Library of Congress

British Library Cataloguing-in-Publication Data
A catalogue record for this book is available from the British Library

ISBN: 978-0-12-801365-6

Publisher: Nikki Levy
Acquisition Editor: Patricia Osborn
Editorial Project Manager: Karen Miller
Production Project Manager: Susan Li
Designer: Matthew Limbert

Typeset by SPi Global, India

Contents

Contributors

P. Bourke
Dublin Institute of Technology, Dublin, Ireland

S. Bußler
Leibniz Institute for Agricultural Engineering Potsdam-Bornim, Potsdam, Germany

J. Christopher Whitehead
The University of Manchester, Manchester, United Kingdom

P.J. Cullen
Dublin Institute of Technology, Dublin, Ireland; University of New South Wales, Sydney, NSW, Australia

K. Ishikawa
Nagoya University, Nagoya, Japan

D.D. Jayasena
Uva Wellassa University, Badulla, Sri Lanka

B. Jiang
Qingdao University Technology; China University of Petroleum, Qingdao, P.R. China

C. Jo
Seoul National University, Seoul, Republic of Korea

K.M. Keener
Iowa State University, Ames, IA, USA

H.-J. Kim
National Institute of Crop Science, RDA, Suwon, Republic of Korea

P. Lu
BioPlasma Group, Dublin Institute of Technology, Dublin, Ireland

N.N. Misra
GTECH, Research & Development, General Mills India Pvt Ltd, Mumbai, India

T. Ohta
Meijo University, Nagoya, Japan

K. Ostrikov
Queensland University of Technology (QUT), Brisbane, QLD; CSIRO – QUT Joint Sustainable Materials and Devices Laboratory, Lindfield, NSW, Australia

S.K. Pankaj
Dublin Institute of Technology, Dublin, Ireland

S. Patil
Dublin Institute of Technology, Dublin, Ireland

O. Schlüter
Leibniz-Institute for Agricultural Engineering Potsdam-Bornim, Potsdam, Germany

O.K. Schlüter
Leibniz Institute for Agricultural Engineering Potsdam-Bornim, Potsdam, Germany

B. Surowsky
Technische Universität Berlin, Berlin, Germany

S. Thomas
Mahatma Gandhi University, Kerala, India

M. Turner
National Centre for Plasma Science and Technology, Dublin City University, Dublin, Ireland

M. Wu
China University of Petroleum, Qingdao, P.R. China

H.I. Yong
Seoul National University, Seoul, Republic of Korea

J. Zheng
China University of Petroleum, Qingdao, P.R. China

D. Ziuzina
Dublin Institute of Technology, Dublin, Ireland

Foreword

When we introduced high-pressure technology for gentler processing of foods than conventional thermal processing concepts, we called pressure the "third dimension" (in addition to the conventional process variables, temperature and time). Now we enter the era of working with the "fourth state of matter." As with other so-called nonthermal technologies, such as high pressure and pulsed electric fields, with plasma adding temperature as an additional process variable, combinations of different processes may occur, but, for the time being, it is very timely to have a summary and key reference work of the current state of the art of "cold" plasma technology as it applies to food production, preservation, and modification.

It also is exciting to see all the other scientific disciplines where plasma research and development are applied. This can open many new ways of interaction, knowledge, and technology transfer, and thus possibly result in completely new fields of applications. Again, as experienced with high pressure, pulsed electric fields, ultrasound, etc. using a given technology as local point may become a new way of expanding food-science knowledge beyond its existing borders.

The editors managed to take on an excellent consortium of experts in the field to provide their knowledge in the two key parts of the book, "Theory and Mechanisms" and "Food and Agriculture Applications." The individual chapters dealing with physics and chemistry, with plasma sources and diagnostics, as well as with antimicrobial mechanisms and interactions with food constituents, offer an in-depth source of information regarding the basics of plasma science and technology.

Conventional food processes of the past have been developed empirically when the mechanisms of action were poorly understood. Pasteurization is the prime example, with industrial-scale processes available long before Louis Pasteur showed the interaction between microorganisms and food spoilage. This may be the first time in the history of food science that we are able to develop a science-driven processing technology using mechanisms and kinetics.

The second part of this valuable book covers more or less the entire food chain from agriculture, food safety, treated plants, and animal-originated foods, as well as packaging applications and effluent and wastewater treatment. The **closing** chapter **of the book** deals with an outlook into the future of plasma in food and agriculture comprising regulatory aspects, plasma source design and process control, future innovations and a discussion about consumer confidence.

The editors and authors have to be congratulated for this first book about plasma in food and agriculture. From a food-science perspective, one of the key requirements for successful adoption of plasma technology is the agreed classification of the operational conditions of "cold plasma," particularly with respect to temperature. Further discussions and clarification around the use of terms, including ambient, atmospheric, nonthermal and even cold need to be completed with a view to ensuring future consumer understanding and acceptance.

Prof. Dr. Dipl.-Ing. Dietrich Knorr
Professor Emeritus of Technische Universität Berlin
President European Federation of Food Science & Technology
President Elect International Union of Food Science and Technology

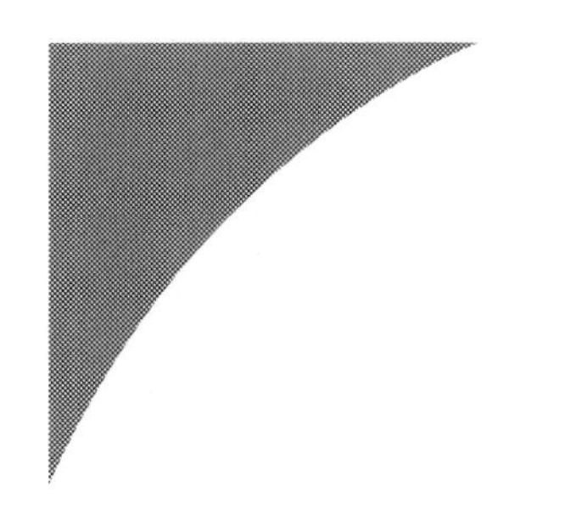

Chapter

Plasma in Food and Agriculture

N.N. Misra*, O. Schlüter†, P.J. Cullen‡,§

*GTECH, Research & Development, General Mills India Pvt Ltd, Mumbai, India, †Leibniz-Institute for Agricultural Engineering Potsdam-Bornim, Potsdam, Germany, ‡Dublin Institute of Technology, Dublin, Ireland, §University of New South Wales, Sydney, NSW, Australia

1 Challenges and Trends in Food Production

The food industry, from farm to fork, must continually adapt to meet the demands of a growing population, both in terms of nutrition and consumer expectations. This must be achieved within the confines of the resources available and regulatory requirements. Innovation with regard to food production and processing is required to meet the emerging challenges of global food security and the complexities of the modern food chain. The key drivers of novel processing technologies, such as cold plasma, within the food and agricultural industries are discussed in the following sections.

1.1 Food Security

With the planet's population projected to reach almost 10 billion by 2050, innovative approaches to both food production and processing will be required to meet food demands. One of the greatest challenges remains the global provision of safe food,

Cold Plasma in Food and Agriculture. http://dx.doi.org/10.1016/B978-0-12-801365-6.00001-9

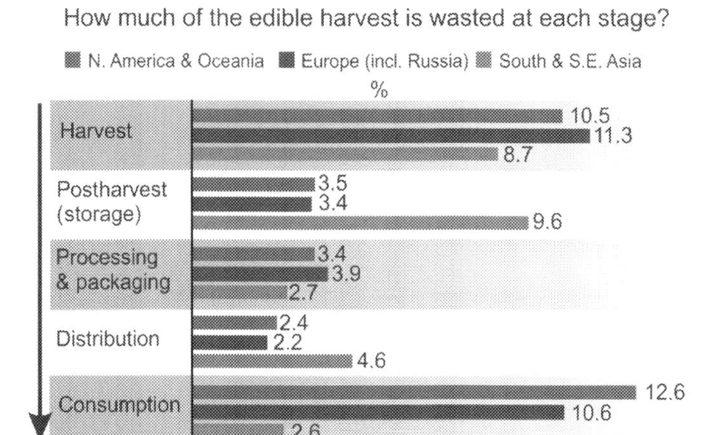

Fig. 1 Food wastage encountered between the harvest, supply chain, and consumption stages. *(From UN FAO.)*

which meets consumer demands of nutritional intake. Almost one-third of worldwide food production for human consumption is lost or wasted (Galanakis, 2015) along the food production, processing, and supply chain as outlined in Fig. 1 (Gustavsson et al., 2013). It is imperative that novel and smart solutions are developed for sustainable food-consumption patterns and global food security. Environmentally friendly intervention strategies that protect food crops or food products from decay or pests, which lead to reduced losses and/or extension of shelf life, are a key component in addressing global food security.

1.2 Food Safety

In the food production chain, food safety remains a major challenge; this is compounded by the emergence of pathogens with low infectious doses and increased virulence. New food-safety intervention strategies are required to manage food safety across the increasingly complex global supply chain. Efficient strategies are required to reduce the microbiological safety risks of food products, while maintaining product quality characteristics with the combined goal of shelf-life extension. Given the vast array of crops and food commodities produced, along with their associated pathogens, no single universal technology is likely to meet all requirements. Consequently, it is important to provide the industry with variable options to meet its specific needs.

1.3 Minimal Processing

Two of the most noticeable developments in food processing over the past 25 years are the rise of *minimally processed foods* and the more recent *nonthermal processing technologies*. The former arose as a consequence of the consumer demands for "fresh-like" fruits and vegetables requiring minimum effort and time for preparation. The latter, a class of ambient temperature technologies, evolved in search of alternatives to conventional thermal processing. Nonthermal processing technologies for food preservation have the potential to address the demands of the consumer and deliver high-quality processed foods with an extended shelf life that are additive-free and have not been subjected to extensive heat treatment. Because of the relatively mild conditions of most nonthermal processes compared with heat pasteurization, consumers are often satisfied by the more fresh-like characteristics, minimized degradation of nutrients, and the perception of high quality.

1.4 Consumer and Regulatory Acceptance

Consumers are not only concerned about the ingredients within the foods they consume, but also the processes which are employed along the "farm to fork" food chain. Paradoxically, consumers are demanding foods which are minimally processed, meet their nutritional and taste desires, yet require minimal preparation. Understanding and addressing consumer issues related to novel food processes is one of the most important challenges facing the developers of innovative food products and processing aids. Research suggests that acceptance of new technologies is based to a great extent on public perceptions of the associated risks, and that perceptions of risk are influenced by trust in information and the source which provides it. Several consumer research studies have consistently shown that consumers have poor knowledge and awareness levels toward most novel food-processing techniques, which serves as a major impediment to their acceptance. Consumer awareness of plasma is in general low, and especially so with regard to food-treatment applications. There are both potentially negative and positive aspects with regard to the "acceptability" of plasma for agricultural and food applications. The association of terms including ionization, radiation, radicals, and reactive species may be associated by the public with food irradiation, leading to potentially negative impressions of the approach. Conversely the proposal to use common "activated" gases like air to treat foods, instead of approaches which leave chemical residues (pesticides, fumigants, chlorine, etc.), would be an attractive solution. It is also likely that the proposed application may also influence consumer acceptance, with differences expected between agricultural applications, waste-water treatments, food packaging, and the various food commodities, for example.

2 The Emergence of Nonthermal Solutions

Fig. 2 displays a timeline marking some of the historic milestones in the development of thermal and nonthermal food processing. One of the major developments in the history of food science was the invention of canning by Nicholas Appert in 1809–10, which remains one of the most widely used methods of food preservation. Similarly the discovery of pasteurization by Louis Pasteur in 1864 revolutionized food safety and preservation and the intervention remains a cornerstone of the food industry. Despite the fact that thermal processes such as canning and pasteurization effectively inactivated microorganisms and enzymes, their detrimental impact on color, flavor, and nutritional quality of foods has not always met consumer demands.

Consequently the food industry has sought alternative or synergistic approaches to provide the treatment objectives. Nonthermal technologies have been designed to meet the required food-product safety or shelf-life demands, while minimizing the effects on the nutritional and quality attributes (Cullen et al., 2012). Nonthermal technologies can be defined as preservation treatments that are effective at ambient or sublethal temperatures, thereby minimizing negative thermal effects on nutritional and quality parameters of foods. The nonthermal processing technologies to receive most attention to date include: high-pressure processing (HPP), irradiation, ultrasound, ozonation, and electrical methods, such as pulsed electric fields (PEF), light pulses, electrolyzed oxidizing water, and oscillating magnetic fields. Most of these topics have been well researched and a wealth of information is available (Knorr et al., 2011), including some recent books (Cullen et al., 2012; Koutchma et al., 2010; O'Donnell et al., 2012; Zhang et al., 2010).

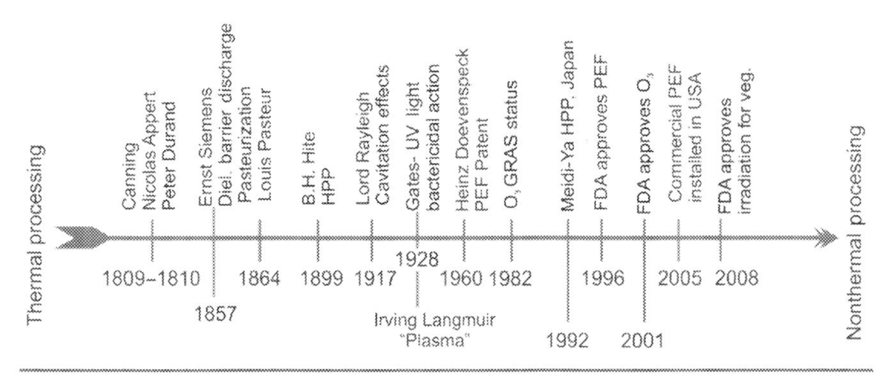

Fig. 2 Timeline depicting milestones in development of nonthermal food-processing and nonthermal plasma technologies.

2.1 Related Nonthermal Technologies

There are a number of nonthermal technologies which are directly relevant to plasma for food applications given their commonalities including:

2.1.1 Pulsed Electric Field Processing

The use of electricity in food processing was introduced in the early 1900s and was first applied for the pasteurization of milk by ohmic heating. In 1960, Heinz Doevenspeck, an engineer in Germany, patented PEF equipment (Doevenspeck, 1960). PEF processing involves the application of pulses of high voltage (typically 20–80 kV/cm) to foods placed between two electrodes. In 1995, the CoolPure® PEF process developed by PurePulse Technologies (4241, Ponderosa Ave., San Diego, CA, 92123, United States) was approved by the US Food and Drug Administration (FDA) for the treatment of pumpable foods. Despite few successful industrial applications in the area of tuber crops and meat preprocessing (ELEA, 2016), PEF is commonly suitable for pumpable liquid foods, and similar to HPP, the high equipment cost could be a concern depending on the desired energy input.

2.1.2 Pulsed Ultraviolet-Light Processing

The bactericidal effect of ultraviolet light was first demonstrated by Gates (1928). Pulsed light technology involves the application of a series of very short, high-power pulses of broad spectrum light to the foods. The use of pulsed light from a xenon lamp with emission between 200 and 1000 nm wavelengths, with a pulse width not more than 2 ms and the cumulative level of the treatment not exceeding 12 J/cm^2, is permitted for food decontamination by the FDA. Pulsed technology is still an emerging technology; one of the only known commercial applications in the food industry includes the decontamination of bottle caps. Sample heating, light penetration issues, shadowing effects, and the absolute necessity for contact between microorganisms and photons are some of the major challenges that have limited the widespread use of pulsed light technology.

2.1.3 Ozone Processing

Interest in ozone has expanded in recent years in response to consumer demands for "greener" food additives, regulatory approval, and the increasing acceptance that ozone is an environmentally friendly technology (O'Donnell et al., 2012). The multifunctionality of ozone makes it a promising food-processing agent. Excess ozone auto decomposes rapidly to produce oxygen, thus leaving no residue in foods from its decomposition. The US FDA's ruling on ozone usage in food as an antimicrobial additive for direct contact with foods of all types (FDA, 2001) has resulted in increased interest in potential food applications worldwide.

2.1.4 General Remarks

While the list of nonthermal technologies extends beyond what is discussed in this section, based on the examples provided, it is fair to remark that there is currently no ideal method to achieve sterilization at ambient temperature. Under this scenario, nonthermal plasma (NTP) technology has emerged as a promising method and has recently gained considerable attention of food scientists and researchers. It has shown potential for solid as well as liquid foods. This is evident from Fig. 3 which shows the increase in the number of scientific publications concerning developments in cold plasma science and its applications toward decontamination in general, and foods in particular. A list of selected research groups active in study of cold plasma technology for food applications can be found in Table 1.

Most nonthermal technologies were originally developed and employed in other fields, and were later extended for food-processing applications. For example, HPP was originally employed for processing of ceramics and materials, while ultrasound

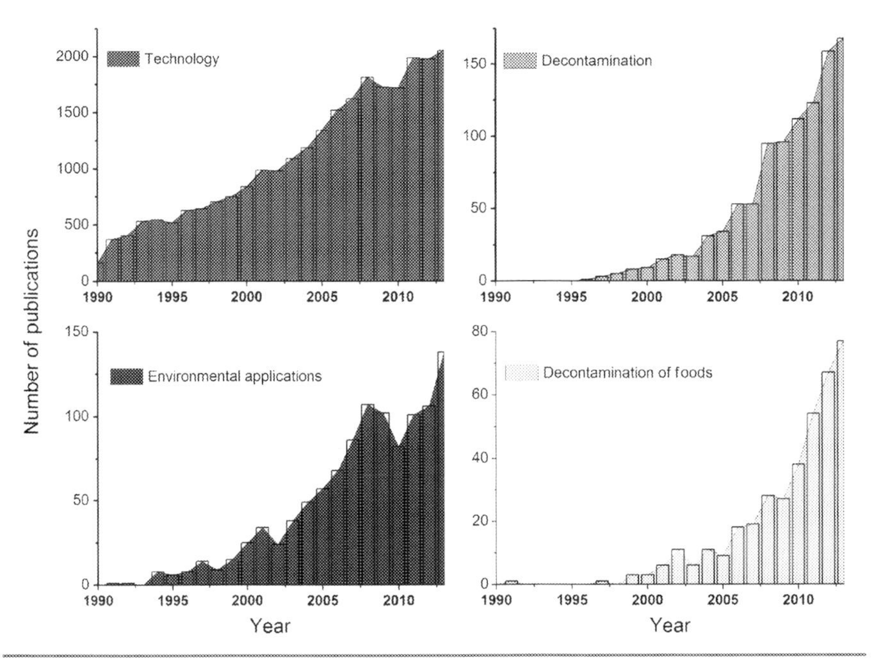

Fig. 3 Number of publications dealing with plasma technology, plasma-based microbial inactivation (in general) and plasma-based microbial inactivation in foods during the last two decades. *(Data accessed from Web of Science on Apr. 2014.)*

TABLE 1 Investigators Exploring Applications of Cold Plasma Technology in Food Industry (by Alphabetical Order of Name)

Names	Affiliation	Topics	References
Annapure and Deshmukh	Institute of Chemical Technology, Mumbai, India	Food grain processing	Sarangapani et al. (2016) and Thirumdas et al. (2015)
Berardinelli and Ragni	University of Bologna, Italy	Resistive barrier discharges for decontamination of eggs and fresh produce	Berardinelli et al. (2016), Pasquali et al. (2016), Ragni et al. (2010), and Tappi et al. (2014, 2016)
Bourke and Cullen	Dublin Institute of Technology, Ireland	Decontamination of fresh produce; property modification of foods and food packaging polymers	Misra et al. (2015), Misra et al. (2014a,b), and Pankaj et al. (2014)
Cheorun Jo	Seoul National University, Seoul, Republic of Korea	Decontamination of meat and meat products, and dairy foods	Kim et al. (2013, 2015) and Yong et al. (2015)
Ehlbeck	Leibniz Inst. for Plasma Research and Technology (INP), Greifswald, Germany	Plasma sources for decontamination of food materials	Ehlbeck et al. (2011) and Schnabel et al. (2015)
Ishikawa and Hori	Nagoya University, Japan	Decontamination of meat and citrus fruits	Ishikawa et al. (2012, 2014) and Ishikawa and Hori (2014)
Keener	Purdue University, Indiana, USA	Fresh produce and egg decontamination	Donner and Keener (2011), Keener et al. (2012), and Klockow and Keener (2009)
Mastwijk, Matser, and Nierop Groot	Wageningen University & Research, The Netherlands	Decontamination of food packaging	van Bokhorst-van de Veen et al. (2015)
Matan	Walailak University, Thailand	Synergistic use of cold plasma and essential oils for food decontamination	Matan et al. (2014, 2015)
Niemira	United States Department of Agriculture, PA, USA	Decontamination of whole fresh fruits	Lacombe et al. (2015), Niemira (2012), and Niemira and Sites (2008)
Potts and Diver	Glasgow University, UK; Anacail Ltd., UK	Decontamination of fresh foods and bakery products	http://www.anacail.com/
Schlüter	Leibniz Institute for Agricultural Engineering (ATB), Potsdam, Germany	Plasma jet and cooled microwave plasma gas decontamination of fresh produce and spices; property modification of protein fractions	Bußler et al. (2015) and Hertwig et al. (2015)

Continued

TABLE 1 Investigators Exploring Applications of Cold Plasma Technology in Food Industry (by Alphabetical Order of Name)—cont'd

Names	Affiliation	Topics	References
Shama	Loughborough University, UK	Decontamination of fresh fruits	Perni et al. (2008a,b) and Shaw et al. (2015)
Shiratani	Kyushu University, Japan	Effect on seed sprouting; decontamination of seeds	Hayashi et al. (2014) and Kitazaki et al. (2014)

for diagnostics and noninvasive monitoring. Likewise, plasma has a long history of applications in semiconductor processing and electronics, and has recently spread to biological and food applications.

3 What Is Cold Plasma?

Plasma is often referred to as the fourth state of matter, according to a scheme expressing an increase in the energy level from solid to liquid to gas, and ultimately to an ionized state of the gas plasma, which exhibits unique properties. The hierarchy of increasing energy levels is illustrated in Fig. 4. Thus, any source of energy which can ionize a gas can be employed for generation of plasma. Plasma is comprised of several excited atomic, molecular, ionic, and radical species, coexisting with numerous reactive species, including electrons, positive and negative ions, free radicals, gas atoms, molecules in the ground or excited state, and quanta of electromagnetic radiation (UV photons and visible light). The free electric charges—electrons and ions—make plasma electrically conductive, internally interactive, and strongly responsive to electromagnetic fields (Fridman, 2008). Most active chemical species of plasma are often characterized by very efficient antimicrobial action.

Plasmas can be subdivided into equilibrium (thermal) and nonequilibrium (low-temperature) plasma. If a gas is heated to sufficiently high temperature (typically in

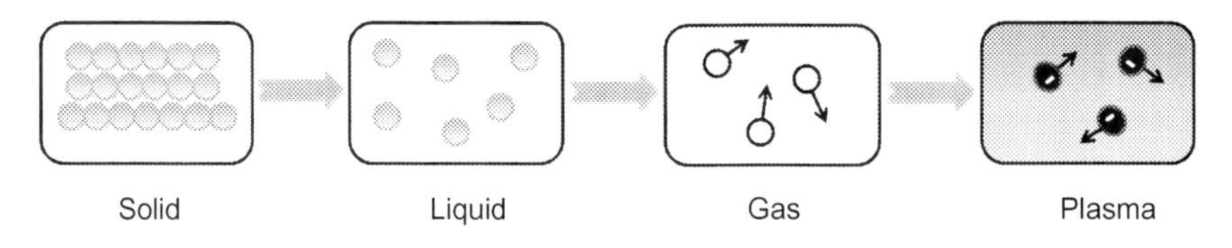

By adding enough energy to any material, we can eventually produce a gas of electrons and ions. This fourth state of matter is referred to as "plasma."

Fig. 4 Pictorial representations of the four states of matter.

the order of 20,000 K) for achieving the ionization of the gas, such plasma would be referred to as "thermal plasma." In thermal plasma, all the constituent chemical species, electrons, and ions exist in thermodynamic temperature equilibrium. The low-temperature plasma can be further branched into quasiequilibrium plasma (typically 100–150°C) and nonequilibrium plasma (<60°C). In the former type, a *local* thermodynamic equilibrium among the species exists, whereas in the latter, cooling of ions and uncharged molecules is more effective than energy transfer from electrons, and the gas remains at low temperature; for this reason nonequilibrium plasma is also called NTP or cold plasma. Nonequilibrium plasmas are typically obtained by means of electrical discharges in gases.

Within the physics and engineering domains, the descriptors of cold plasmas may operate at temperatures of hundreds or thousands of degrees above ambient. Consequently, the term "cold plasma" has recently been employed to distinguish one-atmosphere, near room-temperature plasma discharges from other NTPs. The generation of spatially uniform, well-controlled cold plasma at atmospheric pressures has now become a reality, thereby creating an opportunity to safely and controllably apply plasma to foods and biological surfaces, including medical applications.

Cold plasma is obtained at atmospheric or reduced pressures (vacuum) and requires less power input. Cold plasma can be generated by an electric discharge in a gas at lower pressure or by using microwaves. Typical illustrations for plasma generation at atmospheric pressure include corona discharge, dielectric barrier discharge (DBD), radio-frequency plasma, and the gliding arc discharge. In contrast, thermal plasmas are generated at higher pressures and require high-power inputs.

4 History

Plasmas in nature, such as cloud-to-ground lightning and polar lights, have always intrigued people. The earliest investigations in electrical discharges were conducted by Ernst Siemens, who in 1857 reported about the phenomena of DBDs. In 1928, the American scientist Irving Langmuir proposed that the electrons, ions, and neutrals in an ionized gas could be considered as corpuscular material entrained in some kind of fluid medium. He termed this entraining medium plasma, similar to the plasma (meaning formed or molded in Greek) introduced by the Czech physiologist Jan Purkinje to denote the clear fluid which remains after removal of all the corpuscular material in blood. However, it emerged that there was no fluid medium entraining the electrons, ions, and neutrals in an ionized gas. Nevertheless the name prevailed and the term "plasma" now refers to any system with electrons and ions, where charged particles determine the properties of the system.

Following Langmuir's seminal work, *plasma physics* emerged as an important research field. Townsend was the first to describe the flow of current through a gas, by describing the principle of self-consistency due to the ionization balance during the gas discharge process (Townsend, 1915, 1925). Plasma processing has been used since the 1970s for etching semiconductor materials (Manos and Flamm, 1989). The application of plasmas within the evolving computer industry started in the 1980s, particularly for the fabrication of miniaturized circuits. Since the last decade of 20th century, the development of and advancements in atmospheric pressure plasma has eliminated the need for expensive vacuum chambers and pumping systems. In subsequent years, several other applications of plasma have emerged including plasma medicine, water treatment, and food preservation. More recently, cold plasmas have also been generated inside sealed plastic packages in various configurations, and this has been referred to as the "in-package plasma technology" (Misra et al., 2014a; Patil et al., 2014).

5 Cold Plasma in Food Processing—A Paradigm Shift

An overview of the applications of gas plasma technologies in various areas of science and technology is presented in Fig. 5. It is evident that the range of applications covers many aspects of everyday life and almost all major industries. However, some of the first applications, as mentioned earlier, include the use of plasmas in electronics, especially for material processing, such as the etching of semiconductor surfaces, and later the plasma chemical vapor deposition process as well. DBD-based plasma televisions are probably the best well-known application of plasmas. Other material processing applications include the use of plasma in the polymer and textile industries for surface modification. Cold plasmas are gaining increasing importance in nanotechnology, especially for the synthesis of nanoparticles following the well-established "bottom-up approach." Cold plasmas also represent an alternative technology for gas phase depollution of volatile organic compounds emitted by various industries and are being tested for liquid-phase destruction of pollutants in industrial effluents (Misra, 2015). The latter follows the success of ozone application, which is a predecessor to cold plasma for decontamination processes, typically generated using corona discharges. Apart from various industrial uses, plasmas in general are useful tools in analytical chemistry, especially for ionization of molecules and atoms for optical spectroscopy and mass spectrometry (Kadam et al., 2016). A critical analysis of the developments within plasma science and technology spanning the last decade clearly reveals plasma applications in biology as one of the most exciting and multidisciplinary fields. This marks the shift from treatment of inanimate objects to living or cellular objects. Such applications include the treatment of foods, plant materials, and in animal and human medicine. With the emergence of plasma

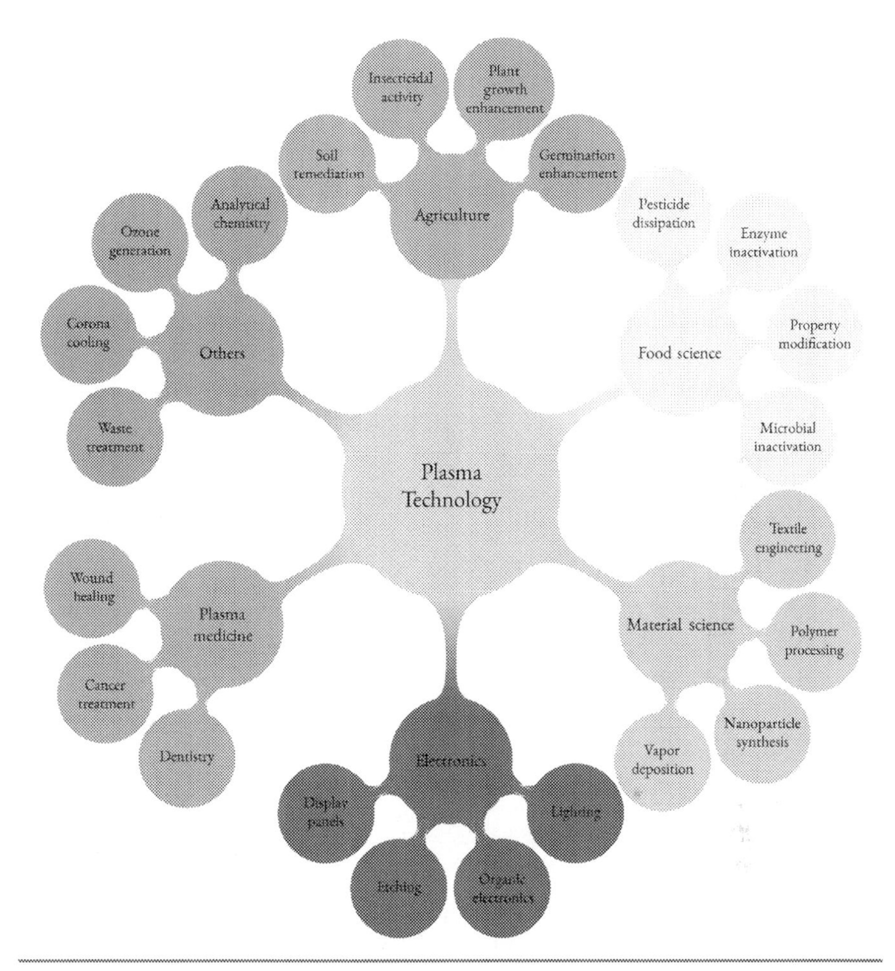

Fig. 5 An overview of applications of nonthermal plasma in various areas of science and technology.

medicine, research in the use of cold plasmas for wound healing, skin treatment, cancer treatment, and bone growth have seen an upsurge (von Woedtke et al., 2013).

Cold plasma as a food technology is a newcomer to this field. In order to appreciate the potential opportunities that cold plasma presents for the food industry, one can compare the limitations of nonthermal technologies and the advantages of cold plasma. Based on the scientific, nonscientific, and patent literature, the described advantages of cold plasma treatment for food preservation can be summarized as follows:

1. Cold plasma offers high microbial inactivation efficiency at low temperatures (generally <50°C). This allows it to extend shelf life, thereby improving the efficiency of the supply chain.

2. Almost all plasma sources available until now allow in situ production of the acting agents, just on demand, and in a range of gases. Therefore, cold plasma is compatible with most existing packaging and modified atmospheres.
3. The active chemical species of plasma are characterized by high diffusivity and therefore act rapidly and access the entire food surface in most cases.
4. Cold plasma is seemingly benign to many food products, if not all, and generally has negligible impact on the product matrix. In addition, it could also reduce preservative use.
5. The application of the cold plasma technology is free of water or solvent; thus, it is also considered environmentally friendly.
6. In general, cold plasma leaves no residues, given sufficient time is provided for the recombination reactions to proceed. However, this may not be universally true, and requires comprehensive validation studies.
7. Most cold plasma sources require only a low energy input; therefore, cold plasma technology is energy efficient.

Another important point to be noted here is that the technology is applicable for both solid as well as liquid foods. The various configurations of plasma sources enable generating plasmas in different gas atmospheres (to treat a range of produce), and also underwater (or liquid) discharges. While we summarized the advantages of cold plasma technologies, we also wish to mention that it is not a universal technology for decontamination of all classes of foods.

Some of the limitations of cold plasma technology include:

1. At present one of the major problems associated with cold plasma treatment of foods is associated with precisely controlling the chemistry of the gas plasma reactions, especially due to the varying levels of humidity introduced by foods.
2. Cold plasma in oxygen containing gas mixture may not be suitable for treating high-fat foods, as the reactive oxygen species formed could lead to oxidation.
3. The cost of the plasma processing is largely dictated by the cost of the gas or gas mixture in which the plasma is generated. The overall process could turn out to be expensive if operated using noble gases.
4. The plasma generation, when carried out using very high voltages, requires additional safety measures. Appropriate measures for destruction and exhaust of the gases are also required.

6 Objective of the Book

The primary objective of this book is to provide insights into the current state of the art and review the emerging applications of cold plasma technology in the food and agricultural industries. The fundamentals of plasma science—its physics and chemistry, process diagnostics, microbial inactivation principles, the effect on microorganisms in various food matrices, and the retention of nutritional and physico-chemical quality of plasma-treated foods and agricultural products—are detailed. A separate chapter is dedicated to discuss the future research needs and the plausible future applications of cold plasma technology in the food industry.

This book is aimed at researchers, students, and industry personnel interested in the area of nonthermal food technology. It is primarily intended for food scientists and food engineers interested in understanding the theory and application of cold plasma for food applications. On the other hand, the book will serve as a good reference for plasma physicists interested in applying novel plasma sources for treatment of foods or biological materials. The contents of the book are self-contained, and explain the fundamentals of plasma physics, chemistry, and technology, before moving on to applications in food processing.

Acknowledgments

We are grateful to many people who contributed to this book, both directly and indirectly. We thank all the authors for their contributions and excellent cooperation. We would also like to thank the many staff members of Elsevier, particularly, Patricia Osborn, Marisa LaFleur, and Karen Miller. Our special thanks to Karen for her patience and constant encouragement during the preparation of this book.

References

Berardinelli, A., Pasquali, F., Cevoli, C., Trevisani, M., Ragni, L., Mancusi, R., Manfreda, G., 2016. Sanitisation of fresh-cut celery and radicchio by gas plasma treatments in water medium. Postharvest Biol. Technol. 111, 297–304.

Bußler, S., Herppich, W.B., Neugart, S., Schreiner, M., Ehlbeck, J., Rohn, S., Schlüter, O., 2015. Impact of cold atmospheric pressure plasma on physiology and flavonol glycoside profile of peas (*Pisum sativum* 'Salamanca'). Food Res. Int. 76, 132–141.

Cullen, P.J., Tiwari, B.K., Valdramidis, V., 2012. Novel Thermal and Non-Thermal Technologies for Fluid Foods. Academic Press, San Diego, CA.

Doevenspeck, H., 1960. Verfahren und Vorrichtung zur Gewinnung der einzelnen Phasen aus dispersen Systemen, German Patent, DE 1237541.

Donner, A., Keener, K.M., 2011. Investigation of in-package ionisation. J. Purdue Undergrad. Res. 1, 10–15.

Ehlbeck, J., Schnabel, U., Polak, M., Winter, J., Th von, W., Brandenburg, R., Hagen, T.V.D., Weltmann, K.D., 2011. Low temperature atmospheric pressure plasma sources for microbial decontamination. J. Phys. D. Appl. Phys. 44, 013002.

ELEA, 2016. Elea Pulsed Electric Field Technology. Elea, Germany.

FDA, 2001. Secondary direct food additives permitted in food for human consumption. Fed. Regist. 66 (123), 33829–33830.

Fridman, A., 2008. Plasma Chemistry. Cambridge University Press, New York.

Galanakis, C., 2015. Food Waste Recovery-Processing Technologies and Industrial Techniques. Academic Press, San Diego, CA.

Gates, F.L., 1928. On nuclear derivatives and the lethal action of ultra-violet light. Science 68, 479–480.

Gustavsson, J., Cederberg, C., Sonesson, C., Emanuelsson, A., 2013. The methodology of the FAO study: global food losses and food waste- extent, causes and prevention. SIK.

Hayashi, N., Yagyu, Y., Yonesu, A., Shiratani, M., 2014. Sterilization characteristics of the surfaces of agricultural products using active oxygen species generated by atmospheric plasma and UV light. Jpn. J. Appl. Phys. 53, 05FR03.

Hertwig, C., Reineke, K., Ehlbeck, J., Knorr, D., Schlüter, O., 2015. Decontamination of whole black pepper using different cold atmospheric pressure plasma applications. Food Control 55, 221–229.

Ishikawa, K., Hashizume, H., Ohta, T., Ito, M., Takeda, K., Tanaka, H., Kondo, H., Sekine, M., Hori, M., 2014. Plasma-biological surface interaction for food hygiene. In: 24th Annual Symposium of the Materials Research Society of Japan, Yokohama, Japan, MRS-J, vol. A-I11-005.

Ishikawa, K., Hori, M., 2014. Diagnostics of plasma-biological surface interactions in low pressure and atmospheric pressure plasmas. Int. J. Mod. Phys.: Conf. Ser. 32, 1460318.

Ishikawa, K., Mizuno, H., Tanaka, H., Tamiya, K., Hashizume, H., Ohta, T., Ito, M., Iseki, S., Takeda, K., Kondo, H., Sekine, M., Hori, M., 2012. Real-time in situ electron spin resonance measurements on fungal spores of *Penicillium digitatum* during exposure of oxygen plasmas. Appl. Phys. Lett. 101, 013704.

Kadam, S.U., Misra, N.N., Zaima, N., 2016. Mass spectrometry based chemical imaging of foods. RSC Adv. 40, 33537–33546.

Keener, K.M., Jensen, J., Valdramidis, V., Byrne, E., Connolly, J., Mosnier, J., Cullen, P., 2012. Decontamination of *Bacillus subtilis* spores in a sealed package using a non-thermal plasma system. In: Hensel, K., Machala, Z. (Eds.), NATO Advanced Research Workshop: Plasma for Bio-Decontamination. Medicine and Food Security, Jasná, pp. 445–455.

Kim, H.-J., Yong, H.I., Park, S., Kim, K., Bae, Y.S., Choe, W., Oh, M.H., Jo, C., 2013. Effect of inactivating *Salmonella typhimurium* in raw chicken breast and pork loin using an atmospheric pressure plasma jet. J. Anim. Sci. Technol. 55, 545–549.

Kim, H.-J., Yong, H.I., Park, S., Kim, K., Choe, W., Jo, C., 2015. Microbial safety and quality attributes of milk following treatment with atmospheric pressure encapsulated dielectric barrier discharge plasma. Food Control 47, 451–456.

Kitazaki, S., Sarinont, T., Koga, K., Hayashi, N., Shiratani, M., 2014. Plasma induced long-term growth enhancement of *Raphanus sativus* L. using combinatorial atmospheric air dielectric barrier discharge plasmas. Curr. Appl. Phys. 14, S149–S153.

Klockow, P.A., Keener, K.M., 2009. Safety and quality assessment of packaged spinach treated with a novel ozone-generation system. LWT Food Sci. Technol. 42, 1047–1053.

Knorr, D., Froehling, A., Jaeger, H., Reineke, K., Schlueter, O., Schoessler, K., 2011. Emerging technologies in food processing. Annu. Rev. Food. Sci. Technol. 2, 203–235.

Koutchma, T., Forney, L.J., Moraru, C.I., 2010. Ultraviolet Light in Food Technology: Principles and Applications. CRC Press, Boca Raton, FL.

Lacombe, A., Niemira, B.A., Gurtler, J.B., Fan, X., Sites, J., Boyd, G., Chen, H., 2015. Atmospheric cold plasma inactivation of aerobic microorganisms on blueberries and effects on quality attributes. Food Microbiol. 46, 479–484.

Manos, D.M., Flamm, D.L., 1989. Plasma Etching: An Introduction. Elsevier Academic Press, San Diego, CA.

Matan, N., Nisoa, M., Matan, N., Aewsiri, T., 2014. Effect of cold atmospheric plasma on antifungal activities of clove oil and eugenol against molds on areca palm (*Areca catechu*) leaf sheath. Int. Biodeterior. Biodegrad. 86, 196–201.

Matan, N., Puangjinda, K., Phothisuwan, S., Nisoa, M., 2015. Combined antibacterial activity of green tea extract with atmospheric radio-frequency plasma against pathogens on fresh-cut dragon fruit. Food Control 50, 291–296.

Misra, N.N., 2015. The contribution of non-thermal and advanced oxidation technologies towards dissipation of pesticide residues. Trends Food Sci. Technol. 45, 229–244.

Misra, N.N., Kaur, S., Tiwari, B.K., Kaur, A., Singh, N., Cullen, P.J., 2015. Atmospheric pressure cold plasma (ACP) treatment of wheat flour. Food Hydrocoll. 44, 115–121.

Misra, N.N., Patil, S., Moiseev, T., Bourke, P., Mosnier, J.P., Keener, K.M., Cullen, P.J., 2014a. In-package atmospheric pressure cold plasma treatment of strawberries. J. Food Eng. 125, 131–138.

Misra, N.N., Sullivan, C., Pankaj, S.K., Alvarez-Jubete, L., Cama, R., Jacoby, F., Cullen, P.J., 2014b. Enhancement of oil spreadability of biscuit surface by nonthermal barrier discharge plasma. Innovative Food Sci. Emerg. Technol. 26, 456–461.

Niemira, B.A., 2012. Cold plasma reduction of *Salmonella* and *Escherichia coli* O157:H7 on almonds using ambient pressure gases. J. Food Sci. 77, M171–M175.

Niemira, B.A., Sites, J., 2008. Cold plasma inactivates *Salmonella Stanley* and *Escherichia coli* O157: H7 inoculated on golden delicious apples. J. Food Prot. 71, 1357–1365.

O'Donnell, C., Tiwari, B.K., Cullen, P., Rice, R.G., 2012. Ozone in Food Processing. Wiley, Oxford.

Pankaj, S.K., Bueno-Ferrer, C., Misra, N.N., O'Neill, L., Jiménez, A., Bourke, P., Cullen, P.J., 2014. Characterization of polylactic acid films for food packaging as affected by dielectric barrier discharge atmospheric plasma. Innovative Food Sci. Emerg. Technol. 21, 107–113.

Pasquali, F., Stratakos, A.C., Koidis, A., Berardinelli, A., Cevoli, C., Ragni, L., Mancusi, R., Manfreda, G., Trevisani, M., 2016. Atmospheric cold plasma process for vegetable leaf decontamination: a feasibility study on radicchio (red chicory, *Cichorium intybus* L.). Food Control 60, 552–559.

Patil, S., Moiseev, T., Misra, N.N., Cullen, P.J., Mosnier, J.P., Keener, K.M., Bourke, P., 2014. Influence of high voltage atmospheric cold plasma process parameters and role of relative humidity on inactivation of *Bacillus atrophaeus* spores inside a sealed package. J. Hosp. Infect. 88, 162–169.

Perni, S., Liu, D.W., Shama, G., Kong, M.G., 2008a. Cold atmospheric plasma decontamination of the pericarps of fruit. J. Food Prot. 71, 302–308.

Perni, S., Shama, G., Kong, M.G., 2008b. Cold atmospheric plasma disinfection of cut fruit surfaces contaminated with migrating microorganisms. J. Food Prot. 71, 1619–1625.

Ragni, L., Berardinelli, A., Vannini, L., Montanari, C., Sirri, F., Guerzoni, M.E., Guarnieri, A., 2010. Non-thermal atmospheric gas plasma device for surface decontamination of shell eggs. J. Food Eng. 100, 125–132.

Sarangapani, C., Thirumdas, R., Devi, Y., Trimukhe, A., Deshmukh, R.R., Annapure, U.S., 2016. Effect of low-pressure plasma on physico-chemical and functional properties of parboiled rice flour. LWT Food Sci. Technol. 69, 482–489.

Schnabel, U., Niquet, R., Schlüter, O., Gniffke, H., Ehlbeck, J., 2015. Decontamination and sensory properties of microbiologically contaminated fresh fruits and vegetables by microwave plasma processed air (PPA). J. Food Process. Preserv. 39, 653–662.

Shaw, A., Shama, G., Iza, F., 2015. Emerging applications of low temperature gas plasmas in the food industry. Biointerphases 10, 029402.

Tappi, S., Berardinelli, A., Ragni, L., Dalla Rosa, M., Guarnieri, A., Rocculi, P., 2014. Atmospheric gas plasma treatment of fresh-cut apples. Innovative Food Sci. Emerg. Technol. 21, 114–122.

Tappi, S., Gozzi, G., Vannini, L., Berardinelli, A., Romani, S., Ragni, L., Rocculi, P., 2016. Cold plasma treatment for fresh-cut melon stabilization. Innovative Food Sci. Emerg. Technol. 33, 225–233.

Thirumdas, R., Deshmukh, R.R., Annapure, U.S., 2015. Effect of low temperature plasma processing on physicochemical properties and cooking quality of basmati rice. Innovative Food Sci. Emerg. Technol. 31, 83–90.

Townsend, J.S., 1915. Electricity in Gases. Clarendon Press, Oxford.

Townsend, J.S., 1925. Motion of Electrons in Gases. Clarendron Press, Oxford.

van Bokhorst-van de Veen, H., Xie, H., Esveld, E., Abee, T., Mastwijk, H., Nierop Groot, M., 2015. Inactivation of chemical and heat-resistant spores of *Bacillus* and *Geobacillus* by nitrogen cold atmospheric plasma evokes distinct changes in morphology and integrity of spores. Food Microbiol. 45, 26–33.

von Woedtke, T., Reuter, S., Masur, K., Weltmann, K.D., 2013. Plasmas for medicine. Phys. Rep. 530, 291–320.

Yong, H.I., Kim, H.-J., Park, S., Kim, K., Choe, W., Yoo, S.J., Jo, C., 2015. Pathogen inactivation and quality changes in sliced cheddar cheese treated using flexible thin-layer dielectric barrier discharge plasma. Food Res. Int. 69, 57–63.

Zhang, H.Q., Barbosa-Cánovas, G.V., Balasubramaniam, V.M., Patrick Dunne, C., Farkas, D.F., Yuan, J.T.C., 2010. Nonthermal Processing Technologies for Food. Wiley-Blackwell, Oxford.

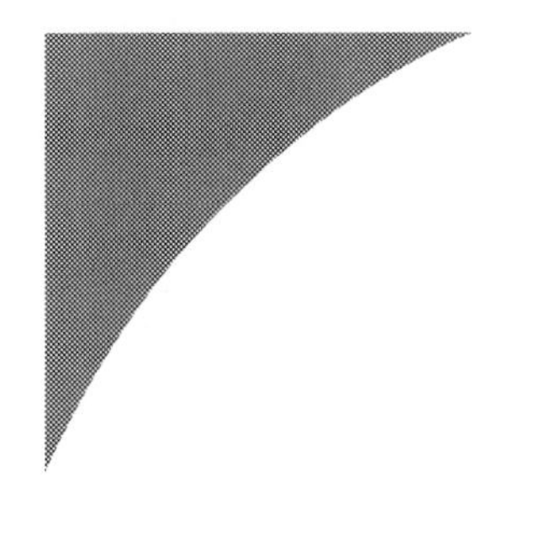

Chapter 2

Physics of Cold Plasma

M. Turner
National Centre for Plasma Science and Technology, Dublin City University, Dublin, Ireland

1 Introduction

Although the atmosphere around us consists almost entirely of neutral atoms and molecules, there is always a small density of charged ions and free electrons present. This ionization is due predominantly to cosmic rays, which produce a density of charged particles of about $10^9 m^{-3}$ at sea level (Hulburt, 1931). If an electric field is present, these charged particles will experience a force and will be accelerated. Of course, the motion along the field direction that this acceleration produces will occasionally be interrupted by collisions with neutral gas molecules. The behavior of electrons and ions in these collisions is crucially different, however. Elementary mechanical considerations show that when two particles with distinct masses m_1 and m_2 collide, the exchange of energy between them depends on the ratio of their masses, m_1/m_2. Consequently, for ions, with $m_1/m_2 \approx 1$, there is a complete sharing of energy with a neutral collision partner, and an ion essentially stops after every collision. For electrons, $m_1/m_2 \ll 1$, and only a small fraction of the available energy can be exchanged. Electrons, therefore, are heated much more than ions, by any given electric field. Consequently, in the presence of an electric field, the electron

Cold Plasma in Food and Agriculture. http://dx.doi.org/10.1016/B978-0-12-801365-6.00002-0

temperature readily becomes much larger than the gas temperature. A ratio of a 100 or more is not unusual. This nonequilibrium behavior is the basis of almost all low-temperature plasma applications, because the hot electrons channel electrical energy efficiently into producing reactive radicals (and not heating the gas, for example). A customary unit for measuring the energy of particles under such conditions is the electron volt (or eV), which is the energy transferred to a particle with one elementary charge in passing through a potential difference of 1 V. In these units, the translational energy of a particle at room temperature is some 0.04 eV. This is to be compared with the energy required to ionize an atom or molecule, which is typically 10–20 eV. So one immediately sees why thermal ionization at room temperature is negligible compared to the effect produced by cosmic rays. At energies between room temperature and the ionization limit, various other atomic and molecular processes can occur, including the production of electronically excited states and dissociation to give reactive radicals. For example, molecular species such as oxygen are dissociated in electron impact processes:

$$e + O_2 \rightarrow e + 2O. \tag{1}$$

Atomic oxygen is an important radical in its own right, and also a precursor for other radical species, including ozone formed by an association reaction involving any third species, denoted by M:

$$O + O_2 + M \rightarrow O_3 + M, \tag{2}$$

and oxides of nitrogen formed by displacement reactions:

$$O + N_2 \rightarrow NO + N, \tag{3}$$

These and other radical species are responsible for many kinds of chemical activity, and are the important agents in most of the applications discussed later in this book. Ozone, in particular, is a powerful antimicrobial agent. Reactions such as (1) are allowed only for electrons with energies of at least several electron volts, which, of course, is much more than is available at room temperature. Consequently, a basic step in the development of such applications is the process of transferring energy from some electrical power source into the plasma electrons, and hence to the reactive radicals that produce chemical activity. These chemical effects are not necessarily the only important phenomena. For instance, in some cases electromagnetic radiation from the plasma, or charged species, may also have a role to play. However, as the predominant source of this radiation is also an electronically excited species produced by the plasma, electrons are still the essential vector through which the process is carried out. Consequently, any kind of basic understanding of plasma applications begins with electron kinetics, and we will therefore consider this aspect in some detail in Section 2. But we need to consider how the products of electron impact processes are chemically transformed into reactive species of interest in applications. This topic will be discussed in Section 3. We will also be discussing some practical features of plasma sources, including the process ("breakdown") by which a plasma is formed, in Section 4. In Section 5, we will briefly describe some popular plasma sources. An important topic here is the development of complicated spatial structures, which to some degree is an almost universal characteristic of plasmas of any description. These structures generally increase the difficulty of designing an optimal plasma source, as we will see. This issue especially affects atmospheric pressure sources intended to operate in ambient air. Since these structures essentially elude analytical mathematical descriptions, we will conclude by surveying the state of the numerical tools that can be used to describe these complex plasmas.

This chapter therefore offers a survey of basic concepts associated with operating gas discharges, with an emphasis on topics likely to be relevant to applications of atmospheric pressure discharges to food processing. Of necessity, many complex topics are treated briefly. Readers are referred to a variety of other sources for more detailed discussion. These include classic extended treatments with broad coverage (Cherrington, 2014; Francis, 1956; Llewellyn-Jones, 1966; Raether, 1964; Raizer, 2011), and relevant papers in the recent literature discussing particular aspects in detail, which are cited in the appropriate places below.

2 Electron Kinetics

In thermal equilibrium at a certain temperature T, the average energy of any kind of particles is $\frac{3}{2}k_B T$, where k_B is the fundamental constant named after Boltzmann. Of course, at the microscopic level, not every particle has exactly this mean energy. Rather, there exists a distribution of particle energies. In thermal equilibrium, this is the Maxwell-Boltzmann distribution. There are various ways of writing the Maxwell-Boltzmann distribution. For instance, we can often conveniently imagine that the velocity vector **v** of each particle is expressed using three components (v_x, v_y, v_z). In equilibrium, each velocity component has the same distribution, which can be written:

$$f(v_x) = \sqrt{\frac{m}{2\pi k_B T}} \exp\left(-\frac{m v_x^2}{k_B T}\right), \tag{4}$$

where

$$\int_{-\infty}^{\infty} f(v_x)\, dv_x = 1. \tag{5}$$

This is a probability distribution, meaning that the chance of finding a particular particle with a velocity between v_x and $v_x + dv_x$ is $f(v_x)dv_x$. Often, we are more interested in the distribution of particle energies than the velocities. If the kinetic energy of a particular particle is $\epsilon = \frac{1}{2}m\left(v_x^2 + v_y^2 + v_z^2\right)$, and each of the three velocity components is distributed according to Eq. (4), then

$$f(\epsilon) = 2\sqrt{\frac{\epsilon}{\pi}}\left(\frac{1}{k_{\mathrm{B}}T}\right)^{3/2} \exp\left(-\frac{\epsilon}{k_{\mathrm{B}}T}\right), \tag{6}$$

where

$$\int_0^\infty f(\epsilon)\, d\epsilon = \int_0^\infty \sqrt{\epsilon} g(\epsilon)\, d\epsilon = 1. \tag{7}$$

Note that either of the functions $f(\epsilon)$ or $g(\epsilon)$ can be used to describe the energy distribution, and there is no generally accepted terminology or notation that distinguishes between them. Care is therefore needed to establish which form is in use in any particular context.

Important information can be derived from the energy distribution. For instance, the rates of chemical reactions depend on the particle energy distribution. The strength of the interaction between any pair of particles is characterized by the collision cross-section, which in general is a function of the relative velocity of the particles, and hence denoted by $\sigma(\epsilon)$. If we are concerned with a collision between an electron and some neutral species, then in most cases this relative velocity is practically equal to the electron velocity, because of the smaller mass and usually higher temperature of the electrons. So the collision cross-section may be taken to be a function of the electron energy, and the velocity distribution of the neutral species can be ignored. Then the rate at which electrons with energies in the range ϵ to $\epsilon + d\epsilon$ collide with a neutral species can be written:

$$dS = nNv(\epsilon)\sigma(\epsilon)f(\epsilon)d\epsilon, \tag{8}$$

where n is the electron density, N is the neutral density, $\sigma(\epsilon)$ is the collision cross-section (units of m^{-2}), and the units of dS are $\mathrm{m}^{-3}\ \mathrm{s}^{-1}$. Hence

$$S = Nn \int_0^\infty v(\epsilon)\sigma(\epsilon)f(\epsilon)\, d\epsilon \tag{9}$$

$$= kNn, \tag{10}$$

where k is called the rate constant.

When an electron collides with a neutral atom or molecule, a number of outcomes are possible. The simplest is elastic scattering, in which the electron transfers a small fraction of its energy to the neutral collision partner. This has the effect of heating the gas. A larger fraction of the electron's energy may be lost when there is a change in the internal state of the collision partner, such as the production of an electronically excited state. In general, then, the total collision cross-section is subdivided into a number of smaller cross-sections, one for each of the possible outcomes of a collision. The number of allowed processes is in principle large, because the number of excited states that can be produced is large. However, a convenient practice is to gather together groups of excited states into a small number of effective cross-sections. For example, when dealing with atomic species such as the noble gases, a common choice is to represent excitation of the entire manifold of excited states by one or two effective cross-sections. This gives a set of four effective cross-sections, one for elastic scattering, two excitation cross-sections, and an ionization cross-section. A set of cross-sections of this kind is shown in Fig. 1, in this case for helium. For molecular species, a considerably wider range of processes can occur, since apart from rotational and vibrational excitation, such species can also be dissociated into various neutral or charged fragments. Consequently, sets of cross-sections for molecular species commonly contain many more processes (see Pitchford et al. (1981), for example).

The partition of power between the various electron impact processes is usually a matter of interest, because this essentially determines the chemical character of the effluent from the discharge. For processes involving internal energy changes in the neutral collision partner, the collision channel must be closed to electrons with energies less than some threshold energy. Consequently, the cross-section must be zero for electron energies less than this threshold. For the simplest case, when the cross-section is a constant above the threshold energy ϵ_0, denoted by σ_0, we have:

$$k(T) = \langle \sigma(\epsilon) v(\epsilon) \rangle \tag{11}$$

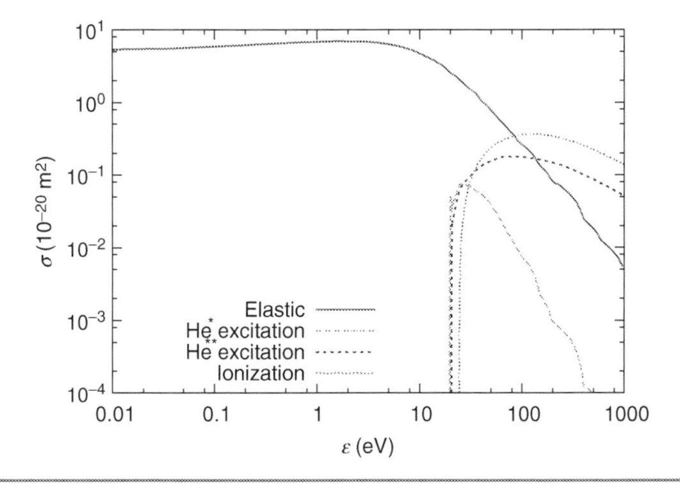

Fig. 1 Electron impact cross-sections for helium.

$$= \int_0^{\infty} v(\epsilon)\sigma(\epsilon)f(\epsilon)\,d\epsilon \tag{12}$$

$$= \sigma_0\sqrt{\frac{8k_BT}{\pi m}}\left(1+\frac{\epsilon_0}{k_BT}\right)\exp\left(-\frac{\epsilon_0}{k_BT}\right). \tag{13}$$

When $\epsilon_0 \gg k_BT$, which is the case for many processes involving electronic excitation or ionization, the exponential factor in Eq. (13) makes the rate constant a strong function of T. This has several consequences. One is that the electron temperature strongly affects the chemical character of a plasma, because the flow of energy into the various collision channels is strongly influenced by the associated rate constants. Another, less immediately obvious, is that the rate constants depend strongly on the shape of the distribution function, $f(\epsilon)$. We have so far assumed that this takes the Maxwell-Boltzmann form given by Eq. (6), but in fact this is the normal case only in thermal equilibrium, and is hardly ever observed in a low-temperature plasma. This implies that if we seek a quantitative understanding of the electron kinetics in a low-temperature plasma, we must know the electron energy distribution function accurately. To find this function, we must solve a complex equation known as the Boltzmann equation, and to this topic we now turn.

In atmospheric pressure discharges, we can usually assume, at least as a first approximation, that the electron energy distribution function is determined by the electric field at each point in space. This is known as the local field approximation. In this case, any influence of spatial gradients is ignored, and time dependence is assumed to arise only through the time dependence of the electric field. The problem is still difficult, because the velocity distributions along each of the three coordinate axes cannot be quite the same, if only because the electrons drift in the direction of the electric field. So the velocity distribution along the axis parallel to the field direction cannot be the same as the velocity distributions along the axes normal to the field direction. However, the difference between the three velocity component distributions is usually small. For instance, if only elastic collisions are significant, then the acceleration along the field between collisions must on average balance the energy loss during a collision. We have seen already that this energy loss depends on the ratio of the electron and neutral masses, δ, and is therefore a small fraction of the electron energy. Since the electron velocity distribution immediately after a collision is the same in all directions, the difference between the electron velocity component parallel to the field and those perpendicular to the field is only due to acceleration between collisions. Since this acceleration can add only a small fraction to the electron energy, it follows that the electron velocity distributions are almost the same on all three axes. This is expressed mathematically by writing the distribution function in the form (Cherrington, 2014; Shkarofsky et al., 1966):

$$g(\epsilon,\theta) = g_0(\epsilon) + g_1(\epsilon)\cos\theta, \tag{14}$$

where θ is a polar angle measured relative to the electric field, and $g_1(\epsilon)$ represents a small perturbation of the isotropic distribution function, $g_0(\epsilon)$. Technically, this approximation is obtained by expressing the distribution function as an expansion in spherical harmonic functions, and truncating the expansion after two terms. The "two-term expansion" is accurate provided that the elastic collision frequency is much larger than the inelastic collision frequency, which is commonly the case, but not invariably so. Even then, the inaccuracy introduced by the two-term expansion is rarely gross, so that this approximation is useful for all but the most detailed and accurate calculations (Pitchford et al., 1981).

The Boltzmann equation determines the electron energy distribution function. Introducing the two-term expansion given by Eq. (14), and assuming that the distribution function is determined by the local electric field, are important assumptions that permit substantial simplifications of the Boltzmann equation, which is in general complex (Cherrington, 2014; Shkarofsky et al., 1966). We will not discuss the rather intricate derivation of this simplified Boltzmann equation from the general form. When only elastic collisions need to be considered, the simplified equation is (Cherrington, 2014)

$$\frac{d}{d\epsilon}\left[\frac{e^2E^2\epsilon}{3N^2\sigma_m(\epsilon)}\frac{dg_0}{d\epsilon}\right]+\frac{2m}{M}\frac{d}{d\epsilon}\left[\epsilon^2\sigma_m(\epsilon)\left(g_0+k_BT\frac{dg_0}{d\epsilon}\right)\right]=0, \quad (15)$$

which can also be expressed in terms of the speed distribution

$$\frac{1}{v^2}\frac{d}{dv}\left[\frac{e^2E^2v}{3m^2N^2\sigma_m(\epsilon)}\frac{dg_0}{dv}\right]+\frac{1}{2v^2}\frac{2m}{M}\frac{d}{dv}\left[v^3\sigma_m(v)\left(vg_0+k_BT\frac{dg_0}{dv}\right)\right]=0. \quad (16)$$

These equations treat elastic collisions as a continuous energy loss process, which is appropriate when $\delta = 2m/M \ll 1$, so that the fractional energy loss in a collision is small. Eq. (16) can be integrated twice to obtain:

$$g_0(v)=C\exp\left[-\int_0^v\frac{mv\,dv}{k_BT\left(1+\frac{2}{3}\frac{1}{\delta}\frac{e^2E^2}{k_BTN^2\sigma_m(v)^2v^2m}\right)}\right], \quad (17)$$

where C is an arbitrary constant. Eqs. (15)–(17) display an important feature, namely that the distribution function g_0 depends not on the gas density N and the electric field E separately, but on the ratio E/N, which is called the reduced electric field. Physically, this follows from the aforementioned consideration that a crucial parameter is the gain of energy from the field between collisions, which also depends on the reduced field, and not on the gas density or the electric field separately. The reduced electric field is such an important parameter that there is an associated unit, the

Townsend, such that 1 Td $= 10^{-21}$ V m^{-2}. Of course, this choice means that experimentally typical reduced fields are expressed as convenient values, usually in the range 1–1000 Td.

Eq. (17) generally does not have an exact solution. If $E = 0$, then the solution is a Maxwellian with temperature T. In this case, the electrons are in thermal equilibrium with the neutral gas. If we assume that the collision frequency is a constant, ie, $\nu_m = v\sigma(v)N$, and $E \neq 0$, then the solution is again a Maxwellian, but with temperature

$$T_e = T + \frac{2}{3}\frac{1}{\delta}\frac{e^2E^2}{k_B m \nu_m^2}, \tag{18}$$

showing that in the presence of an electric field, the electrons have a temperature greater than the gas temperature. If the mean free path, λ, is assumed to be a constant, such that $\nu_m = \lambda/v$, then

$$g_0(v) = C\exp\left(-\int_0^v \frac{mvdv}{k_BT + \frac{2}{3}\frac{1}{\delta v^2}\frac{\lambda^2e^2E^2}{m}}\right). \tag{19}$$

If the electric field is weak, or the collision frequency is large, such that $k_BT \gg \frac{2}{3}\frac{1}{\delta v^2}\frac{\lambda^2e^2E^2}{m}$, then $f_0(v)$ is again a Maxwellian with temperature T. In the opposite limit, we find

$$g_0(v) = C\exp\left(-\frac{3}{8}\frac{\delta m^2v^4}{\lambda^2e^2E^2}\right), \tag{20}$$

which is the Druyvesteyn distribution function (Druyvesteyn and Penning, 1940). This function decreases much more rapidly than the Maxwellian at larger speeds or energies. Consequently, there is a large difference between the rates that are computed using Maxwellian or Druyvesteyn distributions with the same mean energy, especially for processes with threshold energies much greater than the mean energy, which are usually the ones of central interest. Since the assumptions that are required to obtain either of these analytic forms are usually not satisfied in typical experimental contexts, this observation mainly highlights the importance of knowing the distribution function accurately. This means that we must include inelastic processes in the model.

Since we assume that inelastic processes have threshold energies that are comparable to or larger than the mean energy, inelastic collisions are modeled not as causing a continuous flow of particles through energy or velocity space, as in Eq. (15), but as additional processes producing a discontinuous movement of particles from one velocity or energy coordinate to another:

$$\frac{d}{d\epsilon}\left[\frac{e^2E^2\epsilon}{3N^2\sigma_m(\epsilon)}\frac{dg_0}{d\epsilon}\right]+\frac{2m}{M}\frac{d}{d\epsilon}\left[\epsilon^2\sigma_m(\epsilon)\left(g_0+k_BT\frac{dg_0}{d\epsilon}\right)\right]$$
$$+\Sigma_j\left[(\epsilon+\epsilon_j)g_0(\epsilon+\epsilon_j)N\sigma_j(\epsilon+\epsilon_j)-\epsilon f(\epsilon)\sigma_j(\epsilon)\right]=0, \tag{21}$$

where the summation is over the set of inelastic processes to be considered, each with cross-section σ_j and threshold energy ϵ_j. Analytic solutions of Eq. (21) are usually not attempted, because in general the set of inelastic cross-sections is large, and the individual cross-sections are not easily represented by tractable analytic functions. Fortunately, numerical solution is a relatively straightforward procedure, which can be accomplished using readily available tools and cross-sections (Hagelaar and Pitchford, 2005; Pitchford, 2013). The two-term expansion on which Eq. (21) depends is almost always satisfactory, as we already noted.

Fig. 2 shows energy distributions obtained by a numerical solutions of Eq. (21) using a mixture of gases corresponding to humid air. Data are shown for values of *E*/*N* of 0.1, 1, 10, 100, and 1000 Td. This range of values of *E*/*N* roughly spans that likely to be found experimentally, where a self-sustained discharge usually operates in the range 10–100 Td, and lower or higher values may appear during various transient phenomena. On the axes shown in Fig. 2, a Maxwellian distribution would appear as a straight line with a gradient of $-1/k_BT$. Evidently, these distributions

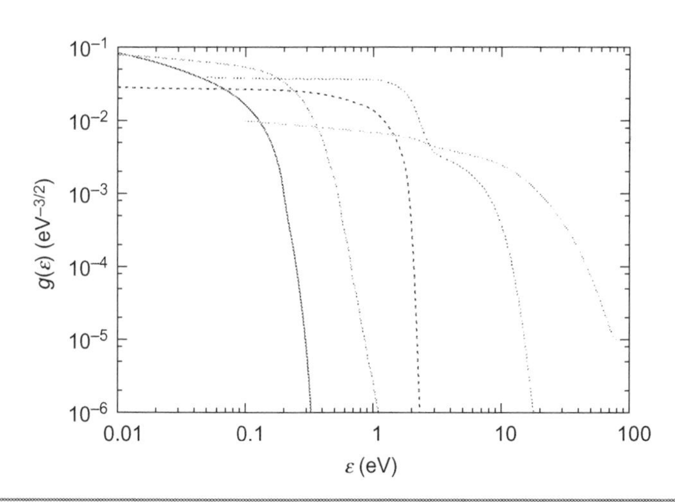

Fig. 2 Energy distributions for values of *E*/*N* of 0.1, 1, 10, 100, and 1000 Td (ordered from left to right by intercept on the horizontal axis). On these axes, a Maxwell-Boltzmann distribution appears as a straight line with gradient $-1/k_BT$. These calculations refer to humid air, here supposed to be a mixture of N_2, O_2, and H_2O in the ratio 80:20:1. Of course, air usually contains traces of other gases as well. These data were obtained by solving Eq. (21) using a numerical procedure. *(Data from Rockwood, S., 1973. Elastic and inelastic cross sections for electron-Hg scattering from Hg transport data. Phys. Rev. A 8 (5), 2348–2358.)*

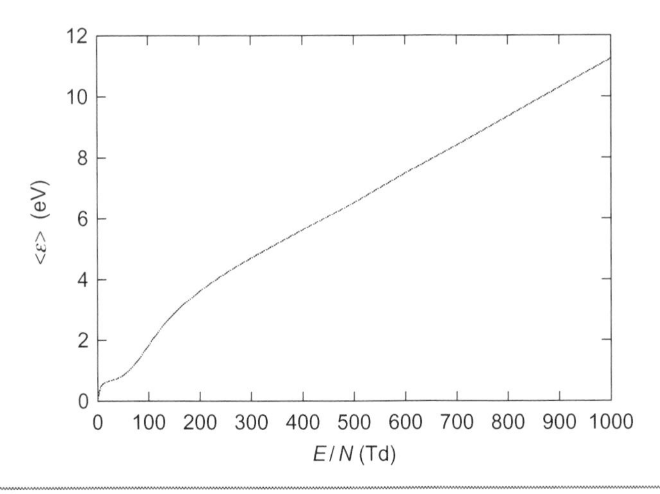

Fig. 3 The mean electron energy as a function of the reduced electric field, for the conditions of Fig. 2.

are grossly different from a Maxwellian. Moreover, the form of the distribution functions changes with *E/N*, so a simple parameterization is not possible. Of course, the mean energy varies as a function of the reduced field, as shown in Fig. 3, and the rate constants for electron impact processes are a strong function of both the reduced field and the mean energy, as shown in Fig. 4.

Other coefficients of interest can be defined in terms of the electron energy distribution function. These include transport coefficients such as the electron mobility

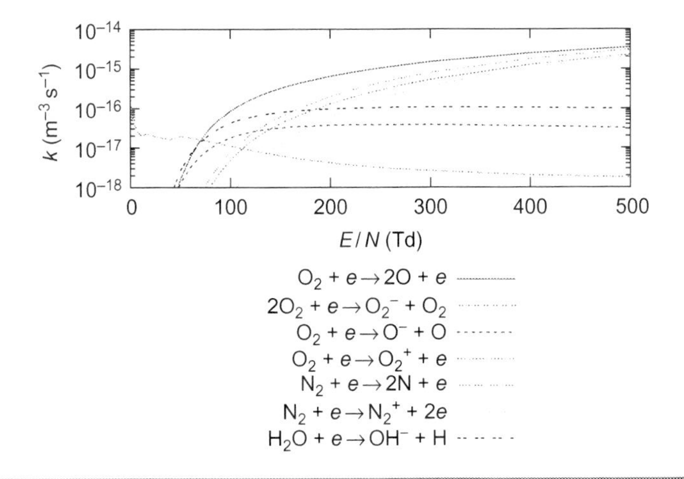

Fig. 4 A selection of representative electron impact rate constants in humid air, for the conditions of Fig. 2.

and diffusion coefficient (Cherrington, 2014; Shkarofsky et al., 1966), and derived coefficients of less obvious significance, such as the first Townsend ionization coefficient, which is important in the discussion of breakdown (see Section 4). This quantity is defined as

$$\alpha(E/N) = k_{\mathrm{i}}(E/N)\, N \,/\, v_{\mathrm{D}}(E/N), \tag{22}$$

where k_{i} is the ionization rate and v_{D} is the drift velocity. The first Townsend coefficient describes the spatial growth of the electron density $n_{\mathrm{e}}(x)$ by ionization:

$$n_{\mathrm{e}}(x) = n_0 \exp(\alpha x). \tag{23}$$

These calculations show that the electron mean energy, and as a result, the rate constants of electron impact processes, are strong functions of the applied electric field. So one might think that the chemical character of the plasma will be freely manipulated by choosing a suitable electric field. Generally speaking, however, this is not so. This is because, if the plasma is in a stationary state, or nearly so, then a plasma balance condition must be satisfied, ie,

$$\frac{dn_{\mathrm{e}}}{dt} \approx 0. \tag{24}$$

For example, a common case in atmospheric pressure plasmas is that the electron density is determined by a balance between ionization and recombination. In a simple model of a pure helium plasma, for instance, the relevant reactions are:

$$e + \mathrm{He} \rightarrow 2e + \mathrm{He}^{+} \tag{25}$$

$$\mathrm{He}^{+} + 2\mathrm{He} \rightarrow \mathrm{He}_2^{+} + \mathrm{He} \tag{26}$$

$$e + \mathrm{He}_2^{+} \rightarrow 2\mathrm{He}, \tag{27}$$

so that

$$\frac{dn_{\mathrm{e}}}{dt} = k_{\mathrm{i}} N n_{\mathrm{e}} - k_{\mathrm{R}} n_{\mathrm{d}} n_{\mathrm{e}} \approx 0, \tag{28}$$

where n_{D} is the dimer ion density and k_{R} is the rate of reaction (27). This implies that

$$k_{\mathrm{i}} - k_{\mathrm{R}} n_{\mathrm{D}} \approx 0. \tag{29}$$

Since both the rate constants k_{i} and k_{R} depend on the electron energy distribution function, and the energy distribution function depends on the electric field, we cannot simply impose a chosen electric field. Rather, detailed processes within the plasma will establish an electric field such that the condition (29) is satisfied. The nature of these processes, and the character of the plasma balance equation, depend in detail on the method chosen to produce the plasma, and the nature of the feedstock gas. But in general, there is always a plasma balance condition, so that the electric field, and the energy distribution function, are essentially fixed by this condition, and not by easily manipulated external parameters, such as an applied voltage.

These considerations imply that the electron mean energy, or effective temperature, in a self-sustaining plasma, cannot vary over a wide range, and is usually found to be a few electron volt. The importance of knowing accurate energy distribution functions is not to determine the mean energy, but rather the distribution of power into various accessible channels. The plasma balance condition fixes the ionization rate, but the ratio of the ionization rate to the rate of dissociation, for instance, may be critically dependent on the exact shape of the distribution function. The chemical character of the discharge may be crucially affected by such ratios. Consequently, calculating the distribution function is often an informative exercise, especially because experimental measurement at atmospheric pressure is difficult, and may yield only an estimate of the gross mean energy (which probably could have been guessed with equal or better precision).

3 Plasma Chemistry

Chemical processes in plasmas are usually initiated by the electron impact processes discussed in the previous section. However, the electronically or vibrationally excited states that are produced by electronic processes are often precursors for more or less complex chains of reactions. The rates of these reactions are also characterized by rate constants, but because there are relatively small differences between the masses of nonelectronic species, we can usually assume that the energy distributions of these species are of Maxwell-Boltzmann form, and have a common temperature. This means that we need not consider elaborate calculations to find these distributions, and the rate constants can generally be expressed in the parametric form:

$$k(T) = AT^{B} \exp\left(-\frac{C}{T}\right), \tag{30}$$

where the coefficients A, B, and C are determined from experiments. Such experiments are of variable reliability. For example, reactions that are important in the chemistry of the atmosphere have often been the subject of careful study, and the rate constants are known with relatively high accuracy, meaning with error bars of a few tens of percent or less (Sander et al., 2011). However, there are many reactions likely to be of significance in atmospheric pressure plasmas that have been much less investigated, and for which the rate constants are correspondingly less well known (Turner, 2015). The range of uncertainty in some such cases may be an order of magnitude or more. For this reason, if one proposes to develop a model for the chemistry of some plasma, a prudent approach should include some attention to the likely errors in the rate constants, and the consequent implications for the uncertainty of prediction (Turner, 2015).

For any given mixture of gases, a chemistry model is usually expressed as a list of reactions, each with a rate constant in the form of Eq. (30) (Table 1). A complete chemistry model for a mixture of molecular gases is typically complex, and may involve dozens of chemical species, and hundreds or thousands of reactions (Van Gaens and Bogaerts, 2013; Murakami et al., 2013). For example, a subset of reactions relevant to neutral chemistry in oxygen is listed in Table 2. This table follows the practice advocated in Turner (2015), in that each rate constant is accompanied by a category label and a dimensionless uncertainty. Category A rate constants have been measured directly, while category C reactions are assigned by analogy with similar processes. The dimensionless uncertainty is the ratio of the experimental error bar to the value of the rate constant. This ratio varies between 0.1 and 5 for the reactions listed in Table 2. A chemistry model for humid air would be vastly more complex than the scheme shown in Table 2. If one aims to predict the densities of the species included in the chemistry scheme, then in principle one first solves the Boltzmann equation as described previously to obtain the rates of electron impact processes. These processes normally initiate chains of chemical processes leading to the radical species of practical interest. Then, one constructs a set of ordinary differential equations governing the densities of the plasma species. For instance, the differential equation for the density of ozone (which is relatively simple) might be written

TABLE 1 Species Included in the Neutral Oxygen Chemistry System Shown in Table 2

O
$O(^1D)$
$O(^1S)$
O_2
$O_2(a^1\Delta_u)$
$O_2(a^1\Delta_u,\nu)$
$O_2(b^1\Sigma_u^+)$
$O_2(b^1\Sigma_u^+,\nu)$
$O_2(\nu)$
O_3
$O_3(\nu)$

The label "ν" indicates a state of vibrational excitation.

TABLE 2 Reactions and Rate Constants for the Neutral Chemistry in Oxygen

Neutral Chemistry of Oxygen						
1	$3O \rightarrow O + O_2$	0.0	$3.8 \times 10^{-44} \left(\frac{300}{T_g}\right) \exp\left(-\frac{170}{T_g}\right)$	1.0	A	Johnston (1968)
2	$3O \rightarrow O + O_2(b^1\Sigma_u^+)$	0.0	$1.4 \times 10^{-42} \exp\left(-\frac{650}{T_g}\right)$	2.0	A	Baurer and Bortner (1978, p. 24–42)
3	$2O + O_2 \rightarrow O + O_3$	0.0	$4.2 \times 10^{-47} \exp\left(\frac{1056}{T_g}\right)$	1.5	A	Rawlins et al. (1987) and Johnston (1968)
4	$2O + O_2 \rightarrow O + O_3(\nu)$	0.0	$9.8 \times 10^{-47} \exp\left(\frac{1056}{T_g}\right)$	1.5	A	Rawlins et al. (1987) and Johnston (1968)
5	$2O + O_2 \rightarrow O_2(a^1\Delta_u) + O_2$	0.0	$6.5 \times 10^{-45} \left(\frac{300}{T_g}\right) \exp\left(-\frac{170}{T_g}\right)$	1.0	A	Slanger and Copeland (2003) and Johnston (1968)
6	$2O + O_2 \rightarrow O_2(b^1\Sigma_u^+) + O_2$	0.0	$6.5 \times 10^{-45} \left(\frac{300}{T_g}\right) \exp\left(-\frac{170}{T_g}\right)$	1.0	A	Slanger and Copeland (2003) and Johnston (1968)
7	$O + O(^1D) \rightarrow 2O$	0.0	2.0×10^{-18}	0.6	A	Sobral et al. (1993)
8	$O + O(^1S) \rightarrow 2O$	0.0	$2.5 \times 10^{-17} \exp\left(-\frac{300}{T_g}\right)$	1.0	A	Schofield (1978)
9	$O + O(^1S) \rightarrow O + O(^1D)$	0.0	$2.5 \times 10^{-17} \exp\left(-\frac{300}{T_g}\right)$	1.0	A	Schofield (1978)
10	$O + 2O_2 \rightarrow O_2 + O_3$	0.0	$1.8 \times 10^{-46} \left(\frac{300}{T_g}\right)^{2.6}$	0.1	A	Sander et al. (2011, p. 2–4), Atkinson et al. (2004, p. 1473), and Rawlins et al. (1987)
11	$O + 2O_2 \rightarrow O_2 + O_3(\nu)$	0.0	$4.2 \times 10^{-46} \left(\frac{300}{T_g}\right)^{2.6}$	0.1	A	Sander et al. (2011, p. 2–4), Atkinson et al. (2004, p. 1473), and Rawlins et al. (1987)

12	$O + O_2 + O_2(a^1\Delta_u) \rightarrow O + 2O_2$	0.0	1.1×10^{-44}	5.0	A	Azyazov et al. (2009) and Braginskiy et al. (2005)
13	$O + O_2 + O_3 \rightarrow 2O_3$	0.0	$1.4 \times 10^{-47} \exp\left(-\frac{1050}{T_g}\right)$	0.3	A	Johnston (1968) and Rawlins et al. (1987)
14	$O + O_2 + O_3 \rightarrow O_3 + O_3(\nu)$	0.0	$3.27 \times 10^{-47} \exp\left(-\frac{1050}{T_g}\right)$	0.3	A	Johnston (1968) and Rawlins et al. (1987)
15	$O + O_2(a^1\Delta_u) \rightarrow O + O_2$	-2.14	1.0×10^{-22}	1.0	A	Sander et al. (2011, p. 1–8)
16	$O + O_2(b^1\Sigma_u^+) \rightarrow O + O_2(a^1\Delta_u)$	-0.65	8.0×10^{-20}	1.0	A	Sander et al. (2011, p. 1–9) and Atkinson et al. (2004, p. 1486)
17	$O + O_3 \rightarrow 2O + O_2$	0.0	$1.2 \times 10^{-15} \exp\left(-\frac{11400}{T_g}\right)$	0.3	A	Johnston (1968)
18	$O + O_3 \rightarrow 2O_2(\nu)$	0.0	$8.0 \times 10^{-18} \exp\left(-\frac{2060}{T_g}\right)$	0.2	A	Sander et al. (2011, p. 1–5), Atkinson et al. (2004, p. 1475), Baulch et al. (1984), and Wine et al. (1983)
19	$O + O_3(\nu) \rightarrow 2O_2$	0.0	4.5×10^{-18}	0.5	A	Steinfeld et al. (1987)
20	$O + O_3(\nu) \rightarrow O_3 + O$	0.0	1.05×10^{-17}	0.5	A	Steinfeld et al. (1987)
21	$O(^1D) + O_2 \rightarrow O + O_2(b^1\Sigma_u^+)$	0.0	$2.64 \times 10^{-17} \exp\left(\frac{55}{T_g}\right)$	0.1	A	Sander et al. (2011, p. 1–5)
22	$O(^1D) + O_2 \rightarrow O + O_2(a^1\Delta_u)$	0.0	$6.6 \times 10^{-18} \exp\left(\frac{55}{T_g}\right)$	0.1	A	Sander et al. (2011, p. 1–5)
23	$O(^1D) + O_3 \rightarrow 2O + O_2$	0.0	1.2×10^{-16}	0.2	A	Sander et al. (2011, p. 1–5), Atkinson et al. (2004), and Steinfeld et al. (1987)
24	$O(^1D) + O_3 \rightarrow 2O_2$	0.0	1.2×10^{-16}	0.2	A	Sander et al. (2011, p. 1–5), Atkinson et al. (2004), and Steinfeld et al. (1987)

Continued

TABLE 2 Reactions and Rate Constants for the Neutral Chemistry in Oxygen—cont'd

Neutral Chemistry of Oxygen						
25	$O(^1S)+O_2 \rightarrow O+O_2$	0.0	$3.0\times10^{-18}\exp\left(-\frac{850}{T_g}\right)$	1.0	A	Baurer and Bortner (1978, p. 24–52) and Slanger and Black (1978)
26	$O(^1S)+O_2 \rightarrow O(^1D)+O_2$	0.0	$1.3\times10^{-18}\exp\left(-\frac{850}{T_g}\right)$	1.0	A	Baurer and Bortner (1978, p. 24–52) and Slanger and Black (1978)
27	$O(^1S)+O_2(a^1\Delta_u) \rightarrow 3O$	0.0	3.2×10^{-17}	0.3	A	Slanger and Black (1981a), Slanger and Black (1981b), and Kenner and Ogryzlo (1982)
28	$O(^1S)+O_2(a^1\Delta_u) \rightarrow O+O_2(b^1\Sigma_u^+)$	0.0	1.3×10^{-16}	0.3	A	Slanger and Black (1981a), Slanger and Black (1981b), and Kenner and Ogryzlo (1982)
29	$O(^1S)+O_2(a^1\Delta_u) \rightarrow O(^1D)+O_2$	0.0	3.6×10^{-17}	0.3	A	Slanger and Black (1981a) and Slanger and Black (1981b)
30	$O(^1S)+O_3 \rightarrow O+O(^1D)+O_2$	0.0	1.93×10^{-16}	0.6	A	Steinfeld et al. (1987)
31	$O(^1S)+O_3 \rightarrow 2O_2$	0.0	1.93×10^{-16}	0.6	A	Steinfeld et al. (1987)
32	$O(^1S)+O_3 \rightarrow 2O+O_2$	0.0	1.93×10^{-16}	0.6	A	Steinfeld et al. (1987)
33	$2O_2 \rightarrow 2O+O_2$	0.0	$6.6\times10^{-15}\left(\frac{300}{T_g}\right)^{1.5}\exp\left(-\frac{59000}{T_g}\right)$	0.5	A	Baurer and Bortner (1978, p. 24–44)
34	$O_2+O_2(\nu) \rightarrow 2O_2$	-0.19	2.0×10^{-20}	0.5	A	Park and Slanger (1994)
35	$O_2+O_2(a^1\Delta_u) \rightarrow 2O_2$	0.0	$3.6\times10^{-24}\exp\left(-\frac{220}{T_g}\right)$	0.2	A	Sander et al. (2011, p. 1–8)
36	$O_2+O_2(a^1\Delta_u,\nu) \rightarrow O_2(\nu)+O_2(a^1\Delta_u)$	-0.19	5.0×10^{-17}	0.5	A	Antonov et al. (2003)

37	$O_2 + O_2(b^1\Sigma_u^+) \rightarrow O_2 + O_2(a^1\Delta_u)$	0.0	3.9×10^{-23}	0.5	A	Sander et al. (2011, p. 1–9), Kebabian and Freedman (1997), and Knickelbein et al. (1987)
38	$O_2 + O_2(b^1\Sigma_u^+, \nu) \rightarrow O_2(\nu) + O_2(b^1\Sigma_u^+)$	-0.19	1.5×10^{-17}	0.1	A	Kalogerakis et al. (2002)
39	$O_2 + O_3 \rightarrow O + 2O_2$	0.0	$7.26 \times 10^{-16} \exp\left(-\frac{11435}{T_g}\right)$	0.3	A	Steinfeld et al. (1987)
40	$O_2 + O_3(\nu) \rightarrow O_2 + O_3$	0.0	4.0×10^{-20}	0.5	A	Steinfeld et al. (1987)
41	$2O_2(a^1\Delta_u) \rightarrow O_2 + O_2(b^1\Sigma_u^+)$	0.0	2.7×10^{-23}	0.2	A	Lilenfeld et al. (1984)
42	$O_2(a^1\Delta_u) + O_2(b^1\Sigma_u^+) \rightarrow O_2 + O_2(b^1\Sigma_u^+)$	0.0	2.7×10^{-23}	2.0	C	
43	$2O_2(b^1\Sigma_u^+) \rightarrow O_2 + O_2(b^1\Sigma_u^+)$	0.0	2.7×10^{-23}	2.0	C	
44	$O_2(a^1\Delta_u) + O_3 \rightarrow O + 2O_2$	0.0	$5.2 \times 10^{-17} \exp\left(-\frac{2840}{T_g}\right)$	0.2	A	Sander et al. (2011, p. 1–8) and Steinfeld et al. (1987)
45	$O_2(a^1\Delta_u) + O_3(\nu) \rightarrow O_2 + O_3$	0.0	5.0×10^{-17}	5.0	A	Steinfeld et al. (1987)
46	$O_2(b^1\Sigma_u^+) + O_3 \rightarrow O + 2O_2$	0.0	$2.4 \times 10^{-17} \exp\left(-\frac{135}{T_g}\right)$	0.2	A	Sander et al. (2011, p. 1–9) and Atkinson et al. (2004)
47	$O_2(b^1\Sigma_u^+) + O_3 \rightarrow O_2 + O_3$	0.0	$5.5 \times 10^{-18} \exp\left(-\frac{135}{T_g}\right)$	0.2	A	Sander et al. (2011, p. 1–9) and Atkinson et al. (2004)
48	$O_2(b^1\Sigma_u^+) + O_3 \rightarrow O_2(a^1\Delta_u) + O_3$	0.0	$5.5 \times 10^{-18} \exp\left(-\frac{135}{T_g}\right)$	0.2	A	Sander et al. (2011, p. 1–9) and Atkinson et al. (2004)
49	$2O_3 \rightarrow O + O_2 + O_3$	0.0	$1.65 \times 10^{-15} \exp\left(-\frac{11435}{T_g}\right)$	0.3	A	Steinfeld et al. (1987)
50	$O_3 + O_3(\nu) \rightarrow 2O_3$	0.0	1.0×10^{-19}	0.5	A	Steinfeld et al. (1987)

The rubric for this table is as follows: Column 1 contains a reference number for each reaction; column 2 specifies the reaction; column 3 gives the threshold energy of the reaction, expressed in electron volt; column 4 gives an expression for the rate constant in MKS units, with the electron temperature expressed in electron volt and the neutral temperature in Kelvin; column 5 is the dimensionless uncertainty discussed in the text; and column 6 gives references.

$$\frac{d[O_3]}{dt} \approx k_3[\mathrm{O}][\mathrm{O}_2][M] - k_{46}[\mathrm{O}_3][\mathrm{O}_2(b^1\Sigma_u^+)], \tag{31}$$

where the subscripts on the rate constants refer to reactions listed in Table 2. The typical behavior of a chemistry scheme, especially for a plasma that is pulsed or otherwise transiently excited, is that a large number of rather short-lived radicals decay quickly into a smaller number of relatively long-lived species, and these long-lived species are responsible for chemical activity of practical interest. For example, in a plasma with oxygen as the main reagent, these long-lived species are atomic oxygen, ozone, and the metastable $O_2(a^1\Delta_g)$ state. These species are strongly oxidizing, and thus often especially important in treatments of food, especially those rich in vitamin C. Examples of the typical behavior of these species are shown in Figs. 5 and 6. Of course, a more complex chemistry will involve a larger set of long-lived radicals.

The process of modeling plasma chemistry appears straightforward, because there are many models already published in the literature, and there are freely available tools for solving and analyzing the associated system of differential equations (Pancheshnyi et al., 2015; Markosyan et al., 2014). However, many published models are of indifferent quality, and the rate data that is available is also of variable accuracy, even when properly used, which is not always the case (Turner, 2015). Some physicochemical processes are probably not fully taken into account. For instance, the rate constants of certain reactions involving molecular species often

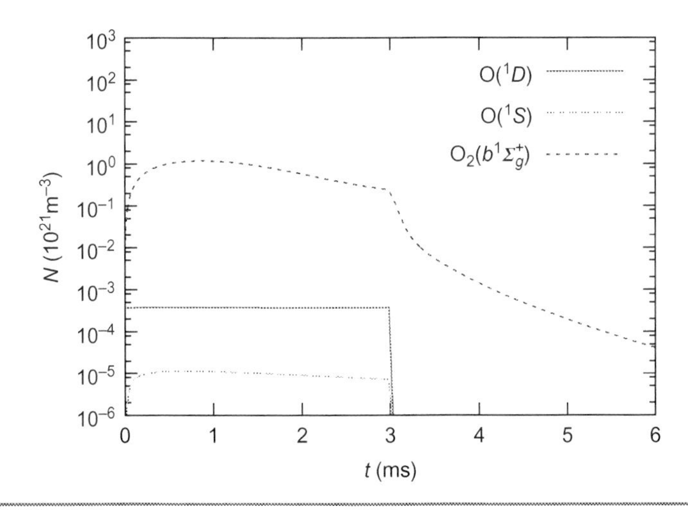

Fig. 5 Time dependent behavior of transient excited state densities in a mixture of helium and oxygen. In this model calculation, power is applied for the first three milliseconds only, so the for the time between 3 and 6 ms, the plasma is unpowered and is decaying. As can be seen, the species depicted in this figure disappear more or less rapidly in this period.

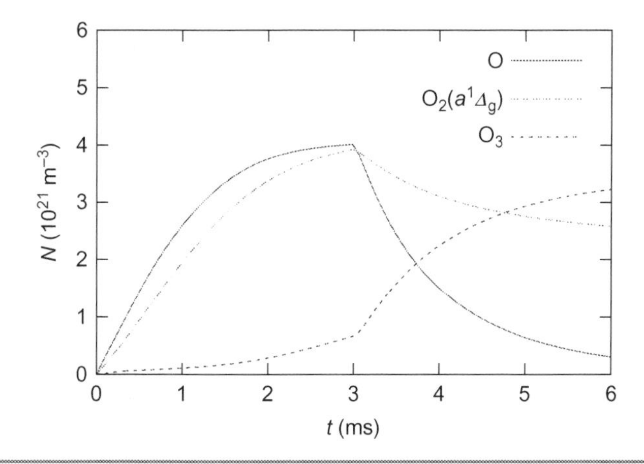

Fig. 6 Time dependent behavior of long-lived excited states under the same conditions as in Fig. 5. These species persist for a relatively long time after the power is turned off, implying, for example, that they could be transported to nearby surfaces and otherwise play an important role in a process.

depend strongly on the vibrational quantum number in ways that have not yet been fully quantified (Marinov et al., 2013). Consequently, plasma chemistry modeling as presently practiced is not usually regarded as a powerful predictive tool, and should be used with caution.

In general, direct measurement of the density of radical species is not straightforward either (Dilecce et al., 2015; Große-Kreul et al., 2015). Detecting transient species is never easy, especially when there is a complex spatial structure to complicate the problem, as is often the case (see Section 5).

4 Breakdown Processes

Breakdown (Llewellyn-Jones, 1966) is the process by which a neutral gas with the background level of ionization discussed in Section 1 develops into a plasma. Breakdown is induced by applying an electric field of sufficient strength to cause multiplication of the background electrons, so the question of how strong of an electric field is needed to produce this effect is of central interest. (Although we are now discussing plasma applications, and therefore producing a plasma is our concern, studies of breakdown have often been motivated by a desire for prevention, for example in many high-voltage engineering contexts.) An empirical answer was given by Paschen in the 19th century (Llewellyn-Jones, 1966; Raizer, 2011). He found that when two electrodes are separated by a distance d, the minimum voltage that must be applied to initiate a breakdown is give by

$$V_{\mathrm{B}} = \frac{Apd}{\ln(Bpd) + C}, \tag{32}$$

where V_{B} is the breakdown voltage, p is the pressure in Pascal, and d is expressed in meters. A, B, and C are coefficients that depend on both the gas and the electrode material. Examples of Paschen curves are shown in Fig. 7. A feature of all of these curves is the existence of a minimum breakdown voltage, given by:

$$V_{\mathrm{B,min}} = \exp(1)\frac{AC}{B}, \tag{33}$$

such that

$$(pd)_{\mathrm{min}} = \exp(1)\frac{C}{A}. \tag{34}$$

The minimum breakdown voltage is usually in the range 100–500 V, while the value of pd where this occurs is about 1 Pa m. At atmospheric pressure, this means that the electrode gap at which the minimum breakdown voltage occurs is about 10 μm. Most of the devices used to produce plasma at atmospheric pressure have characteristic dimensions larger than this, and so are said to operate on the right hand side of the Paschen curve. For instance, a device with electrode separation 1 mm will operate with $pd \approx 100$ Pa m and will have V_{B} of a few kilovolts.

Two basic physical mechanisms determine the form of the Paschen curve. These are electron impact ionization in the gas phase, and so-called secondary emission from electrodes. In the latter process, an ion striking the cathode has a certain probability of causing the emission of an electron. This probability, conventionally

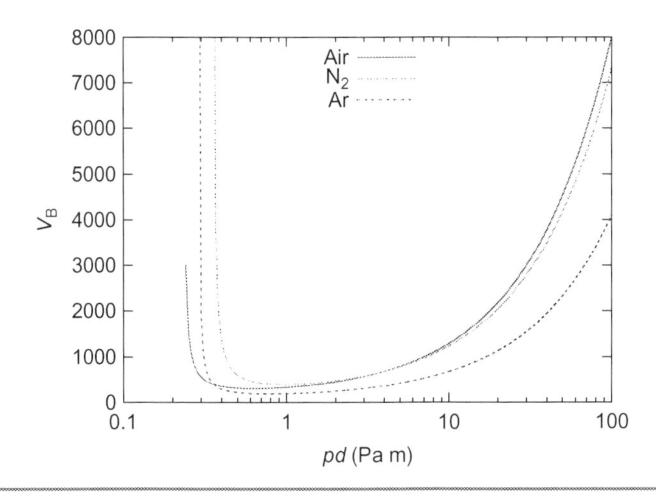

Fig. 7 Paschen curves for argon, nitrogen, and dry air.

denoted by γ, depends on the electrode material, but is often small, ie, $\gamma \lesssim 0.01$. Each secondary electron drifts across the electrode gap, and is ultimately absorbed at the anode. A breakdown can begin only if these processes produce a net increase in the number of charged particles. This breakdown condition requires that for every electron emitted from the cathode, at least $1/\gamma$ ions must be produced by electron impact ionization in the space between the electrodes. The general shape of the Paschen curve depends on the relationship between the electron mean free path and the electrode gap. To the left of the Paschen minimum, the mean free path may be larger than the gap. In this case, the electron collision frequency is low, and a large electric field is needed to ensure that every colliding electron has enough energy to produce ionization. To the right of the Paschen minimum, there are many electron mean free paths across the gap. In this regime, one needs a certain potential difference per mean free path to maintain an appreciable ionization frequency, so that the breakdown voltage is an approximately linear function of the *pd* product. This combination of gas phase ionization and secondary emission from the cathode is known as the Townsend breakdown process.

An assumption of the Townsend mechanism of breakdown is that the charges associated with the onset of ionization processes do not distort the electric field. If instead the space-charge field is comparable with the applied field, then different mechanisms become important. When breakdown is initiated from a single electron-ion pair, the electron drifts toward the cathode, and causes ionization as it travels. As ionization takes place, the primary electron develops into a cloud of electrons. The motion of these electrons is a combination of diffusion, due to the random motion of individual electrons, and drift due to the electric field, which is the same for every electron. Consequently, the electron cloud forms an expanding sphere that drifts toward the anode. The ions, of course, drift much more slowly toward the cathode. This process produces a separation of space charge, with negative charge in the drifting electron cloud, and positive charge filling the region through which the electron cloud has passed. This space-charge distribution sets up an electric field, and if this electric field becomes comparable with the applied electric field, then the Townsend mechanism ceases to apply (Fig. 8). In particular, secondary breakdowns can occur, in which the electrons drift toward regions of positive space charge, rather than the anode. The onset of these mechanisms is determined by the Raether-Meek criterion (Meek, 1940; Raether, 1964), which is

$$\alpha(E/N)d \approx 20, \tag{35}$$

where α is the first Townsend ionization coefficient defined by Eq. (22). Above this threshold, the electric field is a function not just of the applied voltage, but also of the distribution of charges in the gap. This does not greatly affect the breakdown voltage, but the spatial structure of the developing plasma is radically changed. A Townsend breakdown produces a diffuse plasma that rather uniformly

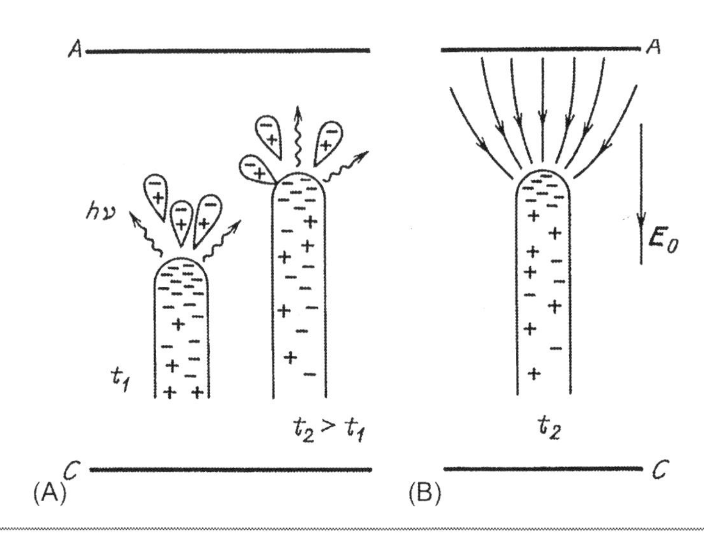

Fig. 8 The space charge structure of an electron avalanche when the Raether-Meek criterion is met. The anode (A) is at the top, and the cathode (C) at the bottom. The left panel (A) shows the avalanche structure, and the right panel (B) emphasizes the electric field distribution. Negative space charge associated with electrons is concentrated in the streamer head, and less-dense positive space charge fills the streamer body. When this separation of space charges produces an electric field comparable with the applied field, secondary avalanches (which may be initiated by cosmic ray events, or photons emitted from the primary avalanche) become directed toward the primary streamer, producing the characteristic branching streamer structure shown in Fig. 9. *(Source: Reproduced from Raizer, Y., 2011. Gas Discharge Physics. Springer, New York. ISBN 978-3-642-64760-4.)*

fills the space between the electrodes, whereas the Raether-Meek mechanism leads to a so-called streamer breakdown, which may produce a plasma with an intricate spatial structure. This is usually not desirable, because the plasma will fill only small fraction of the available volume, and often this leads to a thermal plasma in which the gas temperature and the electron temperature are nearly equal. For most applications, this is an inefficient outcome, because power is consumed by gas heating, and not in production of the reactive radicals which are wanted. An example of a streamer breakdown leading to a thermal arc discharge is shown in Fig. 9.

There are other differences between the Townsend and the streamer breakdown. Breakdown by the Townsend mechanism tends to be relatively slow, as ions must reach the cathode during the process, and the ions move rather slowly. Moreover, as many cosmic ray events usually occur during the breakdown process, statistical effects are not important. A streamer breakdown is a much faster process, and may initiate from a single electron-ion pair. If the voltage across the gap is applied

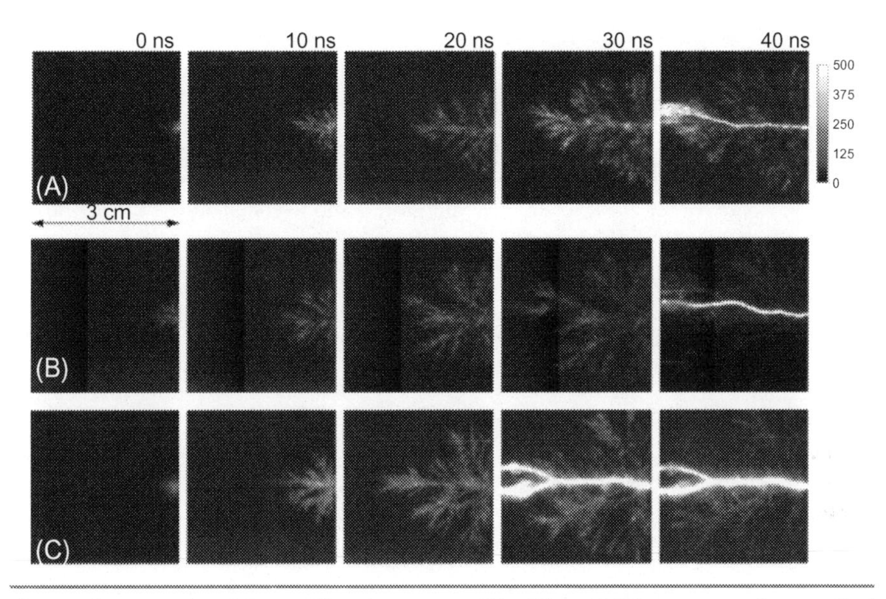

Fig. 9 Streamer breakdown. The three cases (A), (B), and (C) correspond to applied voltages of 18, 20, and 22 kV. The breakdown is initiated from a needle-like anode at the left of the figure, and the 3 cm region between the anode and the cathode is filled with argon at a pressure of 1 atm. Since $pd \sim 300$ Pa m, this breakdown is occurring far out on the right hand side of the Paschen curve. In all cases, after 40 ns a single intense streamer has appeared, which will develop into an arc. *(Source: Reproduced from Takahashi, E., Kato, S., Furutani, H., Sasaki, A., Kishimoto, Y., Takada, K., Matsumura, S., Sasaki, H., 2011. Single-shot observation of growing streamers using an ultrafast camera. J. Phys. D: Appl. Phys. 44 (30), 302001. ISSN 0022-3727.)*

sufficiently fast, a statistical time lag may be observed between the application of the voltage and the onset of breakdown, depending on the time before the next background ionization event (Raether, 1964).

5 Plasma Sources

The preceding sections were concerned with the basic processes that take place during the operation of a glow discharge. Little has been said of the spatial structure of a stationary glow discharge, and yet at all operating pressures, such discharges do exhibit pronounced spatial structure. We are going to discuss first the structure of the classical glow discharge, which appears when the *pd* product is relatively small. This is of limited direct relevance to most atmospheric pressure discharges, which do not operate in this regime. However, since other types of discharge are difficult to discuss without reference to concepts and terminology that originate with low-pressure discharges, we will begin this section by describing the classical glow discharge, and then proceed to discuss other types of discharge with more relevance to atmospheric pressure applications. Perhaps the most important categories of discharges presently used in food research are coronas (Section 5.3) and dielectric barrier discharges (Section 5.4). Either of these techniques can be employed in a jet configuration (Section 5.5), where flowing gas is used to transport the effluent from the discharge toward the processing area.

5.1 Glow Discharge

In the regime where the Raether-Meek criterion is not satisfied, so that breakdown occurs through the Townsend mechanism, there is a characteristic dependence of the discharge voltage on the current. A typical example is shown in Fig. 10. In the region to the left of point B, the current is not steady, and is characterized by random bursts associated with individual cosmic ray events occurring in the plasma volume. Between points B and C, the current becomes steady, and the discharge is sustained by a voltage equal to V_B, the breakdown voltage that was discussed previously. In

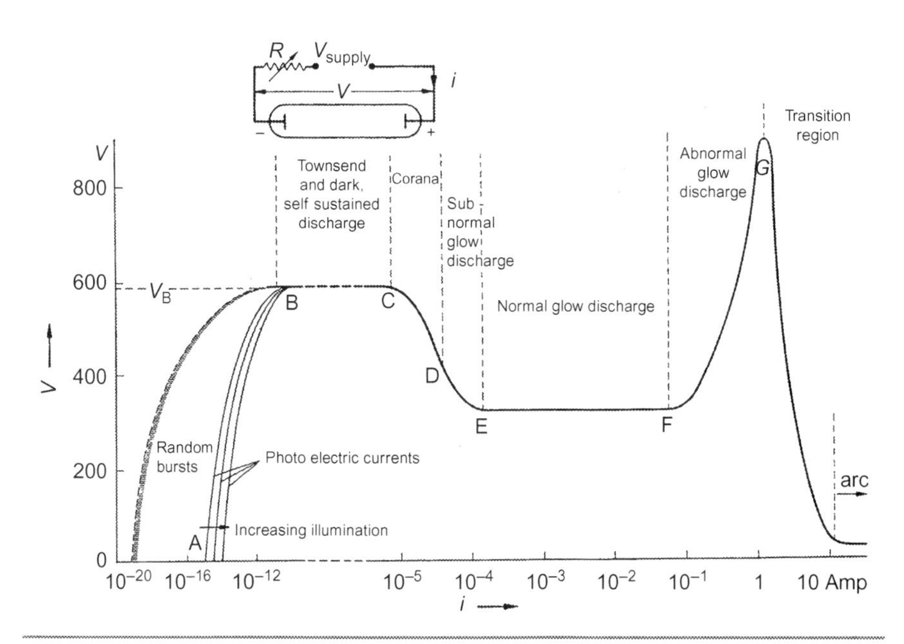

Fig. 10 Current and voltage characteristics of a glow discharge. Note that the horizontal axis is discontinuous, and that the range of currents shown is more than 10^{13}. *(Source: Reproduced from Francis, G., 1956. The glow discharge at low pressure, vol. 4/22. Springer, Berlin, Heidelberg. pp. 53–208. ISBN 978-3-642-45849-1, 978-3-642-45847-7.)*

this region, space-charge effects are not important. A discharge in this range of current density is called a dark or Townsend discharge. In the region to the right of point C, space-charge effects begin to distort the electric field, and once point E has been reached, there develops the characteristic spatial structure of the classical or normal glow discharge. The discharge voltage in this region is considerably less than V_B, and depends weakly on the discharge current. Beyond the point F, the discharge voltage begins to increase again, in the region known as the abnormal glow discharge. Eventually, beyond the point G, the glow discharge develops into an arc, and the voltage falls to a low value. These changes are governed by complex physical processes, which in general are not of high relevance to the topics of this book. We will not, therefore, discuss them in detail. For further information see the classic works (Francis, 1956; Raizer, 2011). We limit our discussion to an outline of the structure of the classical glow discharge in the normal regime.

The existence of spatial structure in a glow discharge is not surprising. A basic physical property of such discharges must be that the electrical current is continuous. If the discharge is stationary, this whole current must be due to the convective motion of charged particles, since there cannot be any displacement current. In regions far from the electrodes, where the densities of electrons and ions are practically equal, almost all of this current will be carried by electrons. However, in the vicinity of the electrodes, complications arise because of the absorption and emission of charged particles at the electrode surfaces. In particular, electrons drift away from the cathode, and are only occasionally emitted from the cathode surface following ion impacts or other similar processes. Consequently, there is a region near the cathode where the electron density is severely depleted. The current in this region must therefore be transported substantially by ions. Since the mobility of ions is much less than that of electrons, there must be a correspondingly larger electric field adjacent to the cathode. This large electric field will accelerate those electrons that are emitted from the cathode to high energies, but this takes a finite amount of space, so that although a large amount of excitation and ionization is associated with these energetic electrons, this takes place at a certain distance from the cathode. Because of this ionization, the electron density increases as one moves away from the cathode, and eventually, at some distance from the cathode, the electron and ion densities become equal. These phenomena largely account for the rather complex spatial structure that a normal glow discharge exhibits. When the discharge is operated at a pressure much less than an atmosphere, this structure is easily seen with the naked eye, as seen in Fig. 11. The principle features (moving from the cathode toward the anode) are: the cathode dark space, the negative glow, the Faraday dark space, the positive column, and the anode glow. The cathode dark space is the region where electrons emitted from the cathode have not yet achieved an energy sufficient to cause excitation leading to light emission. The negative glow is the region of intense ionization and excitation caused by electrons emitted from the cathode once they have reached high energy. The Faraday

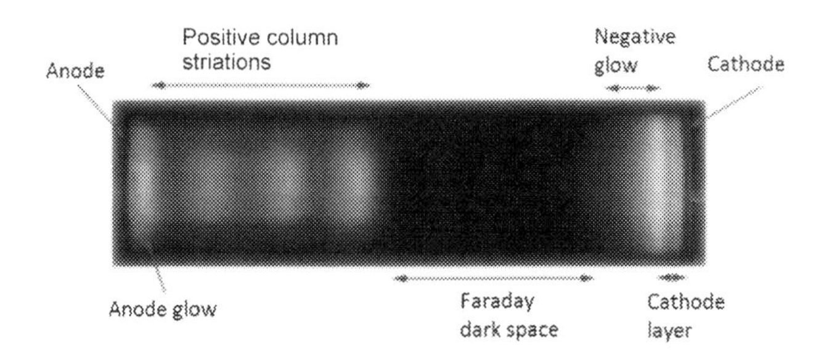

Fig. 11 Discharge structure. *(Reproduced from Lisovskiy, V., Koval, V., Artushenko, E., Yegorenkov, V., 2012. Validating the Goldstein-Wehner law for the stratified positive column of dc discharge in an undergraduate laboratory. Eur. J. Phys. 33 (6), 1537–1545. ISSN 0143-0807, 1361-6404.)*

dark space is a non equilibrium transition region between the negative glow and the positive column. In essence, owing to the intense ionization produced in the negative glow, the plasma in this region need not be self-sustaining (because plasma is transported from the negative glow into the Faraday dark space), and the electron mean energy can fall to a low value, so that light emission also is greatly reduced. The positive column is a region of self-sustaining plasma, in the sense that volume ionization is balanced by either volume recombination or diffusion to the walls. The electron mean energy must therefore be large enough to produce substantial ionization, so that light emission from the positive column is strong, although less intense than from the negative glow. The anode glow is a second region of nonequilibrium behavior, arising because ions are not emitted from the anode.

The length scales of the nonequilibrium regions adjacent to the electrodes are a function of the mean free path of the charged particles, and consequently the size of these regions is approximately inversely proportional to the gas density or pressure. The positive column, however, expands to fill the space available, and may therefore be much larger. This feature is seen in neon signs, for instance, which are almost entirely filled with the positive column of a normal glow discharge. At the other extreme, in a so-called obstructed glow discharge, there may be no space for a positive column, and in that case none appears. Positive columns are subject to a variety of instabilities going under the general rubric of "striations." These are band-like structures that may propagate toward either the cathode or the anode, or less commonly can appear stationary. The mechanisms associated with striations are in general complex, and will not be discussed here (Raizer, 2011).

The features that we have been discussing are readily observed in discharges at pressures much less than one atmosphere, but they exist in glows at all pressures. In particular, they exist at atmospheric pressure, although the length scale of the cathode and anode regions becomes much less than 1 mm, and hence are difficult to observe. A glow discharge operated with alternating current has basically similar properties, in the sense that there are nonequilibrium regions adjacent to each electrode, and a positive column filling the space in between. The nonequilibrium regions are inevitably more complex; however, because of their changing polarities, they become regions of temporal as well as spatial departures from equilibrium.

A normal or obstructed glow discharge is usually the preferred mode of discharge operation for plasma applications. However, for reasons that have already been discussed, at atmospheric pressure this option is limited to plasmas with characteristic dimensions of about 1 mm or less. Otherwise, the Raether-Meek criterion is satisfied, and a streamer breakdown occurs. As the current density in a streamer is high, streamers typically develop into arcs, as shown in Fig. 9. This leads to gas heating, and often melting or other damage to the electrodes, including sputtering of material from the electrodes into the plasma. This can reduce the life of the electrodes and contaminate the process. Consequently, arcing is undesirable, and an important aim when designing a plasma device is to prevent the development of arcs. Preionization can at least delay the onset of an arc discharge when the natural breakdown mechanism involves streamers (Palmer, 1974; Levatter and Lin, 1980). This technique involves increasing the density of seed electrons, such that individual avalanches overlap before space-charge effects begin to distort the field. The increased seed electron density is produced by an electron beam, an X-ray source, or an auxiliary discharge. Preionization is commonly used in gas lasers, but is probably too cumbersome a procedure to be acceptable in most other applications. Therefore, one normally seeks to prevent or limit arcing via the geometrical or electrical configuration of the discharge.

Classical glow discharges are often assumed to be governed by similarity principles. The most important of these is the *pd* product (strictly speaking, the proper parameter is the gas density rather than the pressure, but the use of *pd* is common). For a direct current discharge, this means that if the linear dimensions of the discharge and the gas pressure are varied such that the product *pd* is kept constant, while the discharge voltage is also kept constant, then the electron energy distribution function will be unchanged, and the plasma density will scale inversely with the spatial dimension. Other plasma parameters will also change, but according to simple scaling laws. Consequently, the set of discharges with the same *pd* product and the same voltage can be said to be similar. Of course, the existence of such a scaling principle is closely related to the central role of the reduced electric field in determining the electron energy distribution function. Maintaining a constant value of *pd* keeps the number of mean free paths in the discharge gap constant, and maintaining a constant voltage keeps the potential difference per mean free path constant. For high frequency discharges, a second similarity parameter is the ratio of the excitation frequency to the gas pressure. This similarity principle asserts that the number of collisions per cycle of the excitation frequency should be constant. These similarity principles are powerful ideas, but they have limitations. For instance, any nonlinear phenomena taking place in the plasma volume will not respect the similarity laws. Multistep ionization

$$e + He^* \rightarrow 2e + He, \quad (36)$$

will become more significant for smaller values of d because both the electron density and the excited state density scale inversely as d.

5.2 Microplasmas

The basic idea of a microplasma is to avoid arcing by making the characteristic size sufficiently small (Becker et al., 2006; Iza et al., 2008), so that breakdown occurs by the Townsend mechanism, leading to a diffuse glow discharge. The similarity principles discussed above suggest that one can reproduce the behavior of any particular low pressure discharge at atmospheric pressure by reducing the size of the discharge. In practice, things are not so simple. We have already noticed that the applicability of the similarity principles is limited by nonlinear processes in the plasma. There are other effects. For instance, the absolute value of the electric field increases as the spatial dimensions are reduced, and this can affect the behavior of the electrodes. In particular, one can find that field emission is important in microdischarges, but not in a corresponding macrodischarge, because the absolute value of the electric field is different. More significantly, in the present context, microdischarges are not easily employed in large-volume processes.

5.3 Corona Discharge

A corona discharge is formed when there are pronounced spatial inhomogeneities in the electric field, in particular, when the electric field exceeds the breakdown threshold in a limited spatial region (Raizer, 2011). This commonly occurs when highly asymmetric electrodes are employed, such as a point and a plane. The physical mechanisms involved depend on whether the region of high field is near the anode or the cathode. Since most of the discharge volume is below the critical electric field for breakdown, the plasma in these regions is not self-sustaining, and is produced by secondary processes, such as photoionization or transport of charge carriers from the high-field region. The breakdown in the region of intense field typically produces streamers, but a diffuse plasma, albeit of low density, is formed in the remainder of the region. This low density is a disadvantage, of course, if the aim is to process either the gas or the surface of the larger electrode. Coronas form naturally in such phenomena as St Elmo's fire, and can be a nuisance in electrical power systems, where they cause parasitic power loses. These discharges are used commercially for treating fruit and vegetables with ozone (Guzel-Seydim et al., 2004).

5.4 Dielectric Barrier Discharge

Dielectric barrier discharges, sometimes called silent discharges, are a technique for producing a diffuse plasma in the regime where the natural mode of breakdown involves streamers, without requiring an inhomogeneous electric field (Kogelschatz, 2003) (Fig. 12). The essential idea is to excite the plasma by an oscillating current using a circuit with a large capacitance. This capacitance is introduced by covering one or both of the electrodes with a dielectric material, hence the name. If the Raether-Meek criterion is satisfied, streamers are initiated at essentially random positions between the electrodes as long as the voltage is sufficiently large. However, the presence of the dielectric barrier causes the streamers to be self-extinguishing, as the accumulation of charge on the dielectric surface while current flows through the streamer channel reduces the local electric field. Consequently, the streamers are unable to develop into arcs, and the average effect of many individual streamer breakdowns is to expose the gas between the electrodes to an effectively uniform nonequilibrium plasma. A dielectric barrier is thus a powerful method of producing a plasma in an atmospheric pressure gas with large volume, provided that the instantaneous absence of spatial uniformity is not a concern, which is the case in many applications.

5.5 Jet Sources

A jet source is a category of atmospheric pressure plasma source, that can be adopted in conjunction with any of several plasma excitation schemes, including classical

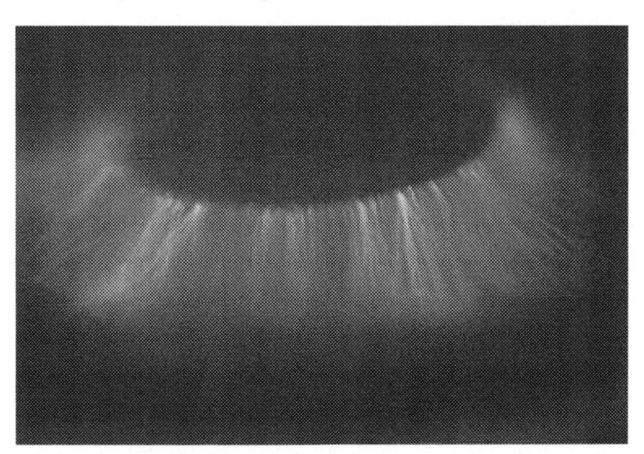

Fig. 12 A typical dielectric barrier discharge, showing the characteristic spatial structure with numerous individual streamers. This discharge is in humid argon. *(Source: Image Courtesy: Maria Prantsidou, University of Manchester.)*

glow discharges, coronas and dielectric barrier discharges. As the name suggests, in a jet source, gas is blown through the discharge region and there exposed to excitation by a plasma. Long-lived radical species produced by interactions with the plasma are carried along in the effluent from the jet, so that objects exposed to the flowing gas can be acted on by the radicals. This approach may be attractive for several reasons. A jet source is a so-called remote source: The power electronics exciting the plasma can be placed at some distance from the processing area, thus avoiding direct exposure of fragile materials to either plasma or electrical systems. Although nonequilibrium plasmas generally do not cause much gas heating, this is also minimized by using a remote source. For these reasons, jet sources are often favored when living tissue is to be treated by a plasma. A well-known jet source is the micro atmospheric pressure plasma jet, or μAPPJ (Schulz-von der Gathen et al., 2007), an example of which is shown in Fig. 13. In common with many jet sources, the μAPPJ operates with predominantly helium gas. This expensive approach is acceptable for some biomedical applications, but is unlikely to be so for food processing, where cost is an important consideration.

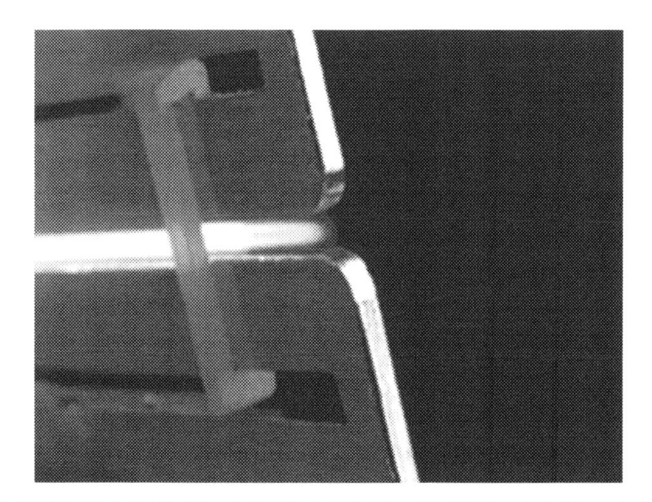

Fig. 13 A μAPPJ source (Schulz-von der Gathen et al., 2007). A dilute mixture of oxygen and helium flows through the channel, where the bright discharge region is clearly visible. The plasma in this case is excited by a radio-frequency voltage applied between the electrodes positioned on two sides of the channel. This source is operating as a classical glow discharge. With an electrode separation $d \approx 1$ mm, this is at the upper limit of the *pd* product where operation in this mode is possible. Both the plasma and the flowing gas are confined by between the electrodes by quartz plates. The effluent from the channel is rich in long-lived oxygen excited states.

6 Modeling Approaches

Our understanding of any physical phenomenon is expressed by our capacity to build models. Such a model is formulated as a mathematical structure, often in terms of partial differential equations. In the case of low-temperature plasma, there is not much doubt as to the nature of the fundamental equations. They are the Boltzmann equation, discussed in simplified form above, and the Maxwell equations describing the electric and magnetic fields that interact with the charged particles. In principle, we need a distinct realization of the Boltzmann equation to describe each species in the discharge, including both the charged and the neutral species. The problem here is that directly solving this system of equations is computationally intractable. The challenge, therefore, is to find simplifying approaches that preserve physical fidelity, while permitting a solution to be found with reasonable computational resources. The central importance of the electron energy distribution means that some solution of the Boltzmann equation for electrons is usually imperative, and sometimes this is extended to other charged species. The most common technique for solving the coupled system of charged particle Boltzmann equations and Maxwell's equations is the particle-in-cell method (Hockney and Eastwood, 1981; Birdsall and Langdon, 1991; Birdsall, 1991). This essentially is a Monte Carlo procedure, in which the large number of physical charged particles are represented by a much smaller number of so-called superparticles. This method is highly accurate, because the basic equations are solved without approximation, but computationally costly, even when limited to charged particles and simplified geometries. Consequently, a common approach is to simplify further, and describe both the charged and neutral particles using continuum equations. These are often called "moment" or "fluid" equations. Formally, one begins with the Boltzmann equation and proceeds to take velocity moments, which leads to a hierarchy of equations involving macroscopic quantities, such as density, momentum, and energy, which are functions only of spatial coordinates and time (Shkarofsky et al., 1966; Raizer, 2011; Cherrington, 2014). So velocity or energy space information is no longer involved, and the resulting equations are readily interpreted as conservation equations. For reasons already discussed, one has to account for the non-Maxwellian character of the electron energy distribution function, which is not now explicitly part of the formulation. The usual approach is to solve a continuity equation for the electron mean energy, and to use this mean energy as a parameter to select transport coefficients and rate constants calculated by solving the two-term approximation. In effect, one uses the equivalent of Fig. 3 to choose an effective value of E/N, and then the equivalent of Fig. 4 to select rate constants, and other parameters that depend on the electron energy distribution function. This hybrid approach (Kushner, 2009) is powerful, because complex chemistries and geometries can be treated with reasonable computational resources.

The downside is that the hybrid coupling of the electron Boltzmann equation with the continuum equations is ad hoc, and necessarily introduces errors that are hard to quantify. These errors are, however, probably at worst comparable with the uncertainty associated with chemical rate coefficients and other data (Turner et al., 2013; Turner, 2015). For discharges that are free of convoluted spatial structure, the hybrid approach permits rather detailed simulations using modest computational resources (Kelly and Turner, 2014a,b).

Presently more difficult is simulation of highly structured discharges such as the one shown in Fig. 12. The challenge is not only that strong spatial gradients are present in highly localized regions, but also that the position of these regions varies strongly in time, as individual filaments develop and decay. Moreover, some of these regions may contain spatial gradients so intense that the basic hybrid approach described above is no longer sufficient, and must be replaced by a more sophisticated approach. How this should best be done is a subject of present research (Li et al., 2012).

7 Summary

The basic concept of most low-temperature plasma applications involves producing energetic electrons by electrical excitation, and then allowing these electrons to transfer their energy to neutral gas atoms in processes that create chemically reactive species. These chemically active species are the main agents of the plasma process. Energetic electrons are readily produced for two reasons: First, because the electrons are charged and thus acted on by electric fields, and second, because the electron mass is small compared to neutral molecules, electrons do not easily lose energy in elastic collisions. Consequently, electrons are rather easily accelerated to energies (several electron volts) at which they can excite, dissociate, or ionize neutral molecules. The energy distribution of electrons is complicated, and not easily described by any simple analytic form. However, modern computational tools allow the distribution to be obtained by straightforward numerical calculation, which permits the rates of electron impact processes to be determined fairly accurately, at least for common gases for which the relevant cross-sections are known. The radical species produced by electron impact processes typically initiate more-or-less complex chains of plasma-chemical reactions, leading to usually a rather small number of relatively long-lived reactive species, which are the agents of the process.

In practice, this rather straightforward understanding of the process is complicated considerably, because the plasma always has a complex spatial structure. This means, for example, that the electron energy distribution function is not characterized simply by the average electric field in the gap, but is modified by the presence of space charges associated with the plasma. At low pressure, these effects are comparatively well understood. However, at atmospheric pressure, and especially in the streamer regime, the spatial structure of the discharge is so intricate that quantitative understanding is presently elusive, even if the general principles have been properly outlined. Moreover, particular applications may involve additional complications not usually considered in the classical discussions of gas discharge physics. These include interactions with liquids and with various types of living organisms. Research in these fields is at an early stage.

References

Antonov, I., Azyazov, V., Ufimtsev, N., 2003. Experimental and theoretical study of distribution of O_2 molecules over vibrational levels in $O_2(a^1\Delta_g)$-I mixture. J. Chem. Phys. 119 (20), 10638–10646. ISSN 0021-9606, 1089-7690.

Atkinson, R., Baulch, D., Cox, R., Crowley, J., Hampson, R., Hynes, R., Jenkin, M., Rossi, M., Troe, J., 2004. Evaluated kinetic and photochemical data for atmospheric chemistry: volume I—gas phase reactions of O_x, HO_x, NO_x and SO_x species. Atmos. Chem. Phys. 1680-7324. 4 (6), 1461–1738.

Azyazov, V., Mikheyev, P., Postell, D., Heaven, M., 2009. $O_2(a^1\Delta)$ quenching in the $O/O_2/O_3$ system. Chem. Phys. Lett. 00092614. 482 (1–3), 56–61.

Baulch, D., Cox, R., Hampson Jr., R., Kerr, J., Troe, J., Watson, R., 1984. Evaluated kinetic and photochemical data for atmospheric chemistry: supplement II : CODATA task group on gas phase chemical kinetics. J. Phys. Chem. Ref. Data 13 (4), 1259–1378.

Baurer, T., Bortner, M.H., 1978. Defense Nuclear Agency Reaction Rate Handbook. Revision Number 7, second ed. General Electric Company, New York. Technical Report DNA-1948H-Rev-7.

Becker, K., Schoenbach, K., Eden, J., 2006. Microplasmas and applications. J. Phys. D: Appl. Phys. 0022-3727. 39 (3), R55.

Birdsall, C., 1991. Particle-in-cell charged-particle simulations, plus Monte Carlo collisions with neutral atoms, PIC-MCC. IEEE Tran. Plasma Sci. 0093-3813. 19 (2), 65–85.

Birdsall, C., Langdon, A., 1991. Plasma Physics Via Computer Simulation. Adam Hilger, London, ISBN: 978-0-7503-0117-6.

Braginskiy, O., Vasilieva, A., Klopovskiy, K., Kovalev, A., Lopaev, D., Proshina, O., Rakhimova, T., Rakhimov, A., 2005. Singlet oxygen generation in O_2 flow excited by RF discharge: I. Homogeneous discharge mode: α-mode. J. Phys. D: Appl. Phys. 38 (19), 3609–3625. ISSN 0022-3727, 1361-6463.

Cherrington, B., 2014. Gaseous Electronics and Gas Lasers. Elsevier, New York, ISBN: 978-1-4832-7896-4.

Dilecce, G., Martini, L., Tosi, P., Scotoni, M., De Benedictis, S., 2015. Laser induced fluorescence in atmospheric pressure discharges. Plasma Sources Sci. Technol. 0963-0252. 24 (3), 034007.

Druyvesteyn, M., Penning, F., 1940. The mechanism of electrical discharges in gases of low pressure. Rev. Mod. Phys. 12 (2), 87.

Francis, G., 1956. The Glow Discharge at Low Pressure, vol. 4/22. Springer, Berlin, Heidelberg. pp. 53–208. ISBN 978-3-642-45849-1, 978-3-642-45847-7.

Große-Kreul, S., Hübner, S., Schneider, S., Ellerweg, D., von Keudell, A., Matejčík, S., Benedikt, J., 2015. Mass spectrometry of atmospheric pressure plasmas. Plasma Sources Sci. Technol. 24 (4), 044008. ISSN 0963-025.

Guzel-Seydim, Z., Greene, A., Seydim, A., 2004. Use of ozone in the food industry. LWT—Food Sci. Technol. 0023-6438. 37 (4), 453–460.

Hagelaar, G., Pitchford, L., 2005. Solving the Boltzmann equation to obtain electron transport coefficients and rate coefficients for fluid models. Plasma Sources Sci. Technol. 14 (4), 722–733. ISSN: 0963-0252, 1361-6595.

Hockney, R., Eastwood, J., 1981. Computer Simulation Using Particles. McGraw-Hill, New York, ISBN: 978-0-07-029108-9.

Hulburt, E., 1931. Atmospheric ionization by cosmic radiation. Phys. Rev. 37 (1), 1–8.

Iza, F., Kim, G., Lee, S., Lee, J., Walsh, J., Zhang, Y., Kong, M., 2008. Microplasmas: Sources, particle kinetics, and biomedical applications. Plasma Process. Polym. 1612-8869. 5 (4), 322–344.

Johnston, H., 1968. Gas phase reaction kinetics of neutral oxygen species. National Bureau of Standards, Technical Report NSRDS-NBS-20.

Kalogerakis, K., Copeland, R., Slanger, T., 2002. Collisional removal of $O_2(b^1\Sigma_g^+,\nu = 2,3)$. J. Chem. Phys. 116 (12), 4877–4885.

Kebabian, P., Freedman, A., 1997. Rare gas quenching of metastable O_2 at 295 K. J. Phys. Chem. A 1089-5639. 101 (42), 7765–7767.

Kelly, S., Turner, M., 2014a. Generation of reactive species by an atmospheric pressure plasma jet. Plasma Sources Sci. Technol. 0963-0252. 23 (6), 065013.

Kelly, S., Turner, M., 2014b. Power modulation in an atmospheric pressure plasma jet. Plasma Sources Sci. Technol. 0963-0252. 23 (6), 065012.

Kenner, R., Ogryzlo, E., 1982. A direct determination of the rate constant for the quenching of $O(^1S)$ by $O_2(a^1\Delta_g)$. J. Photochem. 0047-2670. 18 (4), 379–382.

Knickelbein, M., Marsh, K., Ulrich, O., Busch, G., 1987. Energy transfer kinetics of singlet molecular oxygen: the deactivation channel for $O_2(b^1\Sigma_u^+)$. J. Chem. Phys. 00219606. 87 (4), 2392.

Kogelschatz, U., 2003. Dielectric-barrier discharges: their history, discharge physics, and industrial applications. Plasma Chem. Plasma Process. 0272-4324. 23 (1), 1–46. WOS:000181061100001.

Kushner, M.J., 2009. Hybrid modelling of low temperature plasmas for fundamental investigations and equipment design. J. Phys. D: Appl. Phys. 42 (19), 194013. ISSN 0022-3727, 1361-6463.

Levatter, J., Lin, S., 1980. Necessary conditions for the homogeneous formation of pulsed avalanche discharges at high gas-pressures. J. Appl. Phys. 0021-8979. 51 (1), 210–222. WOS:A1980JF79800034.

Li, C., Teunissen, J., Nool, M., Hundsdorfer, W., Ebert, U., 2012. A comparison of 3D particle, fluid and hybrid simulations for negative streamers. Plasma Sources Sci. Technol. 0963-0252. 21 (5), 055019.

Lilenfeld, H., Carr, P., Hovis, F., 1984. Energy pooling reactions in the oxygen-iodine system. J. Chem. Phys. 0021-9606. 81 (12), 5730–5736. 1089-7690.

Llewellyn-Jones, F., 1966. Ionization and Breakdown in Gases. Methuen, London.

Marinov, D., Guerra, V., Guaitella, O., Booth, J.P., Antoine, R., 2013. Ozone kinetics in low-pressure discharges: vibrationally excited ozone and molecule formation on surfaces. Plasma Sources Sci. Technol. 22 (5), 055018. ISSN 0963-0252, 1361-6595.

Markosyan, A., Luque, A., Gordillo-Vázquez, F., Ebert, U., 2014. Pumpkin: a tool to find principal pathways in plasma chemical models. Comput. Phys. Commun. 0010-4655. 185 (10), 26972702.

Meek, J., 1940. A theory of spark discharge. Phys. Rev. 57 (8), 722–728.

Murakami, T., Niemi, K., Gans, T., O'Connell, D., Graham, W., 2013. Chemical kinetics and reactive species in atmospheric pressure helium-oxygen plasmas with humid-air impurities. Plasma Sources Sci. Technol. 0963-0252. 22 (1), 015003.

Palmer, A., 1974. Physical model on initiation of atmospheric-pressure glow discharges. Appl. Phys. Lett. 0003-6951. 25 (3), 138–140. A1974T569500006.

Pancheshnyi, S., Eisman, B., Hagelaar, G., Pitchford, L., 2015. Computer code ZDPlasKin. University of Toulouse, Toulouse, France.

Park, H., Slanger, T., 1994. $O_2(X, \nu=8–22)$ 300 K quenching rate coefficients for O_2 and N_2, and $O_2(X)$ vibrational distribution from 248 nm O_3 photodissociation. J. Chem. Phys. 100 (1), 287–300. ISSN 0021-9606, 1089-7690.

Pitchford, L., 2013. GEC plasma data exchange project. J. Phys. D: Appl. Phys. 0022-3727. 46 (33), 330301.

Pitchford, L., ONeil, S., Rumble, J., 1981. Extended Boltzmann analysis of electron swarm experiments. Phys. Rev. A 23 (1), 294–304.

Raether, H., 1964. Electron Avalanches and Breakdown in Gases. Butterworths, London.

Raizer, Y., 2011. Gas Discharge Physics. Springer, New York. ISBN: 978-3-642-64760-4.

Rawlins, W., Caledonia, G., Armstrong, R., 1987. Dynamics of vibrationally excited ozone formed by three-body recombination: II. Kinetics and mechanism. J. Chem. Phys. 00219606. 87 (9), 5209.

Sander, S., Freidl, R., Barker, J., Golden, D., Kurylo, M., Wine, P., Abbatt, J., Burkholder, J., Kolb, C., Moortgat, G., Huie, R., Orkin, V., 2011. Chemical kinetics and photochemical data for use in atmospheric studies. Jet Propulsion Laboratory. Technical Report JPL Publication 10-6.

Schofield, K., 1978. Rate constants for the gaseous interaction of $O(2^1D_2)$ and $O(2^1S_0)$—a critical evaluation. J. Photochem. 0047-2670. 9 (1), 55–68.

Schulz-von der Gathen, V., Buck, V., Gans, T., Knake, N., Niemi, K., Reuter, S., Schäper, L., Winter, J., 2007. Optical diagnostics of micro discharge jets. Contrib. Plasma Phys. 47 (7), 510–519. ISSN 08631042, 15213986.

Shkarofsky, I., Johnston, T., Bachynski, M., 1966. The Particle Kinetics of Plasmas. Addison-Wesley Pub. Co., Boston, MA.

Slanger, T., Black, G., 1978. Products of the $O(^1S)$-O_2 interaction. J. Chem. Phys. 68 (3), 998–1000. ISSN 0021-9606, 1089-7690.

Slanger, T., Black, G., 1981a. The product channels in the quenching of $O(^1S)$ by $O_2(a^1\Delta_g)$. J. Chem. Phys. 00219606. 75 (5), 2247.

Slanger, T., Black, G., 1981b. Quenching of $O(^1S)$ by $O_2(a^1\Delta_g)$. Geophys. Res. Lett. 1944-8007. 8 (5), 535–538.

Slanger, T., Copeland, R., 2003. Energetic oxygen in the upper atmosphere and the laboratory. Chem. Rev. 103 (12), 4731–4766.

Sobral, J., Takahashi, H., Abdu, M., Muralikrishna, P., Sahai, Y., Zamlutti, C., de Paula, E., Batista, P., 1993. Determination of the quenching rate of the $O(^1D)$ by $O(^3P)$ from rocket-borne optical (630 nm) and electron density data. J. Geophys. Res. Space Phys. 2156-2202. 98 (A5), 7791–7798.

Steinfeld, J., AdlerGolden, S., Gallagher, J., 1987. Critical survey of data on the spectroscopy and kinetics of ozone in the mesosphere and thermosphere. J. Phys. Chem. Ref. Data 16 (4), 911–951. ISSN 0047-2689, 1529-7845.

Turner, M., 2015. Uncertainty and error in complex plasma chemistry models. Plasma Sources Sci. Technol. 24 (3), 035027. ISSN 0963-0252, 1361-6595.

Turner, M., Derzsi, A., Donkó, Z., Eremin, D., Kelly, S., Lafleur, T., Mussenbrock, T., 2013. Simulation benchmarks for low-pressure plasmas: capacitive discharges. Phys. Plasmas 1070664X. 20 (1), 013507.

Van Gaens, W., Bogaerts, A., 2013. Kinetic modelling for an atmospheric pressure argon plasma jet in humid air. J. Phys. D: Appl. Phys. 0022-3727. 46 (27), 275201.

Wine, P., Nicovich, J., Thompson, R., Ravishankara, A., 1983. Kinetics of $O(^3P_j)$ reactions with hydrogen peroxide and ozone. J. Phys. Chem. 0022-3654. 87 (20), 3948–3954.

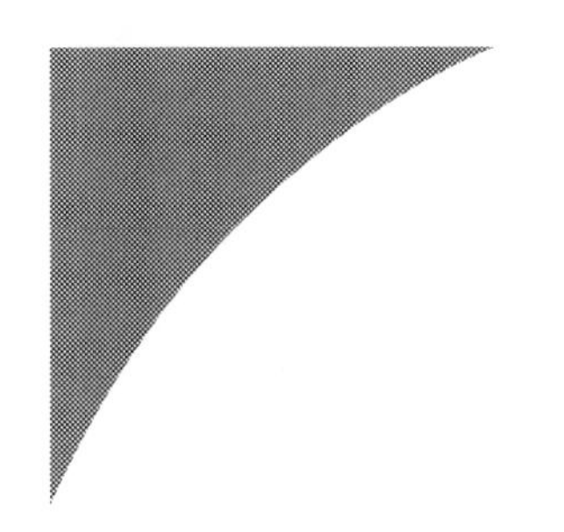

Chapter 3

The Chemistry of Cold Plasma

J. Christopher Whitehead
The University of Manchester, Manchester, United Kingdom

1 Introduction

Plasma technology is a multidisciplinary activity requiring expertise in electrical, chemical, and process engineering, as well as in physics and chemistry. The field of plasma chemistry is probably the most recent to be recognized separately. The book by McTaggart published in 1967 (McTaggart, 1967) can be taken as establishing the discipline; the current state of the field is exemplified in the definitive work by Fridman (2008). To assess the operation and applicability of cold plasma systems in areas such as food and agriculture, it is necessary to identify the range of reactive species that can be produced in a discharge and understand the mechanism of their formation and subsequent activity. This requires a wide range of tools to identify and quantify not just the end-products produced by a plasma but also the transient intermediates that lead to those end-products. This requires sophisticated chemical and spectroscopic instrumentation and leads to the development of robust reaction

Cold Plasma in Food and Agriculture. http://dx.doi.org/10.1016/B978-0-12-801365-6.00003-2

mechanisms that can be incorporated into kinetic simulations and modeling and validated against experimental results (Bogaerts et al., 2010).

The use of cold or nonthermal plasma gives a unique environment for initiating chemistry. In cold plasma, there can be a high degree of nonequilibrium among the various degrees of freedom. In particular, the heavy particles are less energetic than the electrons in the discharge where the heavy particles have kinetic temperatures close to ambient (~300 K), while the electrons are much hotter (~10^5 K). The internal degrees of freedom of the heavy particles—electronic, vibrational, and rotational—also will be much cooler than the electrons but may be somewhat hotter than the translational or kinetic temperature of the gas. The energetic electrons can initiate processes of ionization, excitation, and dissociation that could be achievable only in an equilibrium system at much higher temperatures, such as in flames and combustion, allowing chemistry characteristic of such systems to occur at much lower temperatures that are close to ambient. This gives rise to significant advantages in terms of applicability to systems where temperature stability might be a limitation and in engineering configurations where corrosion and other high temperature problems can cause issues and, in general, results in improved energy efficiency.

In this chapter, we will focus mainly on plasma operating at atmospheric pressure, as these are most likely to form the basis for systems being used in food technology and agriculture and because there are engineering benefits to be gained by operating at ambient pressures. Plasma generation at atmospheric pressure is of interest to the food industry because it does not require extreme process conditions (Misra et al., 2011). The chemical mechanisms that apply at atmospheric pressure can be related to the corresponding processes at reduced pressures but certain reaction steps change their relative importance as the pressure is increased.

A wide variety of cold plasma sources operate at atmospheric pressure, including dielectric barrier discharges (DBD) where a gas passes between two electrodes at least one of which is covered by a layer of dielectric material. Such discharges are characterized by the formation of microdischarges or streamers (Kogelschatz et al., 1999). Similar behavior is found in corona, packed-bed, and surface discharge reactors (Chang, 1993). These devices are driven by pulsed or AC power sources, and their performance can be controlled by the power supplied, its frequency and pulse shape.

Microwave and radio frequency excitation also can be used to generate a discharge generally using an electrode-less configuration (Laroussi and Akan, 2007; Niu et al., 2013). Gliding arc discharges also are finding increasing use for gas processing and are characterized by a warmer discharge than dielectric barrier and corona-like systems (Fridman et al., 1999). Details of nonthermal plasma sources can be found in Chapter 4 and elsewhere (Nijdam et al., 2012).

2 Collisional Processes in Plasma

Plasma is defined as a wholly or partly ionized gas and, as such, we can expect to find a range of charged and neutral species whose collisional behavior will provide the conditions that ultimately determine the chemistry of the plasma. In cold plasma, we usually deal with a partially ionized gas and the chemistry of ionized species does not generally play as important a role as do the neutral species. However, it is the electrons that are produced in the plasma and their subsequent collisions that create the reactive species that drive the plasma chemistry.

2.1 Primary Plasma Processes—Collisions of Electrons

As we have seen in the previous chapter, a high density of electrons is produced in the plasma ($n_e = 10^{11}$–10^{16} cm^{-3}) and these electrons then are accelerated in an electric field through the gaseous medium. This brings about a large number of collisions between the electrons and the atoms and molecules in the gas. The mean free path of the electron in an atmospheric pressure plasma is ~500 nm, meaning that it suffers many collisions when traveling between the typical spacing of ~1 mm between the electrodes in a DBD. These collisions may change the direction and energy of the electrons, but because of the disparity of mass between the electrons and the heavy particles virtually no momentum is transferred in the collisions, which is why the heavy particles stay close to ambient temperature.

A wide range of physical processes can occur when an electron collides with an atom or molecules. These include

Ionization $e + M \rightarrow M^+ + 2e$

Dissociative ionization $e + AB \rightarrow A^+ + B + 2e$

Electron attachment $e + M \rightarrow M^-$

Dissociative electron attachment $e + AB \rightarrow A^- + B$

Excitation $e + M \rightarrow M^* + e$

(This can be electronic excitation in the case of atoms and molecules and also vibrational and rotational excitation for molecules.)

Dissociation $e + AB \rightarrow A + B$

where either A or B, or both, could be excited electronically.

The probability of any of these processes occurring depends on the energy of the electron and the cross-section for the particular process. The electrons have a distribution of energies that is non-Boltzmann in the nonequilibrium conditions of the cold plasma, but it can be calculated using standard programs such as ELENDIF (Morgan and Penetrante, 1990) and BOLSIG (Hagelaar and Pitchford, 2005) and is related to the electron temperature that can be measured experimentally in many cases.

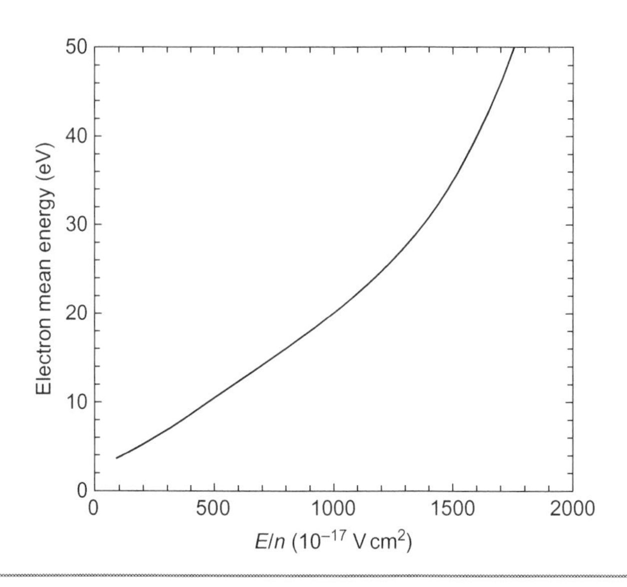

Fig. 1 Mean kinetic energy of the electrons in a discharge of atmospheric-pressure air as a function of the reduced electric field strength, E/n, where $E = V/d$. *(Reproduced with permission from Penetrante, B.M., Bardsley, J.N., Hsiao, M.C., 1997. Kinetic analysis of non-thermal plasmas used for pollution control. Jpn. J. Appl. Phys. 36 (Pt 1), 5007–5017. Copyright 1997 The Japan Society of Applied Physics.)*

A common parameter used to characterize the electron energy distribution function (EEDF) is the electron mean energy. Fig. 1 shows how this is related to the reduced electric field in the discharge ($= (V/d)/n$, where V is the discharge voltage, d is the separation between the electrodes and n the density of the gas in the discharge). The mean electron energy in a nonthermal plasma is generally in the range 2–5 eV, but in the high energy tail of the EEDF there can be electrons with energies in excess of 10 eV. The cross-sections are known for many common atomic and molecular species either through direct measurement of electron-scattering data and ion swarm measurements (Cho et al., 2013; Elford, 1981) or by theoretical calculation. In some cases, the energy range of these cross-sections does not extend to the lowest electron energies found in the plasma and some form of extrapolation or estimation is required.

The form of the EEDF is key to the identity and relative abundance of the species that are produced in the discharge and then initiate the various chemical processes that we will be interested in. It is a strong function of the composition, temperature, and pressure of the gas in the discharge which can be changed to tailor the EEDF to produce different outcomes. In addition, the nature of the discharge, for example, DBD or corona or microwave or gliding arc can affect the EEDF and the electron mean energies. It has been found that in some systems the outcome of the plasma processing is not a function of the nature of the discharge (Penetrante et al., 1995, 1996) but depends only on the specific input energy (SIE = plasma power/gas flow rate) deposited into the plasma. However, it is increasingly being recognized that the mechanical design and electrical engineering of the discharge can be used as parameters that can control the chemistry.

The contribution of different electron-impact processes as the mean electron energy is increased can be seen in Fig. 2 for an air plasma at atmospheric pressure (Penetrante et al., 1997). The electrical power that is supplied to the plasma goes mainly into vibrational excitation of N_2 at the lowest energy. As the electron energy is increased, we successively see dissociation of O_2 and then of N_2, followed by ionization of N_2 and O_2. At still higher energies, electronic excitation of molecular O_2 and N_2 also can occur. More details about the possible dissociation and ionization processes are shown in Fig. 3. The formation of both ionized and dissociated species can be seen clearly with the dissociation of O_2 occurring for lower electron energies than for N_2, reflecting the greater bond strength of molecular nitrogen compared to oxygen. Both positive and negative ions are seen in the case of oxygen. Dissociative electron attachment, which yields O_2^-, is a process that generally is observed only for species with high electron affinities such as halogen-, oxygen-, and sulfur-containing molecules. In addition, it also should be noted that some dissociation processes produce not just ground-state species but also electronically excited states, for

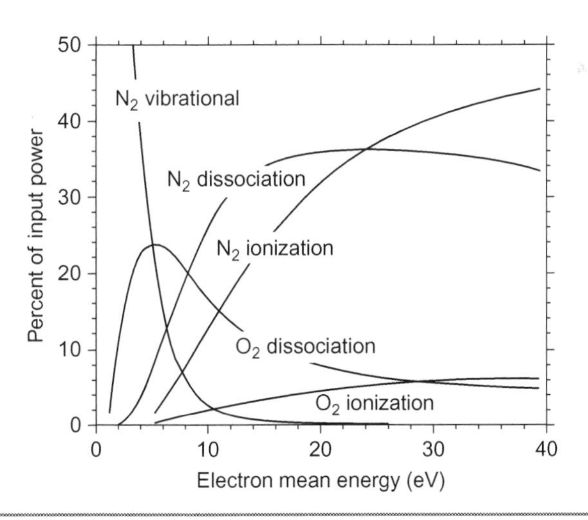

Fig. 2 Power dissipation in an atmospheric-pressure air plasma, showing the percentage of input power consumed in the electron-impact processes leading to vibrational excitation, dissociation, and ionization of O_2 and N_2. *(Reproduced with permission from Penetrante, B.M., Bardsley, J.N., Hsiao, M.C., 1997. Kinetic analysis of non-thermal plasmas used for pollution control. Jpn. J. Appl. Phys. 36 (Pt 1), 5007–5017. Copyright 1997 The Japan Society of Applied Physics.)*

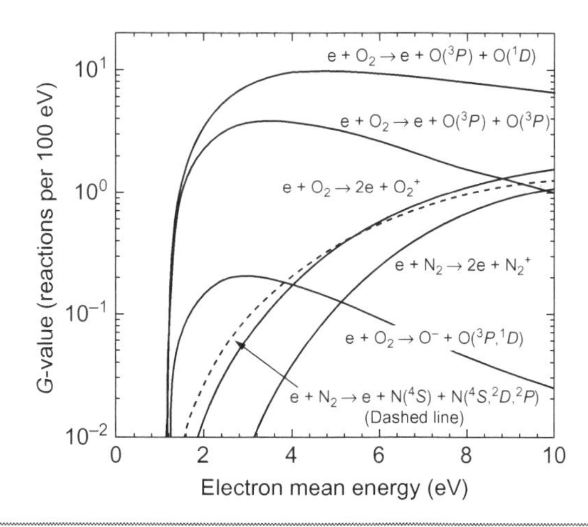

Fig. 3 *G*-values (no. of reactions per 100 eV of plasma input energy) for dissociation and ionization of oxygen and nitrogen in pure air, shown as a function of the electron mean energy in an atmospheric-pressure plasma. *(Reproduced with permission from Penetrante, B.M., Bardsley, J.N., Hsiao, M.C., 1997. Kinetic analysis of non-thermal plasmas used for pollution control. Jpn. J. Appl. Phys. 36 (Pt 1), 5007–5017. Copyright 1997 The Japan Society of Applied Physics.)*

example, $O(^1D)$. The plasma chemistry will be driven by these plasma-produced ions and neutral fragments and the reactivity of the fragments may be different depending on whether they are in their ground or an electronically excited state. These primary processes, directly initiated by electron collisions, then give rise to secondary collisions that involve the heavy particles that have been created by the electrons.

In addition to ions, radicals, and excited states, nonthermal plasma is a source of photons produced from the decay of electronically excited states. This can be of high energy in the vacuum ultraviolet. However, the photon flux from a typical, atmospheric pressure cold plasma is orders of magnitude too small to activate a catalytic material allowing photocatalysis to take place (Kim et al., 2015).

2.2 Secondary Plasma Processes—Collisions of Heavy Particles

It is the species formed by ionization, dissociation, and excitation in the electron collisions with the discharge gases that become the reagents in the next stage of the plasma chemistry. These species will be inherently more reactive than the untreated gases being fed to the discharge as collisions of ions with molecules, atoms with molecules, radicals with molecules and those involving excited states generally are characterized by lower activation energies than collisions solely between molecules, giving rise to large reaction rates. These species can be formed in the discharge by the supplied electrical energy without any heating of the gas. This nonthermal or cold plasma is a unique environment for initiating and promoting chemical reactions.

In addition to the species formed from oxygen and nitrogen, another very important species formed in both primary and secondary collision processes in the plasma is the hydroxyl radical, $OH^{\cdot}$. The $OH^{\cdot}$ radical is a very reactive, strongly oxidizing species that can abstract a hydrogen atom from a range of organic compounds (RH) to generate further radicals, $R^{\cdot}$

$$OH^{\cdot} + RH \rightarrow R^{\cdot} + H_2O \tag{1}$$

This high reactivity makes the hydroxyl radical a useful reagent in advanced oxidation processes for purifying wastewater where dissolved organic pollutants can be destroyed by a sequence of free radical chain reactions initiated and propagated by $OH^{\cdot}$. These pollutants include pesticides, dyes, pharmaceuticals, and nitrogen-containing species (Magureanu et al., 2015; Tijani et al., 2014). Many of these species result from food and agricultural production processes (Preis et al., 2013). $OH^{\cdot}$ also is classed as a reactive oxygen species that potentially can be damaging to living cells. This high reactivity means that its lifetime in an atmospheric discharge is very short (<100 μs; Hibert et al., 1999).

$OH^{\cdot}$ is commonly found in discharges containing water, such as humid air. It can be produced in a primary electron collision process involving dissociation of water

$$H_2O + e \rightarrow H^{\cdot} + OH^{\cdot} + e \tag{2}$$

or by secondary processes involving neutralization of ions

$$H_2O^+ + e \rightarrow H^{\cdot} + OH^{\cdot} + e \tag{3}$$

and by reactions of excited states of atomic oxygen, $O(^1D)$, and of molecular nitrogen, $N_2(A)$, formed in primary steps

$$\begin{aligned} H_2O + O(^1D) &\rightarrow OH^{\cdot} + OH^{\cdot} \\ H_2O + N_2(A) &\rightarrow H^{\cdot} + OH^{\cdot} + N_2 \end{aligned} \tag{4}$$

The electronically excited state of nitrogen, $N_2(A^3\Sigma_u^+)$, is an important species in plasma containing nitrogen, such as air. It is a metastable state with an energy of 6.224 eV and cannot return to the N_2 ground state by emitting a photon without violating electronic spin conservation. This gives it a long radiative lifetime ($\tau_0 = 2.0$ s) and makes it an important energy-containing species capable of dissociating molecules in the gas, in many cases, comparable to or more important than the contribution made by electron-induced dissociation. It can be formed directly by an electron collision with ground-state molecular nitrogen.

$$N_2\left(X^1\sum\nolimits_g^+\right) + e \rightarrow N_2\left(A^3\sum\nolimits_u^+\right) + e \tag{5}$$

Other metastable species (often called "dark" species as they cannot radiate because of spin conservation rules) also can be found in plasma. They have long lifetimes and act as energy reservoirs and can provide energy to enable reactive and dissociative collisions. Examples include atomic and molecular states of oxygen (O ^1D, $E=1.967$ eV, $\tau_0=150$ s: O_2 $a^1\Delta$, $E=0.977$ eV, $\tau_0=64.6$ min) and atomic nitrogen (N ^2D, $E=2.38$ eV, $\tau_0\sim 17$ h: N ^2P, $E=3.576$ eV, $\tau_0\sim 12$ s) (Golde, 1988; Golde and Moyle, 1985). In addition, inert gases such as argon often are added to plasma systems commonly to stabilize the discharge. However, it should be recognized that these gases are not inert in plasma as the discharge creates excited states (eg, Ar($^2P_{2,0}$) at 11.55 and 11.72 eV) that then can go on to collisionally ionize, dissociate or excite the other species in many cases.

As an example of other secondary reactions following electron excitation in a plasma, we can consider the formation of the oxides of nitrogen, NO, NO_2, N_2O, and N_2O_5 which can be formed in nitrogen discharges that also contain oxygen, such as air. The first step is the recombination of nitrogen and oxygen atoms formed in the primary electron collision steps

$$N + O + M \rightarrow NO + M \quad (6)$$

where M is any gas molecule. Recombination generally occurs only by a three-body recombination process of this type because the energy released when the fragments recombine (equivalent here to the NO bond energy) has to be removed to stabilize the product. This can be achieved only by the third body, M, taking away this excess energy. The rate of a three-body recombination process such as this ($=k$[N][O][M]) is then proportional to the concentration of M, which is equivalent to the total pressure of the system. This means that recombination processes will play a much more important role in the reaction mechanisms for plasma chemistry in atmospheric pressure discharges than for those operating at low pressure. In addition, NO can be formed by the reaction of excited state oxygen and nitrogen atoms with molecular nitrogen and oxygen, respectively,

$$O^* + N_2 \rightarrow NO + N \quad (7)$$

$$N^* + O_2 \rightarrow NO + O \quad (8)$$

It can be removed by reaction with nitrogen atoms

$$N^* + NO \rightarrow N_2 + O \quad (9)$$

or oxidized to NO_2 by a three-body recombination reaction with atomic oxygen

$$O + NO + M \rightarrow NO_2 + M \quad (10)$$

or by reaction with ozone, O_3, (whose formation in a discharge will be discussed later)

$$O_3 + NO \rightarrow NO_2 + O_2 \tag{11}$$

In turn, NO_2 can be returned to NO by reaction with an oxygen atom

$$O + NO_2 \rightarrow NO + O_2$$

NO_2 can recombine with an oxygen atom to form a radical species, NO_3, which can recombine with another NO_2 to form nitrogen pentoxide, N_2O_5 (Fitzsimmons et al., 1999)

$$NO_2 + O + M \rightarrow NO_3 + M \tag{12}$$

$$NO_3 + NO_2 + M \rightarrow N_2O_5 + M \tag{13}$$

NO_2 can react with a nitrogen atom to form nitrous oxide, N_2O,

$$NO_2 + N \rightarrow N_2O$$

which also can be formed by reactions of metastable nitrogen, $N_2(A^3\Sigma_u^+)$,

$$N_2\left(A^3\sum\nolimits_u^+\right) + O_2 \rightarrow N_2O + O \tag{14}$$

$$N_2\left(A^3\sum\nolimits_u^+\right) + O + M \rightarrow N_2O + M \tag{15}$$

Fig. 4 shows the time evolution of some oxides of nitrogen formed by a microwave plasma in an air mixture obtained using kinetic modeling of the discharge involving a mechanism of ~450 reactions to describe the relevant intermediate species (ions, atoms, radicals, and molecules), including excited states and the reaction products (Kossyi et al., 1992). The concentrations of the calculated species versus reaction time is plotted using a logarithmic-logarithmic format with the time scale for reaction spanning nine orders of magnitude from 1 μs to 1 ks. The mechanism is a complex set of competing concurrent and consecutive reactions and it can be seen that various species such as O, N, and NO appear early in the sequence, whereas N_2O_5 only appears at about 0.1 s. To describe plasma chemistry, even for a system as simple as air, requires a reaction mechanism involving many species, some created by discharge-produced electrons that much later generate short-lived intermediate species, forming part of a complex set of reactions that eventually yield a range of reaction products. These models can be validated against experiments where end-products are identified and quantified; in some cases, intermediates also are probed.

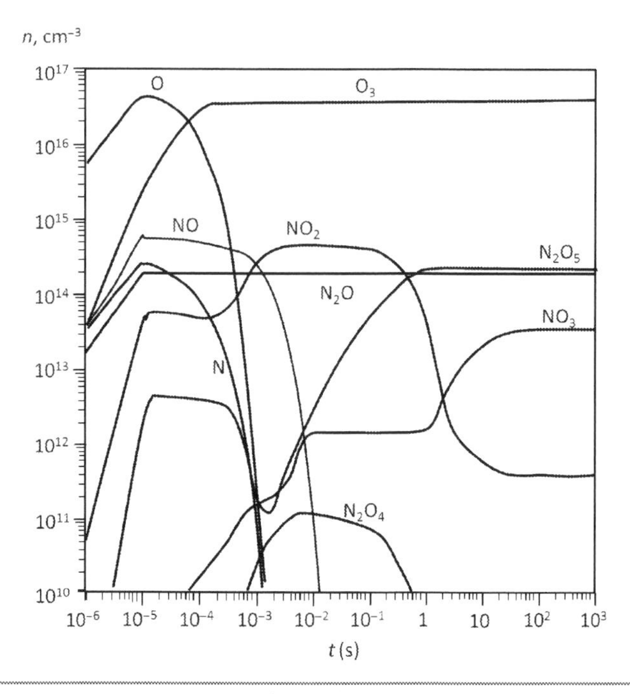

Fig. 4 Modeled time profile from 10^{-6} to 10^{3} s for the concentrations of major species (intermediates and products) in a microwave plasma discharge of a 4:1 N_2/O_2 mixture. *(Reproduced with permission from Kossyi, I.A., Kostinsky, A.Y., Matveyev, A.A., Silakov, V.P., 1992. Kinetic scheme of the non-equilibrium discharge in nitrogen-oxygen mixtures. Plasma Sources Sci. Technol. 1, 207–220. © IOP Publishing. All rights reserved.)*

3 Some Case Studies in Plasma Chemistry of Relevance to Food and Agriculture

To illustrate some of the points made above, we present three case studies involving applications relevant to the use of cold plasma in food and agriculture. These are the production of ozone from oxygen and air, the fixation of nitrogen, and the removal of volatile organic compounds (VOCs) by plasma.

3.1 The Plasma Chemistry of Ozone Formation

Ozone is an important oxidizing agent and has many biocidal uses. In terms of its oxidizing potential, ozone is second only to fluorine (2.07 c/f 2.87 V) and exceeds other oxidizing agents such as hydrogen peroxide, H_2O_2 (1.78 V), Cl_2 (1.36 V), O_2 (1.23 V), and nitric acid (0.94 V). Ozone is being used in the horticultural industry to remove ethylene to delay ripening and for antimicrobial purposes and postharvest treatment of fruit and vegetables. It can remove traces of pesticides from fruit and vegetables when used both in the gaseous state and when dissolved into water (Ebihara et al., 2013; Misra, 2015). Ozone also can be used for treating and sterilizing food packaging (Pankaj et al., 2014), and ozonated water also is used for soil treatment (Mitsugi et al., 2014).

The convenience of using ozone comes from the fact that it is a gas that is unstable with respect to oxygen and has a half-life of $\sim$1 day in clean dry air at room temperature; its lifetime rapidly decreases with increasing temperature or humidity. For this reason, it has to be prepared in situ and is not readily transportable. It is harmful in relatively low concentrations ($<$1 ppm) and is regarded as a particularly potent air pollutant. Although it is a very reactive substance and converts to oxygen, which is safe, care must be taken in its use.

Ozone can be produced from molecular oxygen either by ultraviolet light with a wavelength shorter than 200 nm or by an electrical discharge. In the first case, the O_2 is dissociated by absorption of a photon and in the second case by collision with an electron as described above. The production of ozone by plasma is a mature technology, and nonthermal plasma at atmospheric pressure has been used successfully for more than 150 years, following the work in 1857 by Siemens using a silent discharge (a form of DBD) to produce ozone from air or oxygen (Siemens, 1857). Ozone was used for the purification of municipal drinking water in Nice as early as 1906.

The chemistry in an oxygen plasma is relatively simple. The primary step is electron-impact dissociation of molecular oxygen to give two oxygen atoms

$$O_2 + e \rightarrow O + O \tag{16}$$

followed by recombination of an oxygen atom with an oxygen molecule in a three-body reaction involving an O_2 molecule as the third body, M,

$$O + O_2 + O_2 \rightarrow O_3 + O_2 \tag{17}$$

Two possible competitive processes will decrease the overall efficiency of ozone production:

$$O + O + M \rightarrow O_2 + M \tag{18}$$

$$O + O_3 + M \rightarrow 2O_2 + M \tag{19}$$

which give alternative reactions for the atomic oxygen, one of which also consumes the ozone, placing a limit on the amount of ozone that can be produced. Ions play a negligible role in the formation of ozone (Kogelschatz et al., 1988).

The formation of ozone is strongly affected by temperature because of thermal decomposition of ozone. This process can be represented (Benson and Axworthy, 1957) by the equilibrium

$$M + O_3 \leftrightarrows M + O_2 + O \tag{20}$$

and

$$O + O_3 \rightarrow 2O_2 \tag{21}$$

With increasing temperature, the equilibrium increasingly favors the decomposition of ozone. Compared to production at 24°C, ozone yields at 40°C are reduced by about a factor of two, whereas reducing the temperature to 4°C almost doubles the yield.

A second factor that affects the production of ozone in a discharge is the presence of water because of a catalytic cycle involving the OH radical that is formed in a primary process by electron dissociation of the water as discussed above

$$e + H_2O \rightarrow OH + H \tag{22}$$

The OH reacts rapidly with the ozone to give the peroxy radical, HO_2,

$$OH + O_3 \rightarrow HO_2 + O_2 \tag{23}$$

which can remove a second ozone molecule

$$HO_2 + O_3 \rightarrow OH + 2O_2 \tag{24}$$

The net effect of this cycle is conversion of ozone to oxygen

$$2O_3 \rightarrow 3O_2 \tag{25}$$

For this reason, high efficiency ozone generators are operated with oxygen (or air) that is dried and cooled. Additionally, commercial ozonizers are operated at high frequency where the operating voltage is lower for a given power. This gives less strain on the dielectric materials and allows for higher power densities giving higher ozone production rates and concentrations (Kogelschatz et al., 1999).

Ozone also can be produced by discharges in air, but the chemistry is more complicated because of the presence of nitrogen and the chemistry that the species created from it in the discharge can undergo. In particular, it can generate an additional source of atomic oxygen from the reactions of atomic and electronically excited metastable N_2^* created in primary electron collision processes in the discharge, as described in Section 2.2,

$$N + O_2 \rightarrow NO + O \tag{26}$$

$$N_2^* + O_2 \rightarrow N_2O + O \tag{27}$$

$$N_2^* + O_2 \rightarrow N_2 + 2O \tag{28}$$

It is estimated that about half of the ozone formed in air discharges comes from these processes (Kogelschatz et al., 1999).

Inspection of Fig. 2 shows that at lower electron energies, O_2 dissociation dominates over N_2 dissociation. However, raising the electron energy can favor the dissociation of N_2. The electron energy can be controlled by the choice of plasma operating parameters, in particular the electric field strength (V/d). As previously discussed, this can be changed within limits by modifying the design and construction of the plasma reactor and varies inherently between different types of cold plasma sources such as radio frequency, microwave, corona, and DBD.

Under conditions of high NO_x concentrations (>0.1%; Fridman, 2008), it is possible to inhibit ozone production altogether in an air-fed ozonizer and obtain only NO, NO_2, and N_2O. This is called discharge poisoning and results from a catalyzed decomposition cycle of ozone by NO

$$NO + O_3 \rightarrow NO_2 + O_2 \tag{29}$$

$$NO_2 + O \rightarrow NO + O_2 \tag{30}$$

This has the net effect of removing both ozone and atomic oxygen

$$O_3 + O \rightarrow 2O_2 \tag{31}$$

Interestingly, this mechanism also is responsible for the depletion of ozone by nitrogen oxides in the stratospheric ozone layer (Kogelschatz et al., 1999).

An interesting application of the chemical mechanism outlined above can be seen in the kinetic modeling of NO and ozone densities produced by an RF plasma torch in argon containing 2% air (Van Gaens et al., 2014). These microplasma torches where the plasma exiting from the torch can be applied to a surface can be used for a range of sterilization applications and their output contains a range of oxidizing species such as O, NO, and O_3 and also HO_2 and OH when the source uses humid air or when the output gas stream interacts with moisture in the surrounding air. Fig. 5 shows the contribution made by different chemical reactions to the evolution of atomic oxygen as a function of distance from the nozzle. This shows the current sophistication of kinetic modeling for plasma processes where the evolution of species in three dimensions in complex geometries can be simulated, including the thermodynamics, heat transfer, and fluid dynamics of the gas flow. An example of a commonly used package for this modeling is PLASIMO from the Technical University of Eindhoven (van Dijk et al., 2009) and COMSOL (COMSOL, n.d.).

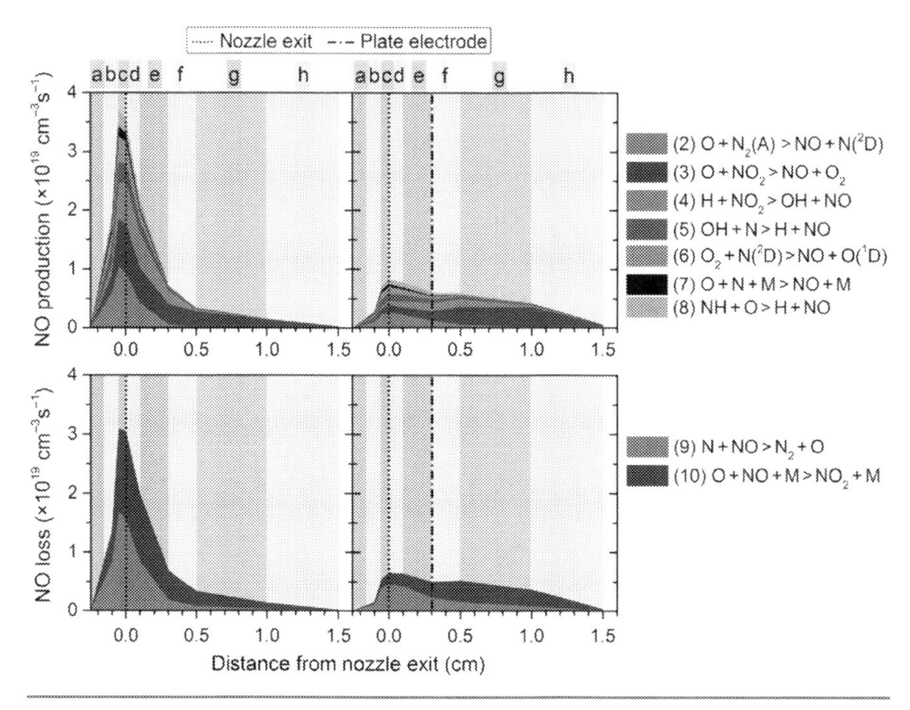

Fig. 5 The contribution of the most important chemical production and loss processes for atomic oxygen, O, throughout the plasma jet, for two different configurations. *(Reproduced with permission from Van Gaens, W., Bruggeman, P.J., Bogaerts, A. 2014. Numerical analysis of the NO and O generation mechanism in a needle-type plasma jet. New J. Phys. 16, 060354. © IOP Publishing. All rights reserved.)*

3.2 Nitrogen Fixation by Cold Plasma

Nitrogen produced from air in the form of nitrogen oxides, nitrates, and with hydrogen to give ammonia is a vital source of nutrients for plants applied in the form of liquid or solid fertilizers. Because of the very large bond strength of nitrogen (945 kJ mol^{-1} or 9.79 eV), the production of ammonia and nitric acid involving molecular nitrogen is extremely energy intensive. Ammonia is made industrially by the Haber-Bosch process at pressures of several hundred atmospheres and temperatures of 400–600°C

$$N_2 + 3H_2 \rightarrow 3NH_3 \tag{32}$$

Ammonia can be converted into nitric acid by the Ostwald process by first burning it in oxygen to produce NO, which is then oxidized to NO_2, and finally dissolved in water to produce nitric acid

$$4NH_3 + 5O_2 \rightarrow 4NO + 6H_2O \tag{33}$$

$$2NO + O_2 \rightarrow 2NO_2 \tag{34}$$

$$3NO_2 + H_2O(l) \rightarrow 2HNO_3(aq) + NO \tag{35}$$

Considerable research effort is being applied to search for more sustainable and energy-efficient methods for nitrogen fixation, and cold plasma is currently being assessed as a promising technology (Cherkasov et al., 2015; Patil et al., 2015b). One advantage that plasma may have is that it could provide a localized, low-volume facility situated at the point of use, drawing its power from renewable sources, and therefore reducing the costs associated with the transportation of fertilizers.

3.2.1 The Plasma Production of Nitrogen Oxides and Nitric Acid

Although the overall formation of nitric oxide from nitrogen and oxygen

$$N_2 + O_2 \rightarrow 2NO \tag{36}$$

is endothermic by 90 kJ mol^{-1} or ~1 eV, we have seen in the preceding Sections 2.2 and 3.1 that it is possible to produce the oxides of nitrogen, NO and NO_2, using cold plasma in air or other mixtures of N_2 and O_2 under conditions of higher electron energy that favor the dissociation of nitrogen over oxygen (see Fig. 2) and hence the formation of nitrogen oxides over ozone. In addition to forming nitric acid by dissolving NO_2 in water as in reaction (35), it also is possible using humidified gases to directly produce nitric acid in the gas phase from the reaction of OH radicals with NO_2

$$OH + NO_2 \rightarrow HNO_3 \tag{37}$$

For many years, researchers have attempted to improve the yield of the nitrogen oxides by combining cold plasma with a catalyst. Plasma activation of a catalyst in plasma processing can sometimes be a synergistic arrangement in which the plasma creates species in the gas phase at low temperature that would be possible on the catalytic surface in conventional thermal catalysis only at much higher temperature. Cavadias and Amouroux (1986) and Gicquel et al. (1986) investigated the use of WoO_3 as a catalyst with a low-pressure radio frequency discharge for the production of NO and found that the overall extent of nitrogen fixation was increased from 8% to 19%. Mutel et al. (1984) also used a low-pressure microwave discharge with a MoO_3 catalyst. The plasma-catalysis mechanism involves the creation of vibrationally excited nitrogen molecules in the plasma (see Fig. 2), which are more efficiently adsorbed onto the surface of the WoO_3 catalyst and then react with free oxygen atoms from the catalyst. This process can be operated at atmospheric pressure using a catalyst of WoO_3/Al_2O_3 (Hessel et al., 2013).

NO also has been produced directly in the gas phase from vibrationally excited nitrogen, $N_2^\dagger$, formed (Azizov et al., 1980) in an electron cyclotron resonance (ECR)-microwave discharge of a nitrogen-oxygen mixture in which molecular nitrogen is excited by collisions with electrons at reduced pressure and low temperatures. This then reacts with the atomic oxygen in the plasma to form nitric oxide, NO,

$$N_2^\dagger + O \rightarrow NO \qquad (38)$$

A recent gas-phase study at atmospheric pressure has used a millimeter-scale gliding arc discharge (Patil et al., 2015a) and investigated the effect of the ratio of nitrogen and oxygen in the gas feed upon the yield of NO and NO_2. The study concluded that the relative yields of the various nitrogen oxides can be controlled by the plasma conditions and the gas composition.

Despite much research activity, the energy consumption of cold-plasma production of nitric acid is not yet competitive with the Ostwald process (Hessel et al., 2013). For the fixation of nitrogen in the form of nitrogen oxides, the energy consumption for nonthermal plasma exceeds that for the industrial Haber-Bosch process by a factor of at least 2–3 and the plasma processes have small conversions to nitrogen oxides ($<10\%$ vol.) (Patil et al., 2015a).

3.2.2 Ammonia Production by Nonthermal Plasma

The direct formation of ammonia from molecular nitrogen and hydrogen via reaction (32) is an exothermic reaction (-46 kJ mol^{-1}), but the rate of reaction is immeasurably slow. The reaction proceeds with reasonable yield only at very high pressures and with the use of an iron catalyst. Several groups have demonstrated that it is possible to produce ammonia using cold plasma from a gas stream of N_2 and H_2 at normal pressure and temperature but generally with energy consumption rates that are too high for commercial implementation (Patil et al., 2015b). Patil et al. (2015b) recently tabulated a summary of the investigations undertaken so far. Many of these involve the use of a catalyst to improve the yield of ammonia and the selectivity toward its production.

In a study of the formation of ammonia from N_2 and H_2 in a microwave plasma performed by Nakajima and Sekiguchi (2008) at relatively high gas temperatures (790–1240 K), it was suggested that the mechanism of ammonia formation begins in the primary electron process with the formation of excited nitrogen atoms, $N(^2D)$, which react with molecular hydrogen to form a NH radical

$$N(^2D) + H_2 \rightarrow NH + H \qquad (39)$$

which then goes on to form NH_3 by successive reactions with H or H_2. An alternative mechanism for the formation of NH comes from the reaction of ionized molecular nitrogen, $N_2{}^+$, with H_2

$$N_2^+ + H_2 \rightarrow N_2H^+ + H \quad (40)$$

followed by neutralization of the N_2H^+ ion

$$N_2H^+ + e \rightarrow NH + H \quad (41)$$

The N_2^+ ion can be formed directly by an ionizing electron collision with N_2 or by the reaction of two metastable electronically excited nitrogen molecules formed in a primary electron process

$$N_2^* + N_2^* \rightarrow N_2^+ + N_2 \quad (42)$$

The authors suggest that only undischarged molecular hydrogen is necessary to produce ammonia and that active nitrogen species, in particular $N(^2D)$, are key for the formation of NH and ultimately NH_3. This may explain their observation that the addition of hydrogen downstream of the discharge is more effective by a factor of 20 in forming ammonia than mixing it with nitrogen before discharging. At more moderate conditions than for the microwave experiment (80–155°C), Bai et al. (2003) used a dielectric barrier microdischarge to produce ammonia with a yield of 12.5%. They proposed that hydrogen atoms, as well as N_2^+ and N atoms, are necessary to form NH_3 from NH radicals adding the following processes to reactions (39) and (41)

$$N + H \rightarrow NH \quad (43)$$

$$NH + H \rightarrow NH_2 \quad (44)$$

$$NH + H_2 \rightarrow NH_3 \quad (45)$$

$$NH + 2H \rightarrow NH_3 \quad (46)$$

$$NH_2 + H \rightarrow NH_3 \quad (47)$$

van Helden et al. (2007) used laser spectroscopy to detect both NH and NH_2 radicals in an expanding thermal plasma that was either of N_2 to which H_2 was added downstream of the discharge or of N_2 and H_2 discharged together. NH radicals were observed in both configurations, but NH_2 was found only for the case where H_2 was added downstream. They conclude that in the expanding discharge of N_2 and H_2, NH is formed by reactions of N atoms with H_2 (reaction 39) and also a reaction of N atoms with NH_2

$$NH_2 + N \rightarrow NH + NH \quad (48)$$

where the NH_2 is formed from

$$H + NH_3 \rightarrow NH_2 + H_2 \quad (49)$$

Reaction (49) is likely to be absent when hydrogen is injected downstream of the nitrogen discharge with the absence of H atoms explaining the absence of NH_2. The consensus is that the NH radical is the precursor of NH_3, formed by reactions between NH and H_2, but ammonia can be decomposed by H atoms in the plasma.

More promising in terms of efficiency are the reports of the production of ammonia at atmospheric pressure from nitrogen and hydrogen using the combination of plasma and a catalyst. A range of catalysts, mainly metals (eg, Mo, Fe, Pt, Al, Ag, Ru), have been used (Patil et al., 2015b). The catalytic mechanism involved in plasma-catalysis differs from that occurring in the conventional thermal catalytic Haber-Bosch process where N_2 and H_2 are dissociatively adsorbed onto the surface of a Fe catalyst as N and H atoms and recombination and reaction then occurs on the surface until NH_3 is formed and desorbs (Somorjai, 1994). Fig. 6 shows a schematic view of the formation of ammonia from nitrogen and hydrogen by plasma-catalysis using a Ru-alumina membrane catalyst in a DBD at atmospheric pressure (Mizushima et al., 2004, 2007). Plasma-excited molecular nitrogen adsorbs dissociatively onto the alumina substrate, as can excited H_2. NH_3 then can be formed from surface reactions producing NH and ultimately NH_3. In an additional process, H_2 can be dissociatively adsorbed onto the Ru as H atoms that then can spill over onto the alumina and react rapidly with N atoms on the surface. Mizushima et al. (2004) found that depositing Ru onto the alumina significantly increases the conversion to ammonia as well as increasing the overall energy efficiency by providing an accelerated route to ammonia formation through the nonplasma dissociation of hydrogen on the ruthenium.

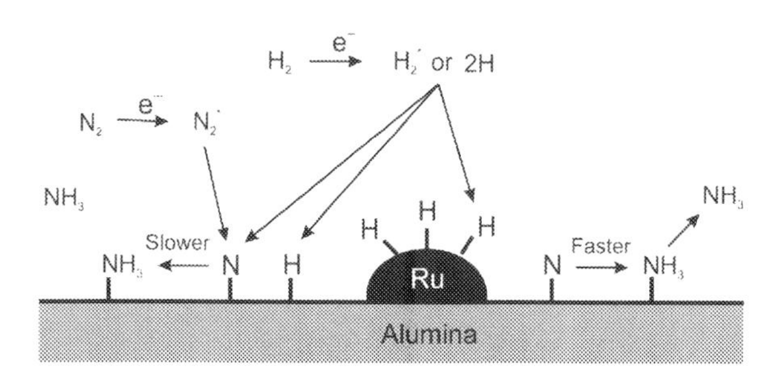

Fig. 6 A schematic account of the plasma-catalytic processes occurring on a Ru/Al_2O_3 membrane catalyst during the synthesis of ammonia from a N_2/H_2 dielectric barrier discharge. *(Reproduced with permission from Mizushima, T., Matsumoto, K., Ohkita, H., Kakuta, N., 2007. Catalytic effects of metal-loaded membrane-like alumina tubes on ammonia synthesis in atmospheric pressure plasma by dielectric barrier discharge. Plasma Chem. Plasma Process. 27, 1–11 with permission from Springer Publishing and Business Media.)*

The University of Minnesota has reported a scheme that uses electricity derived from wind power to generate ammonia from nitrogen and hydrogen in a DBD together with a Ru catalyst under ambient conditions giving a 6% yield (Deng et al., 2007).

Ammonia also has been produced directly in cold plasma of CH_4/N_2 mixtures. Methane is an attractive source of hydrogen because it is an abundant gas with a high greenhouse warming potential. It usually is converted into hydrogen by steam, reforming before reacting with nitrogen in the Haber-Bosch process. The plasma route to ammonia reduces these two steps into one. Fig. 7 shows a schematic of the mechanism of ammonia production deduced from a study of the conversion of an atmospheric pressure stream of dilute methane in nitrogen in a ferroelectric packed-bed plasma reactor (Pringle et al., 2004). It differs from the mechanism for the formation of ammonia from N_2 and H_2 only by the production of H atoms from the dissociation of methane in the plasma either in a primary process by electron collisions or by an abstraction reaction with electronically excited atomic nitrogen formed by primary electron dissociation of N_2

$$e + CH_4 \rightarrow CH_3 + H + e \tag{50}$$

$$N^* + CH_4 \rightarrow CH_3 + H + N \tag{51}$$

Bai et al. (2008) also have studied this CH_4/N_2 system in their microdischarge dielectric barrier system as a function of a range of plasma parameters. In their system, they have identified the role played by the ions N^+ and N_2^+

$$N_2^+ + CH_4 \rightarrow N_2^+H + CH_3 \tag{52}$$

$$N_2^+ + CH_3 \rightarrow N_2^+H + CH_2 \tag{53}$$

$$N^+ + CH_4 \rightarrow CH_3^+ + NH \tag{54}$$

Followed by reaction (41). The yield of ammonia is 9.1% (v/v).

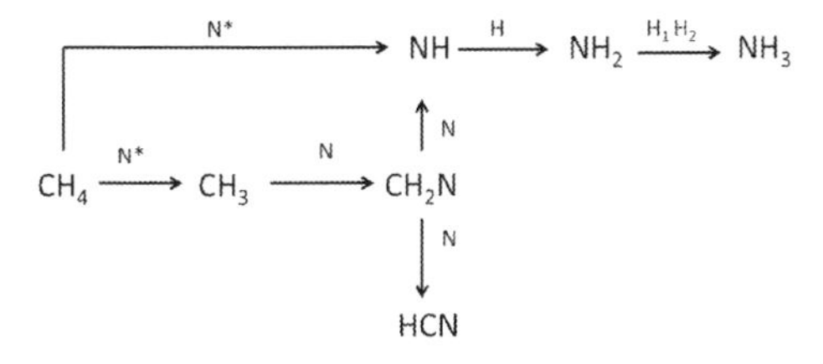

Fig. 7 The chemical mechanism for the production of ammonia by a plasma discharge of methane in nitrogen at atmospheric pressure. *(Reproduced with permission from Pringle, K.J., Whitehead J. C., Wilman J. J., Wu J. 2004. The chemistry of methane remediation by a non-thermal atmospheric pressure plasma. Plasma Chem. Plasma Process. 24, 421–434, with permission from Springer Publishing and Business Media.)*

Another nitrogen-fixing reaction that can be observed in the processing of methane with nitrogen is the formation of hydrogen cyanide, HCN, as is shown in Fig. 7. Rapakoulias and Amouroux (1979, 1980) have investigated the mechanism of HCN formation from CH_4 and N_2 in a low-pressure RF discharge. They find that the key reagent for the formation of HCN is electronically excited N_2^{*+} formed in high vibrational levels in the primary electron collisions. It reacts with CH_4 to ultimately form HCN

$$N_2^{*+} + CH_4 \rightarrow \cdots\cdots \rightarrow HCN + N_2 \tag{55}$$

N_2^* formed in the plasma without vibrational excitation reacts differently with methane to give ethyne, C_2H_2,

$$N_2^* + CH_4 \rightarrow \cdots\cdots \rightarrow C_2H_2 + H_2 + N_2 \tag{56}$$

which is a process with a lower energy threshold than (55) and which can proceed without the extra vibrational excitation of N_2^*. The extent of vibrational excitation of N_2^* can be controlled by the SIE to the plasma, where increasing the SIE will increase the vibrational excitation and the yield of HCN. The use of a metal catalyst (Fe, Cu, W, Ta, or Mo) increases the yield of HCN because chemisorption of N_2^* onto the metal increases its degree of vibrational excitation.

3.3 Cold Plasma Treatment of VOCs

VOCs are common components in the waste from a wide range of processing. Cold-plasma technology is increasingly being employed as a solution for their abatement, particularly using the combination of plasma with catalyst (Whitehead, 2010, 2014). Gases that are emitted from food processing (eg, cooking, drying, and smoking) or from the decay of stored materials may represent health hazards but most commonly are a concern because of the associated odors that are unacceptable in a modern society (Preis et al., 2013). Provision for the elimination of odors has to be incorporated into new processes and environmental health officials can require existing processes to be retrofitted. Most odor treatment methods (with the exception of masking) work either by completely removing the odorous molecules from the gas stream or by converting them into significantly less odorous species that can be emitted in their place.

As an example, we examine the removal of methyl mercaptan, CH_3SH, and dimethyl sulfide, $(CH_3)_2S$, in waste gas streams containing nitrogen or air using a ferroelectric packed-bed plasma reactor (Ahmad et al., 2003). These substances are emitted from a variety of sources, including sewage processing, landfill off-gas, food processing and preparation, animal husbandry, and wood pulp and paper production. The odor detection thresholds for methyl mercaptan and dimethyl sulfide are ~0.001 and 0.02 ppm respectively, much lower than other sulfides such as CS_2 (0.121 ppm) and SO_2 (4.4 ppm) (Anon., 1989). Atmospheric pressure cold plasma technology has considerable advantages to offer in the field of odor control as it is well known (Huang et al., 2001; Krasnoperov et al., 1997; Snyder and Anderson, 1998; Tonkyn et al., 1996; Yamamoto et al., 1992) that plasma destruction becomes significantly more efficient as the concentration of the added species decreases.

Fig. 8 shows a schematic of the mechanism for the removal of methyl mercaptan in nitrogen or air. The main products in pure nitrogen are H_2S, HCN and NH_3,while in the presence of oxygen these products are not formed and SO_2, CH_2O, CO, and CO_2 are the oxidation end products. In pure nitrogen, the important species created in the discharge are electronically excited nitrogen atoms, $N(^2D)$, and metastable triplet molecular nitrogen, $N_2(A^3\Sigma_u^+)$ and the mechanism parallels that for methane

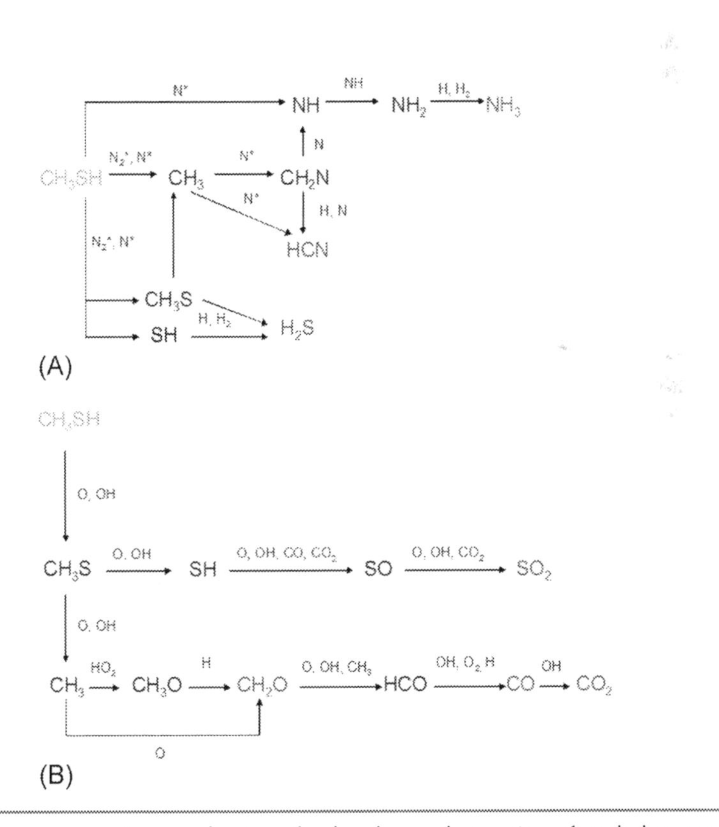

Fig. 8 Schematic reaction mechanisms for the plasma destruction of methyl mercaptan in (A) nitrogen and (B) air. *(Reproduced with permission from Ahmad, I.K., Wallis, A.E., Whitehead, J.C., 2003. The plasma destruction of odorous molecules: organosulphur compounds. High Temp. Mater. Process. 7, 487–499, with permission from De Gruyter.)*

described in the previous section (Fig. 7). In the presence of oxygen, the excited state nitrogen atoms react with O_2 to form NO that is further oxidized to NO_2. The metastable *A* state of nitrogen also reacts much more rapidly with O_2 than with CH_3SH because of the much larger concentration of O_2. The chemistry involving the nitrogen species formed in the discharge is the source of the NO_x that is produced in the air discharges and is largely decoupled from the oxidative chemistry responsible for the destruction of CH_3S. The major reactive species in the air discharge that is responsible for the destruction of CH_3S is atomic oxygen formed in the primary electron collision step, which initially reacts with the mercaptan by abstracting a hydrogen atom generating CH_3S, which is the source of the sulfur-containing end products, and OH radicals, which is more reactive than O atoms and therefore rapidly take over the oxidation process and are regenerated. Oxidation of CH_3S to methyl radicals, CH_3, ultimately yields CO_2 and CO via CH_2O (Fig. 8B).

In general, the initial step in the oxidation of a VOC molecule, RH, by a cold plasma, involves the breaking of the RH bond by an oxygen atom abstracting a H atom

$$O + RH \rightarrow R + OH \tag{57}$$

The OH radical is considerably more reactive than O and takes over the decomposition of the organic molecule

$$OH + RH \rightarrow R + H_2O \tag{58}$$

An additional source of OH can come from electron dissociation of H_2O if there is moisture in the air stream as may commonly occur in waste gases from food processing. In some cases, the primary step might involve dissociation by electron collision, especially when the organic molecule has a large electron affinity

$$e + RH \rightarrow R + H + e \tag{59}$$

or by collision with metastable electronically excited N_2^*

$$N_2^* + RH \rightarrow R + H + N_2^* \tag{60}$$

Collisions with electrons and N_2^* also can result in the dissociation of different bonds in RH, giving a range of possible reaction pathways. Hydrogen atom abstraction from RH also can occur in reaction with N atoms to give NH and R, although the dominant pathways for reactions of N atoms and N_2^* leads to the production of nitrogen oxides, NO and NO_2.

The organic radical, R, combines with O_2 to form an organic peroxy radical

$$R + O_2 + M \rightarrow RO_2 + M \tag{61}$$

The peroxide radical RO_2 is then able to react with another VOC molecule, RH, forming organic peroxide, RO_2H, and propagating a chain mechanism for the oxidation of RH

$$RO_2 + RH \rightarrow RO_2H + R \tag{62}$$

The peroxides RO_2 and RO_2H are further oxidized to CO_2 and H_2O with or without the additional consumption of OH and other plasma-generated oxidizers, giving complete oxidation of the VOC.

As a further example, we show in Fig. 9 the outline mechanism for the cold plasma oxidation of dichloromethane, CH_2Cl_2, a common solvent (Fitzsimmons et al., 2000). In addition to the abstraction of a hydrogen atom in the initial step,

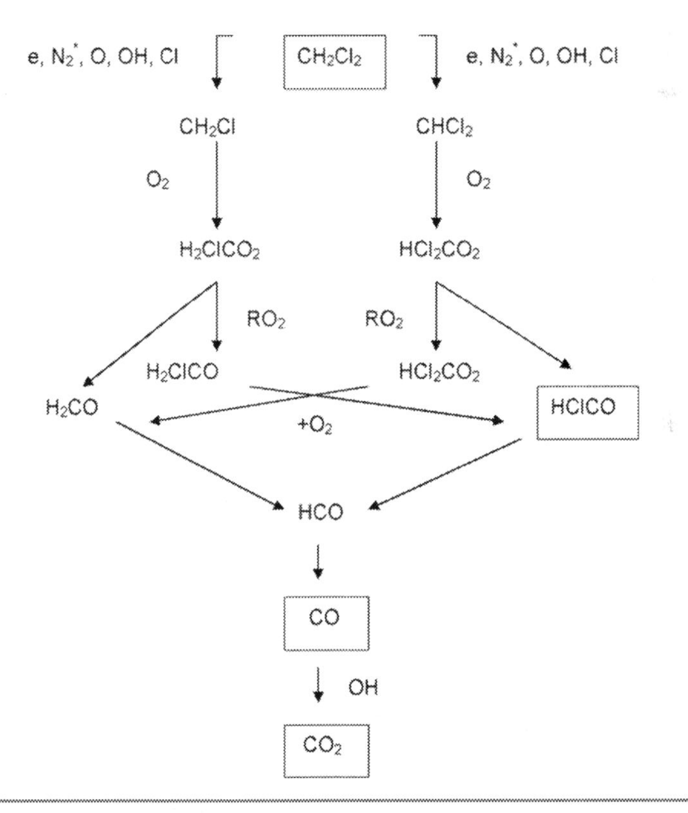

Fig. 9 Schematic reaction mechanisms for the plasma destruction of dichloromethane in air. The detected species are enclosed in textboxes. *(Reproduced with permission from Fitzsimmons, C., Ismail, F., Whitehead, J.C., Wilman, J.J., 2000. The chemistry of dichloromethane destruction in atmospheric pressure gas streams by a dielectric packed-bed plasma reactor. J. Phys. Chem. A 104, 6032–6038, with permission from the American Chemical Society.)*

it is possible that abstraction of a chlorine atom can occur, giving two possible peroxy radicals, H_2ClCO_2 or HCl_2CO_2. As the reaction intermediate, HClCO, is observed, it can be deduced that H-atom abstraction represents the dominant pathway. Atomic chlorine can be generated in the processing by a collision of N_2^* with CH_2Cl_2

$$N_2^* + CH_2Cl_2 \rightarrow CH_2Cl + Cl + N_2 \tag{63}$$

and is very effective at abstracting a H-atom from a CH_2Cl_2 molecule. If the ratio of oxygen to nitrogen is altered, the maximum destruction of dichloromethane is obtained with ~1–3% O_2, indicating that variation of the composition of the treated gas can be an important process variable. It also has been found that adding a catalyst such as TiO_2 or a zeolite into the discharge region can increase the destruction of dichloromethane by up to 33% and also can minimize the NO_x formation that is always present when an air stream is treated by plasma (Wallis et al., 2007).

Metal oxide catalysts are found to be very effective in a plasma-catalyst environment for the removal of a wide range of VOCs. Among the most commonly used are TiO_2 and MnO_2. An effective arrangement for the VOC destruction involves the adsorption of the VOC onto the oxide surface, which then is treated by a stream of ozone generated by cold plasma. The ozone dissociates on the oxide surface creating active oxygen species, commonly atomic oxygen, which then initiate the decomposition of the VOC (Barakat et al., 2014).

4 Concluding Remarks

From the consideration of the mechanisms described in this chapter, we can draw a few generic points of guidance:

- When working at atmospheric pressure in plasma containing a relatively dilute VOC in air or a mixture of air and nitrogen, a good starting point for an appropriate mechanism can be drawn from mechanisms developed by the atmospheric chemistry community for the polluting effect of the VOC in tropospheric air. The major difference between processes in the troposphere and in plasma is the additional contribution from the dissociated and excited nitrogen that is created in the plasma. This will introduce an additional source of nitrogen oxides that may have the potential to react with the organic species (Futamura et al., 2000; Harling et al., 2005).
- In an oxygen-nitrogen mixture at atmospheric pressure, the most common and hence most important recombination processes involving radicals will be the three-body recombination with O_2, which yield a peroxy radical, RO_2. This is because the rate of the three-body recombination process ($=k[R][O_2][M]$) will be large because of the concentration of oxygen, $[O_2]$, which is up to 20% when using atmospheric air.

- Overall, the rates of radical-radical recombination reactions generally will be small because the steady state concentrations of radicals (and atoms) in a typical plasma are much lower than the undissociated components of the gas mixture. This occurs despite of the fact that radical-radical reactions have zero or very small activation energies, giving large rate constants, but this is reduced in the rate expression by their low concentrations. Reactions of atoms or radicals with the undissociated components of the gas will correspondingly be more important.
- The rates of reactions of the common plasma radicals with a particular molecule usually are in the sequence $OH > O \sim H$ (based on data for the hydrogen abstraction reactions with methane; Manion et al., 2015).
- An additional source of chemical mechanisms can be found in the field of flame and combustion chemistry, which often is performed at atmospheric pressure but at much higher temperatures than those in a typical cold plasma. This means that these mechanisms will contain many reactions with high activation energies that become important in the high temperature environment of a flame but will not play any part in the chemistry of a cold plasma.

These generalizations will not generate complete mechanisms but will allow the main pathways to be identified. Modeling often involves mechanisms with hundreds of species and many hundreds of reactions, but the main features of the plasma chemistry often can be described by only a very small number of species and reactions. These simplified mechanisms are important in gaining chemical insight into the important and dominant features of the process. This often can be achieved by a sensitivity analysis of the kinetic modeling that identifies the most important reaction steps and their relative contributions. This was seen in Fig. 5 where the contribution of a few key reactions to the production and destruction of atomic oxygen at different points in the nitrogen-oxygen plasma torch was determined (Van Gaens et al., 2014).

In addition to the insight that we can gain from an identification of the chemical mechanism that determines the plasma chemistry, we also have noted the importance of the primary electron-collision processes that occur in the early stages of the discharge. These will determine the identity and yields of the species that will initiate and drive the chemistry, which will be determined by many plasma parameters, including the design and type of plasma reactor, the gas composition, the plasma energy, the nature of the input energy in terms of frequency, modulation and pulse shape and duration. All of these can be used to control the chemical outcomes, as can the temperature and residence times of the gases in the plasma reactor and the possible use of catalysts. It is a complex multiparameter problem with a high degree of coupling between the parameters, but when understanding comes by employing a range of experiments with different diagnostics supported by realistic modeling and simulation, then it will be possible to design processes *a priori* for different situations with well-defined outcomes.

References

Ahmad, I.K., Wallis, A.E., Whitehead, J.C., 2003. The plasma destruction of odorous molecules: organosulphur compounds. High Temp. Mater. Process. 7, 487–499.

Anon., 1989. Odor Thresholds for Chemicals with Established Occupational Health Standards. American Industrial Hygiene Association, Fairfax, VA.

Azizov, R.I., Zhivotov, V.K., Krotov, M.F., Rusanov, V.D., Tarasov, Y.V., Fridman, A.A., Sholin, G.V., 1980. Synthesis of nitrogen-oxides in a non-equilibrium UHF discharge under electron-cyclotron resonance conditions. High Energy Chem. 14, 275–277.

Bai, M.D., Zhang, Z.T., Bai, X.Y., Bai, M.D., Ning, W., 2003. Plasma synthesis of ammonia with a microgap dielectric barrier discharge at ambient pressure. IEEE Trans. Plasma Sci. 31, 1285–1291.

Bai, M., Zhang, Z., Bai, M., Bai, X., Gao, H., 2008. Synthesis of ammonia using CH_4/N_2 plasmas based on micro-gap discharge under environmentally friendly condition. Plasma Chem. Plasma Process. 28, 405–414.

Barakat, C., Gravejat, P., Guaitella, O., Thevenet, F., Rousseau, A., 2014. Oxidation of isopropanol and acetone adsorbed on TiO_2 under plasma generated ozone flow: gas phase and adsorbed species monitoring. Appl. Catal. B Environ. 147, 302–313.

Benson, S.W., Axworthy, A.E., 1957. Mechanism of the gas phase, thermal decomposition of ozone. J. Chem. Phys. 26, 1718–1726.

Bogaerts, A., De Bie, C., Eckert, M., Georgieva, V., Martens, T., Neyts, E., Tinck, S., 2010. Modeling of the plasma chemistry and plasma-surface interactions in reactive plasmas. Pure Appl. Chem. 82, 1283–1299.

Cavadias, S., Amouroux, J., 1986. Nitrogen-oxides synthesis in plasmas. Bull. Soc. Chim. Fr. 147–158.

Chang, J.-S., 1993. Energetic electron induced plasma processes for reduction of acid and greenhouse gases in combustion flue gas. In: Penetrante, B.M., Schultheis, S.E. (Eds.), Non-Thermal Plasma Techniques for Pollution Control, Part A, vol. 34. Springer-Verlag, Berlin, pp. 1–32.

Cherkasov, N., Ibhadon, A.O., Fitzpatrick, P., 2015. A review of the existing and alternative methods for greener nitrogen fixation. Chem. Eng. Process. 90, 24–33.

Cho, H., Yoon, J.S., Song, M.Y., 2013. Evaluation of total electron scattering cross sections of plasma-relevant molecules. Fusion Sci. Technol. 63, 349–357.

COMSOL (online). Available at www.comsol.com.

Deng, S., Le, Z., Ruan, R., Yu, F., Reese, M., Cuomo, G., Chen, P., 2007. PHYS 360-nonthermal plasma synthesis of ammonia using renewable hydrogen. Abstr. Pap. Am. Chem. Soc. 234.

Ebihara, K., Mitsugi, F., Ikegami, T., Nakamura, N., Hashimoto, Y., Yamashita, Y., Baba, S., Stryczewska, H.D., Pawlat, J., Teii, S., Sung, T.-L., 2013. Ozone-mist spray sterilization for pest control in agricultural management. Eur. Phys. J. Appl. Phys. 61, 24318.

Elford, M.T., 1981. Recent Electron Scattering Cross Sections Derived from Swarm Transport Coefficients. Pergamon, New York, NY, pp. 11–20.

Fitzsimmons, C., Shawcross, J.T., Whitehead, J.C., 1999. Plasma-assisted synthesis of N_2O_5 from NO_2 in air at atmospheric pressure using a dielectric pellet bed reactor. J. Phys. D Appl. Phys. 32, 1136–1141.

Fitzsimmons, C., Ismail, F., Whitehead, J.C., Wilman, J.J., 2000. The chemistry of dichloromethane destruction in atmospheric pressure gas streams by a dielectric packed-bed plasma reactor. J. Phys. Chem. A 104, 6032–6038.

Fridman, A., 2008. Plasma Chemistry. Cambridge University Press, New York, NY.

Fridman, A., Nester, S., Kennedy, L.A., Saveliev, A., Mutaf-Yardimci, O., 1999. Gliding arc gas discharge. Prog. Energy Combust. Sci. 25, 211–231.

Futamura, S., Zhang, A.H., Yamamoto, T., 2000. Behavior of N_2 and nitrogen oxides in nonthermal plasma chemical processing of hazardous air pollutants. IEEE Trans. Ind. Appl. 36, 1507–1514.

Gicquel, C., Cavadias, S., Amouroux, J., 1986. Heterogeneous catalysis in low-pressure plasmas. J. Phys. D. Appl. Phys. 19, 2013–2042.

Golde, M.F., 1988. Reactions of $N_2(A)$. Int. J. Chem. Kinet. 20, 75–92.

Golde, M.F., Moyle, A.M., 1985. Study of the products of the reactions of N_2 (A): the effect of vibrational energy in N_2 (A). Chem. Phys. Lett. 117, 375–380.

Hagelaar, G.J.M., Pitchford, L.C., 2005. Solving the Boltzmann equation to obtain electron transport coefficients and rate coefficients for fluid models. Plasma Sources Sci. Technol. 14, 722–733.

Harling, A.M., Whitehead, J.C., Zhang, K., 2005. NOx formation in the plasma treatment of halomethanes. J. Phys. Chem. A 109, 11255–11260.

Hessel, V., Cravotto, G., Fitzpatrick, P., Patil, B.S., Lang, J., Bonrath, W., 2013. Industrial applications of plasma, microwave and ultrasound techniques: nitrogen-fixation and hydrogenation reactions. Chem. Eng. Process. 71, 19–30.

Hibert, C., Gaurand, I., Motret, O., Pouvesle, J.M., 1999. OH(X) measurements by resonant absorption spectroscopy in a pulsed dielectric barrier discharge. J. Appl. Phys. 85, 7070–7075.

Huang, L., Nakajo, K., Ozawa, S., Matsuda, H., 2001. Decomposition of dichloromethane in a wire-in-tube pulsed corona reactor. Environ. Sci. Technol. 35, 1276–1281.

Kim, H.-H., Teramoto, Y., Negishi, N., Ogata, A., 2015. A multidisciplinary approach to understand the interactions of nonthermal plasma and catalyst: a review. Catal. Today 256, 13–22.

Kogelschatz, U., Eliasson, B., Hirth, M., 1988. Ozone generation from oxygen and air-discharge physics and reaction-mechanisms. Ozone Sci. Eng. 10, 367–377.

Kogelschatz, U., Eliasson, B., Egli, W., 1999. From ozone generators to flat television screens: history and future potential of dielectric-barrier discharges. Pure Appl. Chem. 71, 1819–1828.

Kossyi, I.A., Kostinsky, A.Y., Matveyev, A.A., Silakov, V.P., 1992. Kinetic scheme of the non-equilibrium discharge in nitrogen-oxygen mixtures. Plasma Sources Sci. Technol. 1, 207–220.

Krasnoperov, L.N., Krishtopa, L.G., Bozzelli, J.W., 1997. Study of volatile organic compounds destruction by dielectric barrier corona discharge. J. Adv. Oxid. Technol. 2, 248–256.

Laroussi, M., Akan, T., 2007. Arc-free atmospheric pressure cold plasma jets: a review. Plasma Process. Polym. 4, 777–788.

Magureanu, M., Mandache, N.B., Parvulescu, V.I., 2015. Degradation of pharmaceutical compounds in water by non-thermal plasma treatment. Water Res. 81, 124–136.

Manion, J.A., Huie, R.E., Levin, R.D., Burgess Jr., D.R., Orkin, V.L., Tsang, W., McGivern, W.S., Hudgens, J.W., Knyazev, V.D., Atkinson, D.B., Chai, E., Tereza, A.M., Lin, C.-Y., Allison, T.C., Mallard, W.G., Westley, F., Herron, J.T., Hampson, R.F., Frizzell, D.H., 2015. NIST chemical kinetics database. NIST standard reference database 17, version 7.0 (web version, online). Available at http://kinetics.nist.gov/.

McTaggart, F.K., 1967. Plasma Chemistry in Electrical Discharges. Elsevier, Amsterdam, London.

Misra, N.N., 2015. The contribution of non-thermal and advanced oxidation technologies towards dissipation of pesticide residues. Trends Food Sci. Technol. 45, 229–244.

Misra, N.N., Tiwari, B.K., Raghavarao, K.S.M.S., Cullen, P.J., 2011. Nonthermal plasma inactivation of food-borne pathogens. Food Eng. Rev. 3, 159–170.

Mitsugi, F., Nagatomo, T., Takigawa, K., Sakai, T., Ikegami, T., Nagahama, K., Ebihara, K., Sung, T., Teii, S., 2014. Properties of soil treated with ozone generated by surface discharge. IEEE Trans. Plasma Sci. 42, 3706–3711.

Mizushima, T., Matsumoto, K., Sugoh, J., Ohkita, H., Kakuta, N., 2004. Tubular membrane-like catalyst for reactor with dielectric-barrier-discharge plasma and its performance in ammonia synthesis. Appl. Catal. A Gen. 265, 53–59.

Mizushima, T., Matsumoto, K., Ohkita, H., Kakuta, N., 2007. Catalytic effects of metal-loaded membrane-like alumina tubes on ammonia synthesis in atmospheric pressure plasma by dielectric barrier discharge. Plasma Chem. Plasma Process. 27, 1–11.

Morgan, W.L., Penetrante, B.M., 1990. ELENDIF: a time-dependent Boltzmann solver for partially ionized plasmas. Comput. Phys. Commun. 58, 127–152.

Mutel, B., Dessaux, O., Goudmand, P., 1984. Energy-cost improvement of the nitrogen-oxides synthesis in a low-pressure plasma. Rev. Phys. Appl. 19, 461–464.

Nakajima, J., Sekiguchi, H., 2008. Synthesis of ammonia using microwave discharge at atmospheric pressure. Thin Solid Films 516, 4446–4451.

Nijdam, S., van Veldhuizen, E., Bruggeman, P., Ebert, U., 2012. An Introduction to Nonequilibrium Plasmas at Atmospheric Pressure. Wiley-VCH Verlag GmbH & Co. KGaA, Weinheim, pp. 1–44.

Niu, J., Peng, B., Yang, Q., Cong, Y., Liu, D., Fan, H., 2013. Spectroscopic diagnostics of plasma-assisted catalytic systems for NO removal from $NO/N_2/O_2/C_2H_4$ mixtures. Catal. Today 211, 58–65.

Pankaj, S.K., Bueno-Ferrer, C., Misra, N.N., Milosavljević, V., O'Donnell, C.P., Bourke, P., Keener, K.M., Cullen, P.J., 2014. Applications of cold plasma technology in food packaging. Trends Food Sci. Technol. 35, 5–17.

Patil, B.S., Rovira Palau, J., Hessel, V., Lang, J., Wang, Q., 2015a. Plasma nitrogen oxides synthesis in a milli-scale gliding arc reactor: investigating the electrical and process parameters. Plasma Chem. Plasma Process. 36, 241–257.

Patil, B.S., Wang, Q., Hessel, V., Lang, J., 2015b. Plasma N_2-fixation: 1900–2014. Catal. Today 256, 49–66.

Penetrante, B.M., Hsiao, M.C., Merritt, B.T., Vogtlin, G.E., Wallman, P.H., 1995. Comparison of electrical discharge techniques for nonthermal plasma processing of NO in N_2. IEEE Trans. Plasma Sci. 23, 679–687.

Penetrante, B.M., Hsiao, M.C., Merritt, B.T., Vogtlin, G.E., Wallman, P.H., Neiger, M., Wolf, O., Hammer, T., Broer, S., 1996. Pulsed corona and dielectric-barrier discharge processing of NO in N_2. Appl. Phys. Lett. 68, 3719–3721.

Penetrante, B.M., Bardsley, J.N., Hsiao, M.C., 1997. Kinetic analysis of non-thermal plasmas used for pollution control. Jpn. J. Appl. Phys. 36 (Pt 1), 5007–5017.

Preis, S., Klauson, D., Gregor, A., 2013. Potential of electric discharge plasma methods in abatement of volatile organic compounds originating from the food industry. J. Environ. Manag. 114, 125–138.

Pringle, K.J., Whitehead, J.C., Wilman, J.J., Wu, J., 2004. The chemistry of methane remediation by a non-thermal atmospheric pressure plasma. Plasma Chem. Plasma Process. 24, 421–434.

Rapakoulias, D., Amouroux, J., 1979. Synthesis and quenching reactor in a nonequilibrium plasma—application in C_2H_2 and HCN synthesis. Rev. Phys. Appl. 14, 961–968.

Rapakoulias, D., Amouroux, J., 1980. Catalytic processes in nonequilibrium plasma chemical reactors.1. Nitrogen-fixation in N_2-CH_4 system. Rev. Phys. Appl. 15, 1251–1259.

Siemens, W., 1857. Ueber die elektrostatische Induction und die Verzögerung des Stroms in Flaschendräten. Poggendofls Ann. Phys. Chem. 102, 66.

Snyder, H.R., Anderson, G.K., 1998. Effect of air and oxygen content on the dielectric barrier discharge decomposition of chlorobenzene. IEEE Trans. Plasma Sci. 26, 1695–1699.

Somorjai, G.A., 1994. Introduction to Surface Chemistry and Catalysis. John Wiley & Sons, New York, NY.

Tijani, J.O., Fatoba, O.O., Madzivire, G., Petrik, L.F., 2014. A review of combined advanced oxidation technologies for the removal of organic pollutants from water. Water Air Soil Pollut. 225, 2102.

Tonkyn, R.G., Barlow, S.E., Orlando, T.M., 1996. Destruction of carbon tetrachloride in a dielectric barrier/packed-bed corona reactor. J. Appl. Phys. 80, 4877–4886.

van Dijk, J., Peerenboom, K., Jimenez, M., Mihailova, D., van der Mullen, J., 2009. The plasma modelling toolkit Plasimo. J. Phys. D Appl. Phys. 42, 194012.

Van Gaens, W., Bruggeman, P.J., Bogaerts, A., 2014. Numerical analysis of the NO and O generation mechanism in a needle-type plasma jet. New J. Phys. 16, 060354.

van Helden, J.H., Wagemans, W., Yagci, G., Zijlmans, R.A.B., Schram, D.C., Engeln, R., Lombardi, G., Stancu, G.D., Ropcke, J., 2007. Detailed study of the plasma-activated catalytic generation of ammonia in N_2-H_2 plasmas. J. Appl. Phys. 101, 043305.

Wallis, A.E., Whitehead, J.C., Zhang, K., 2007. The removal of DCM from atmospheric pressure air streams using plasma-assisted catalysis. Appl. Catal. B Environ. 72, 282–288.

Whitehead, J.C., 2010. Plasma catalysis: a solution for environmental problems. Pure Appl. Chem. 82, 1329–1336.

Whitehead, J.C., 2014. Plasma Catalysis for Volatile Organic Compounds Abatement. Imperial College Press, London, pp. 155–172.

Yamamoto, T., Ramanathan, K., Lawless, P.A., Ensor, D.S., Newsome, J.R., Plaks, N., Ramsey, G.H., 1992. Control of volatile organic compounds by an ac energized ferroelectric pellet reactor and a pulsed corona reactor. IEEE Trans. Ind. Appl. 28, 528–533.

Chapter 4

Atmospheric Pressure Nonthermal Plasma Sources

P. Lu*, P.J. Cullen*,†, K. Ostrikov‡,§

*Dublin Institute of Technology, Dublin, Ireland, †University of New South Wales, Sydney, NSW, Australia, ‡Queensland University of Technology (QUT), Brisbane, QLD, Australia, §CSIRO – QUT Joint Sustainable Materials and Devices Laboratory, Lindfield, NSW, Australia

1 Introduction

In recent decades, atmospheric pressure plasma (APP) has attracted increasing attention. Special interest in APP has grown because it maintains similar properties as lower-pressure plasma, and expensive vacuum facilities are not required. Density and temperature are two critical plasma characteristic parameters. Categorized by relative temperature among electrons, ions, and neutrals, APPs are classified into (1) thermal equilibrium, (2) local thermal equilibrium, and (3) nonthermal equilibrium. In nonthermal equilibrium plasma (also named as nonthermal plasma, nonequilibrium plasma, or cold plasma), the electron temperature (T_e) is considerably higher than that of the ions (T_i) and neutrals (T_n), that is, $T_e \gg T_i$, T_n. For the atmospheric pressure molecular gas, the electrons can quickly transfer their energy to molecular rotational and vibrational states because the energy levels of the

Cold Plasma in Food and Agriculture. http://dx.doi.org/10.1016/B978-0-12-801365-6.00004-4

rotational and vibrational states of the molecules are much lower than that of the electrons' excitation and ionization. In this case, one of the factors determining the nonequilibrium state is the difference between the "kinetic" gas temperature and rotational and vibrational temperatures. In molecular gas plasma, most of the excitation energy goes into these rotational and vibrational states, which leads to extreme reactivity of the molecules and radicals. Atmospheric pressure nonthermal plasmas (APNTP) are widespread in many applications, such as exhaust fume treatment, water purification, sterilization, material deposition and synthesis, flow control and so on. Its applications have recently expanded to biological applications which exhibit huge potential.

It is easier to produce nonequilibrium gas discharges at low pressure, with gas temperatures close to room temperature. However, at atmospheric pressure a gas discharge requires a higher electric field to initiate the breakdown. In the case of a large discharge gap, the breakdown voltage is quite high. On the other hand, at atmospheric pressure the gas discharge after breakdown easily develops into a spark or arc. Thus, it is necessary to utilize a special electrode configuration and select a suitable power source and operation gas.

It is known that there are generally two breakdown mechanisms: Townsend breakdown and streamer breakdown. At atmospheric pressure, initiation of a Townsend breakdown is challenging compared to streamer breakdown. In the case of a small discharge gap, a glow discharge following a Townsend breakdown can be generated, such as in microdischarge or special dielectric barrier discharge (DBD). The glow regime can also exit in well-controlled corona discharge. Glow discharge has a relatively low current density and is typical for nonthermal plasma. However, streamer breakdown is the general form at atmospheric pressure. Most atmospheric pressure nonequilibrium discharges are streamer discharge with a contracted filamentary appearance, as found for most corona discharges and DBDs. At atmospheric pressure a streamer discharge tends to transit into a spark discharge, in particular at high current density, which leads to heating of the gas. Thus, it is necessary to avoid the transition of streamer into spark, either by limiting the current density or controlling the streamer development time. The principal approaches to produce APNTP have been summarized in the field of low-temperature plasma science and technology, mainly including (Chu and Lu, 2014):

- Increasing the local electric field strength by using sharp electrodes, as in corona discharges.
- Limiting discharge current by introducing dielectric or resistive barriers.
- Preventing thermalization/equilibrium by using pulsed power supply.
- Improving the heat transfer by forced convection via high gas-flow rate, using a highly thermal conductive gas (eg, helium).

2 Corona Discharge APNTP

2.1 Corona Discharge

An electrical discharge at atmospheric pressure requires a sufficiently high electric field to initiate and develop the ionization process. The local electric field enhancement is an approach to form a local nonuniform discharge. A corona discharge is a well-known, local, nonequilibrium discharge, which occurs only when the field is highly nonuniform (Raizer, 1991). Thus, the electrodes with a small curvature radius, such as a sharp point or a thin wire as shown in Fig. 1 are generally used as one or both electrodes, which are called corona electrodes or active electrodes. The ionization process is confined to a local region in the vicinity of the high voltage corona electrode due to the nonuniformity of the field. The weak luminance from the thin corona region is like a crown, which is the name origin of "corona."

A corona discharge is a local self-sustained discharge near the high-voltage (HV) electrode. It can be classified into positive and negative corona discharge by the polarity of the HV corona electrode. Positive corona indicates the HV electrode is the anode, and in negative corona the cathode is the HV electrode. Electrode polarity determines the different ignition processes of positive and negative corona discharge. In negative corona discharge, the maximum field region is close to cathode. The ignition process of negative corona is similar to the Townsend breakdown, which is based on the secondary electron emission (SEE) of the cathode and the electron avalanche multiplication via impact ionization. In the atmospheric air the attachment effects, due to the existence of electronegative molecules, must be taken into account. The ignition criteria can be described by Eq. (1), in which $\alpha(E)$, $a(E)$, and γ are

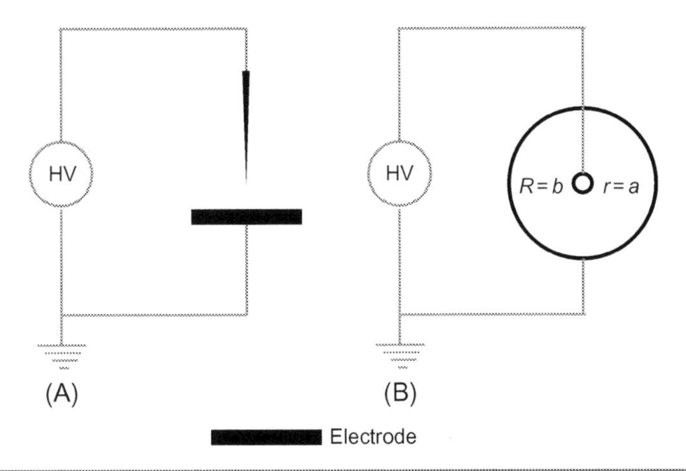

Fig. 1 Basic electrode configurations of corona discharge. (HV is high voltage power supply; R is the radius of outer electrode, r is the radius of inner electrode)

the electron impact ionization coefficient, attachment coefficient, and the SEE coefficient. These coefficients are the function of the E/p, in which E is the effective field of the external field and space charge field. The anode-directed electron avalanche usually extends to only a very short distance because the field strength decreases steeply away from the cathode. This distance is reflected in Eq. (1), where x_1 indicates the maximum distance from the cathode where the condition $\alpha(x_1)=a(x_1)$ is satisfied (Raizer, 1991).

$$\int_0^{x_1}[\alpha(x)-a(x)]dx = \ln\left(1+\gamma^{-1}\right) \tag{1}$$

The cathode SEE effect can be neglected in the positive corona discharge because the external field strength near the cathode is very weak and the initiation of a positive corona is far from the cathode. The positive corona discharge is formed by the photoionization of the gas near the anode, which is the element process of the anode streamer (cathode-directed) theory. The appearance of the positive corona displays multiple streamers running away from the anode region. Thus, the formation condition of the streamer can be used as the ignition criteria of the positive corona (Meek and Graggs, 1953). The field strength of the positive space charge at the avalanche head should reach the same order of magnitude of the external field. However, in the positive corona discharge, $\alpha(E)$ is more sensitive to the local field and the distance from the anode due to the high nonuniformity of the field (Raizer, 1991).

$$\int_0^{x_1}[\alpha - a]dx \approx 18-20 \tag{2}$$

A corona discharge can be driven by direct-current (DC), alternating-current (AC), or pulsed voltage. Traditional research and application on corona discharge generally use a simple DC or low frequency (power frequency) AC power source. The stable driving voltage leads to a stable space charge field, which distorts the external field. Depending on the applied voltage polarity and the electrode configuration, the corona discharge can develop into several different forms after ignition (Chang et al., 1991). In the case of a positive needle or sphere-plate corona discharge, the initial discharge manifests a burst pulse corona, which is not a fully developed streamer. It then proceeds to a streamer corona, glow corona, and spark discharge with the increasing voltage (Fig. 2). The streamer corona is noisy and its current is not steady. In a negative point-plane corona, one can recognize the initial form is the regular Trichel pulse corona, followed by a pulseless corona and again a spark. For a wire-cylinder or wire-plate electrode configuration, the positive corona appears as a narrow sheath around the wire electrode, which is named a "Hermstein's glow layer," or as a series of thin conducting streamers moving away from the wire electrode. For the negative corona, it appears a negative glow or a small concentrated spot.

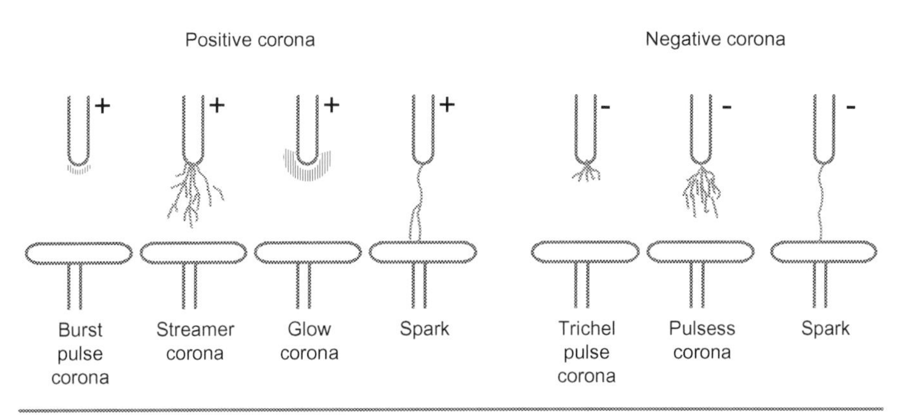

Fig. 2 Schematic type of corona discharge. *(From Chang, J.-S., Lawless, P.A., Yamamoto, T., 1991. Corona discharge processes. IEEE Trans. Plasma Sci. 11 (6), 1152–1166).*

The corona discharge is not a full breakdown in the gap. The current in the outer-gap region of a corona discharge is an nonsustaining discharge. The current of the charge carrier is limited by the space charge of the same sign. The current characteristic $I\ (V)$ of a corona discharge between two coaxial cylinders with radii r and R ($r \ll R$) was derived by Townsend, with the current per unit expressed as (Raizer, 1991):

$$I = \frac{8\pi\varepsilon_0\mu(V - V_S)V}{R^2\ \ln(R/r)} \tag{3}$$

where V and V_S is the applied voltage and the corona ignition voltage and μ is the mobility of the charged particles determining conductivity outside the active corona volume. The same parabolic current-voltage relation is also valid for other corona configurations. For example, an expression for corona generated in atmospheric air between a sharp point cathode with a small curvature radius $r = 3$–$35\ \mu$m and a perpendicular plane anode of distance $d = 4$–16 mm is obtained by experiments:

$$I = \left(\frac{52}{d}\right)V(V - V_S) \tag{4}$$

The corona ignition voltage V_S is 2.3 kV and is independent of d. Increasing the applied voltage can increase the corona current and the corresponding discharge power. However, the increased voltage leads to formation of a streamer corona with an unsteady current. These streamers are the precursors to a spark discharge and increase the potential of a full breakdown in the gap. Thus, applications of the DC coronas should use a current-ballistic resistor or operate under a lower applied voltage to limit the corona current.

2.2 Pulsed Corona Discharge

DC coronas are not continuous steady-state discharges, as the corona current has a pulsed nature in time with different pulse widths depending on the discharge parameters. During the pulse interval, accumulated space charges will disappear by diffusion and drift to initiate the next discharge.

Pulsed corona discharge has been widely studied in recent decades with a view to increasing the corona current by increasing the discharge voltage, but without transition to sparks. Pulsed corona discharge can even be simply called pulsed discharge, although other atmospheric discharges can be driven by HV pulse, such as DBD.

Pulsed corona discharge has been widely applied to many fields and is regarded as one of the most economic and efficient APNTP sources. They are mainly used to generate streamers; thus it is also named, "pulsed streamer discharge." An atmospheric nonthermal plasma is a high impedance load for the pulsed power generator. A very short rising time of the applied HV pulse is essential to initiate the streamer and avoid excessive growth to spark discharge. The streamer velocity is generally in an order of 10^8 cm/s. Therefore, for a gap of several centimeters, the pulse rising time is tens of nanoseconds (ns). Thus, the discharge loop inductance should be dedicatedly designed and reduced as much as possible. With the application of extremely short HV pulses, the applied voltage amplitude can be higher than the DC corona discharge. Consequently, highly nonequilibrium states can be produced in which the electron can require higher energy, while the ion and neutral gas are still cold due to the short energization times of pulsed voltage. The highly nonequilibrium state brings higher energy conversation efficiency from the input electrical energy to the streamer corona reactor (Yan et al., 2001a). Physical and chemical properties and applications of positive and negative streamers at atmospheric pressure driven by HV pulses with different pulse parameters have been systematically investigated by many groups, such as Ebert's group (Briels et al., 2006, 2008; Nijdam et al., 2008, 2009), Van Heesch's group (Van Veldhuizen, 2000; Smulders et al., 1998; Yan et al., 2001b; Winands et al., 2006; Van Heesch et al., 2008), Akiyama's group (Hackam and Akiyama, 2000; Wang et al., 2007; Katsuki et al., 2002; Samaranayake et al., 2000; Namihira et al., 2000; Matsumoto et al., 2011) and Ono's group (Teramoto et al., 2012; Ono et al., 2010, 2011; Ono and Oda, 2003). Fig. 3 shows a streamer driven a very short pulse voltage with only a 5 ns pulse width (Matsumoto et al., 2011).

2.3 Application of Corona APNTP

Applications of corona discharge processes can be dated back to the early 1880s, to the first electrostatic precipitator (ESP). Since then, corona has been extensively used

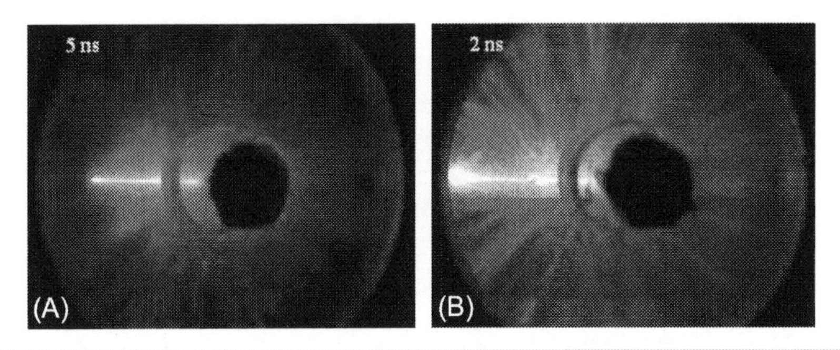

Fig. 3 (A) Five nanosecond pulsed streamer discharge plasma. (B) Two nanosecond pulsed streamer discharge plasma. Exposure time 1/30 s. *(From Matsumoto, T., Wang, D., Namihira, T., Akiyama, H., 2011. Discharge appearances of 2- and 5-ns pulsed power. IEEE Trans. Plasma Sci. 39, 1162–2263.)*

in several commercial ways and is gaining increasing attentions for other emerging applications. Application of corona discharge is principally divided into two kinds: corona ion-based and corona induced plasma-based applications. ESP is a well-known application of corona discharge, which serves as an ion source to charge dust. Other corona ion based applications include: electrophotography, clean room ionizers for static control, and atmospheric pressure ionization sources for mass spectrometry (Chang et al., 1991). The physical and chemical processes of corona induced plasma have been known to be highly effective in promoting oxidation, enhancing molecular dissociation, and producing free radicals. Therefore, corona plasma has widespread applications. Moreover, corona discharge has enabled plasma generation at atmospheric pressure with low gas temperatures. Application of atmospheric pressure corona plasma has extended to cover a wide range of fields, such as ozone synthesis, gas, and liquid cleaning. Corona plasma process depends on their physical construction and energization method. Various corona plasma reactors have been studied including point-plane, multiple points, packed bed, coaxial, and duct. Corona discharge is a broad technology which can be combined with other discharge configurations.

3 Dielectric Barrier Discharge APNTP

3.1 Dielectric Barrier Discharge

The name "DBD" defines its configuration, in which the discharge is blocked by a dielectric barrier layer. In DBD, a dielectric layer covers one or both electrodes, or can be suspended between two electrodes. Thus, the conduction current and charge transfer are limited. DBD is an easy and safe approach to produce nonthermal equilibrium discharge at atmospheric pressure. The design can avoid spark or arc discharges by limiting the discharge current.

DBDs can operate under a wide range of gas pressures (normally 10^4–10^6 Pa). It can be driven by AC voltage with a broad frequency band (50 Hz–1 MHz) and also a pulsed voltage. In some configurations with special dielectric material layout, it can be driven by a DC voltage. Basic DBD configurations are illustrated in Fig. 4. In Fig. 4A, the high-voltage electrode is covered by a dielectric layer. The thermal heat during extended operation can be dissipated through the metallic electrode. In Fig. 4B both electrodes are covered by the dielectric layer. Here electrode erosion can be avoided, in the case of corrosive gases. Using the configuration Fig. 4C, two different plasma composites can be produced separately in upper and lower gaps. DBD geometries can be very flexible. These basic configurations can be planar, coaxial, surface, or other novel combinations. In some applications, such as food packages, the flexible electrodes, and dielectric sheets can be used to form a plasma container. The typical gap distance in DBDs varies from 0.1 mm to several centimeters.

The dielectric material properties determine the stability of the DBD and most discharge parameters. The insulation strength and dielectric constant are important parameters for the dielectric material. In addition, the service temperature and mechanical properties should also be considered. Common materials for the dielectric layer are glass, quartz, ceramics, and polymer layers.

The breakdown of the gap in a DBD at atmospheric pressure is generally streamer breakdown. The breakdown voltage of the DBD gap is the same as other streamer breakdowns with a similar external field distribution. However, the existence of the dielectric barrier alters the gap field strength. The gap field strength is higher than

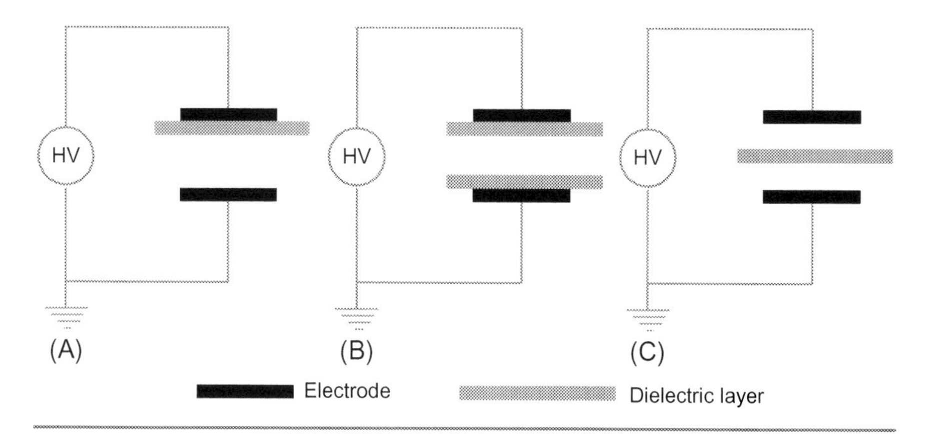

Fig. 4 Basic configurations of DBD (HV is high-voltage power supply).

the average field strength because the dielectric constant of barrier layer is higher than the value of air or other gases. A thicker dielectric layer can provide a sufficient breakdown strength and reliable insulation of discharge current to avoid spark discharges. However, a thicker layer requires a higher applied voltage, so a compromise must be made.

$$\begin{aligned} E_d/E_g &= \varepsilon_d/\varepsilon_g \\ V &= 2l_d E_d + l_g E_g \\ E_d &= \frac{V\varepsilon_g}{2l_d\varepsilon_g + l_g\varepsilon_d} \\ E_g &= \frac{V\varepsilon_d}{2l_d\varepsilon_g + l_g\varepsilon_d} \end{aligned} \tag{5}$$

The average field strength in dielectric layers and in the gap can be calculated from the continuity of the flux density (Eq. 5). For example, in the case of a parallel plate DBD configuration with two glass ($\varepsilon = 4$) layers of 3 mm thickness, for a 4 mm gap and 25 kV applied voltage, the average field strength is 25 kV/cm, but the gap field strength can reach to 45 kV/cm, which is sufficiently higher than the air breakdown threshold 30 kV/cm.

After ignition of the discharge, during each half cycle of the applied AC voltage (or a single pulse of repetitive pulsed voltage), the DBD current consists of a series of irregular microdischarge current pulses. These microcurrent pulses correspond to a thin streamer channel spanning the gap as shown in the schematic of Fig. 5 (Kogelschatz, 2002a). Each microdischarge pulse has a very short lifetime of only several tens of nanoseconds. The typical radius of the microdischarge is about 100–200 μm and the current density is about 100–1000 A/cm^2. The power of DBD can be characterized by the number of microdischarge channels per second.

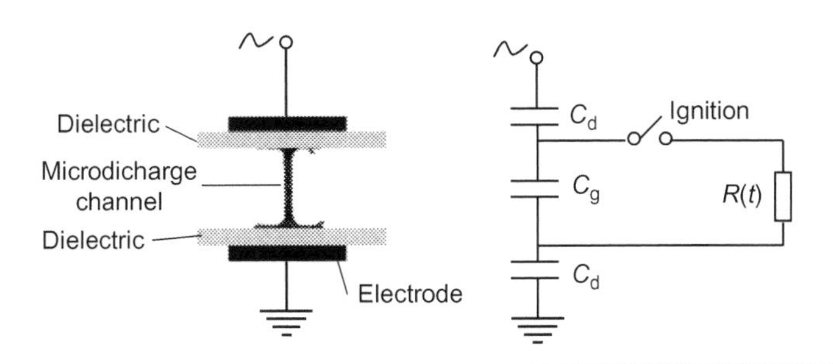

Fig. 5 Sketch of a micro discharge and a simple equivalent circuit in DBD. *(From Kogelschatz, U., 2002. Dielectric-barrier discharges: their history, discharge physics, and industrial applications. Plasma Chem. Plasma Process. 23 (1), 1–46.)*

However, an evaluation of the average DBD power by these microcurrent pulses is complicated.

In many applications, the average DBD power is used. The average DBD power is derived from the charge-voltage (Q-V) relation of the capacitor. In formula (6), f is the frequency of applied AC voltage, V_{op} is the peak value of applied voltage and V_d is the minimum applied voltage which can ignite the discharge. All the parameters in formula (6) can be measured experimentally. For a given DBD configuration, the discharge power is proportional to f, V_{op}, and V_d.

$$P = \frac{4fC_d^2}{C_d + C_g} V_d \left(V_{op} - V_d\right) \tag{6}$$

A more straightforward method to determine the average DBD power is to observe the voltage charge (Q-V) using a Lissajous figure as proposed by Kogelschatz (2002a). By adding a capacitance in series with the DBD configuration, the time-integrated discharge current can be derived by the measured voltage across this measurement capacitor. The average discharge power can be obtained (Eqs. 7, 8) as well. Fig. 6 shows an example of Q-V Lissajous figure (Connolly et al., 2013).

$$I = C_M \frac{dV_M}{dt} \tag{7}$$

$$P = \frac{1}{T}\int_0^T VIdt = \frac{C_M}{T}\int_0^T V\frac{dV}{dt}dt = fC_M \oint VdV_M \tag{8}$$

3.2 Different Patterns of DBD

DBDs at atmospheric pressure are generally nonuniform filamentary discharges as shown in Fig. 7. Formation of avalanches and development of streamers arise from separated local sites of the dielectric surface. Each streamer represents a so-called microdischarge distributed in the gap (Kogelschatz, 2002a). These filaments are unstable and move around erratically. Dynamic distribution of these filaments determines the pattern of the DBD appearance, which is practically very important for specific applications. Conventional applications of these filamentary DBD plasmas takes advantages of these irregular streamer pulses.

In recent decades, many DBDs have been found to exhibit regular structures depending on their electrode configurations, gas properties, and operation parameters. Using periodically arranged electrodes, 1D and 2D DBDs with regular patterns can be acquired (Kogelschatz, 2002b). These discharges are actually a series of regularly spaced microglow discharges which differ significantly from filamentary DBDs, which can even exhibit a so-called self-organized discharge pattern. This self-organized structure is generated with a high aspect ratio structure, in which

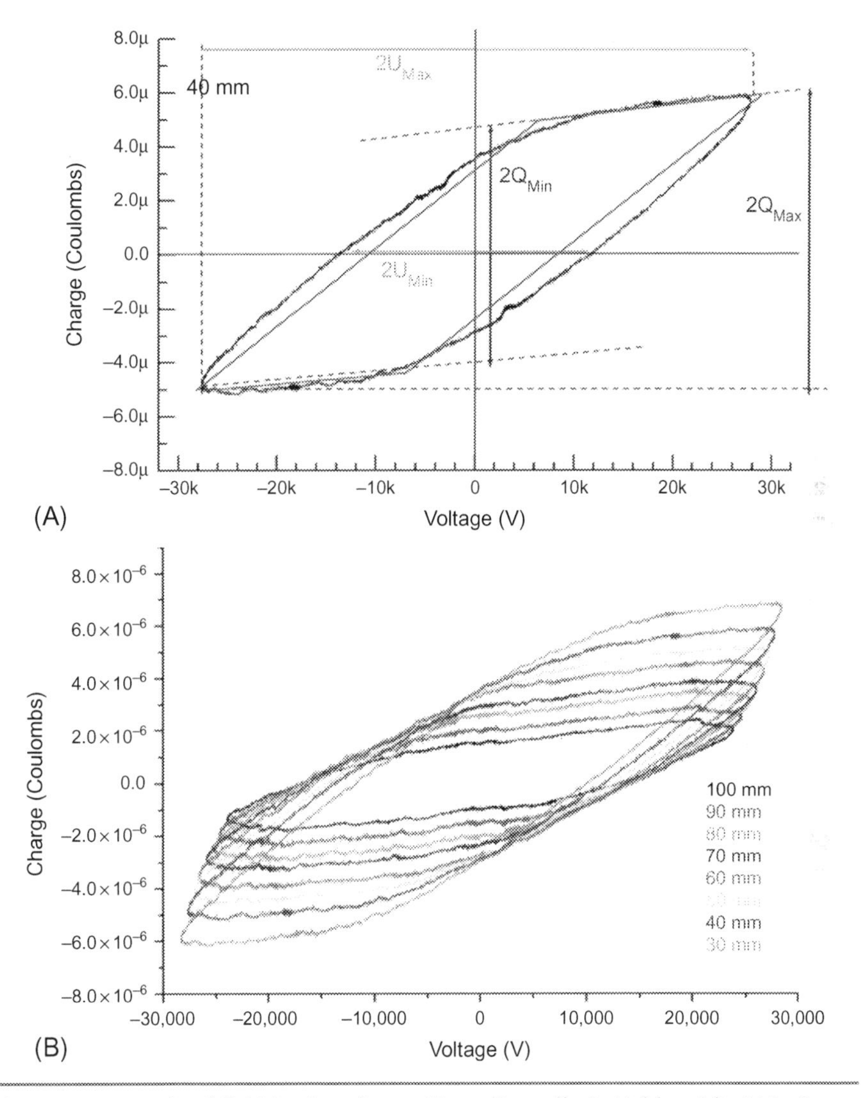

Fig. 6 An example of *Q-V* Lissajous figure. *(From Connolly, J., Valdramidis, V.P., Byrne, E., Karatzas, K.A., Cullen, P.J., Keener, K.M., Mosnier, J.P., 2013. Characterization and antimicrobial efficacy against* E. coli *of a helium/air plasma at atmospheric pressure created in a plastic package. J. Phys. D. Appl. Phys. 46 (3), 035401.)*

the gap distance is much smaller than the lateral electrode (Stollenwerk, 2010). By using a transparent metal or water, electrodes various regular and singular DBD patterns have been observed along the discharge current direction. Fig. 8 shows a series of self-organized patterns for DBDs (Dong et al., 2008). It should be noted that although the conduction current waveform of this self-organized pattern DBD

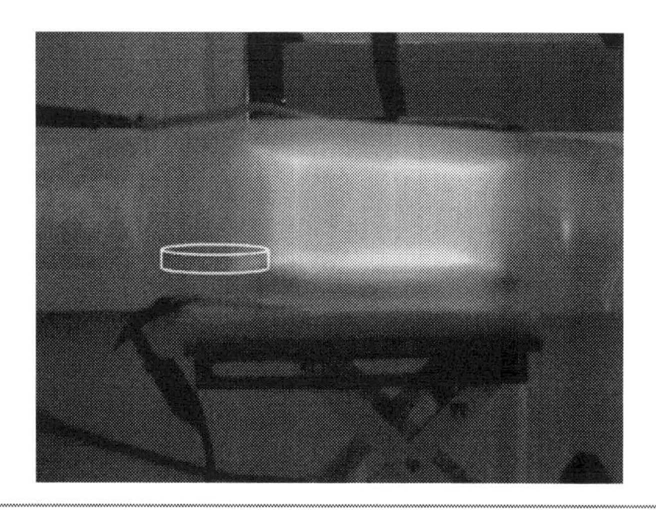

Fig. 7 DBD plasma operated in the filamentary mode. *(From Connolly, J., Valdramidis, V.P., Byrne, E., Karatzas, K.A., Cullen, P.J., Keener, K.M., Mosnier, J.P., 2013. Characterization and antimicrobial efficacy against* E. coli *of a helium/air plasma at atmospheric pressure created in a plastic package. J. Phys. D. Appl. Phys. 46 (3), 035401.)*

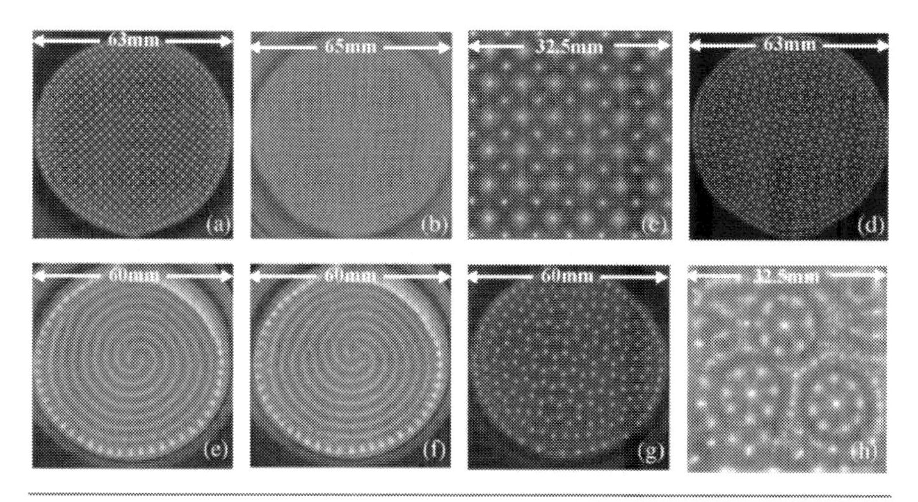

Fig. 8 Samples of diverse patterns obtained in a gas discharge by using a DBD system with two water electrodes. (A) Square lattice, applied voltage $U=2.4$ kV, gas pressure $p=1$ atm, gas gap $d=1.4$ mm, and frequency $f=50$ kHz. (B) Square-texture pattern, $U=4.1$ kV, $p=1$ atm, $d=1.5$ mm, and $f=60$ kHz. (C) Square superlattice pattern, $U=4.2$ kV, $p=1$ atm, $d=1.5$ mm, and $f=62$ kHz. (D) Hexagonal superlattice state, $U=3.6$ kV, $p=70$ kPa, $f=58$ kHz, and $d=1.5$ mm. (E) Two-armed spiral pattern, $U=4.2$ kV, $p=1$ atm, $f=61$ kHz, and $d=1.4$ mm. (F) Three-armed spiral pattern, the parameters are the same to that in (E). (G) Hollow-hexagon pattern, $U=3.7$ kV, $p=1$ atm, $d=1.5$ mm, and $f=55$ kHz. (H) Rotating wheels, $U=3.5$ kV, $p=1$ atm, $f=49$ kHz, and $d=1.5$ mm. The free boundary condition is used in (B), while in the other ones, the confined boundary conditions with respective circular and rectangular geometries are applied. *(From Dong, L., Fan, W., He, Y., Liu, F., 2008. Self-organized gas-discharge patterns in a dielectric-barrier discharge system. IEEE Trans. Plasma Sci. 36 (4), 1356–1357.)*

discharge shows exactly one current peak per half-cycle of the driving voltage, which is the signature of the diffuse glow-like DBD discharge, this discharge is not a homogeneous diffuse glow discharge, but regularly filamentary. However, it is suggested that these filaments are different from spark-like filaments. Experimental observations and theoretical explanations on the self-organized patterns of DBD discharge have been interesting research subjects in recent decades (Stollenwerk, 2010; Dong et al., 2008; Purwins, 2011; Babaeva and Kushner, 2014). Aside from the physic insights of this interesting phenomenon, the applications of these self-organized uniform-like DBD is expected to initiate novel applications.

Filamentary DBDs are easy to produce and the simplest form of a DBD plasma source. In some occasions, where the uniformity and homogeneity of plasma effects are highly required, uniform glow DBD plasma have been developed. Glow discharge is easily realized at low pressure or lower *Pd* values. But, the generation of a stable glow discharge at atmospheric pressure has more practical significance. Under certain circumstances DBDs can be a kind of uniform diffuse or spaced glow discharge. Since about the 1990s, glow-mode DBDs have been extensively studied (Kogelschatz, 2002b). Okazaki et al. (1993) reported the generation of atmospheric pressure glow (APG) using a 50 Hz power source based on fine wire mesh electrodes together with the dielectric plates. This stable glow DBD was generated in different gases, including He, Ar, O_2, N_2, and air. Massines et al. (1997) presented more detailed work on atmospheric pressure glow discharge (APGD) based on dielectric barrier (Massines et al., 1997; Gherardi et al., 2000; Massinesa et al., 2003; Gherardi and Massines, 2001). Fig. 9 shows comparative images of glow and filamentary DBDs (Gherardi et al., 2000). The mechanism on how to obtain a glow DBD and control the transition from glow to filament was explained based on electrical and optical diagnostics. At the same time, (Roth, 2001) proposed a one-atmosphere uniform glow discharge plasma

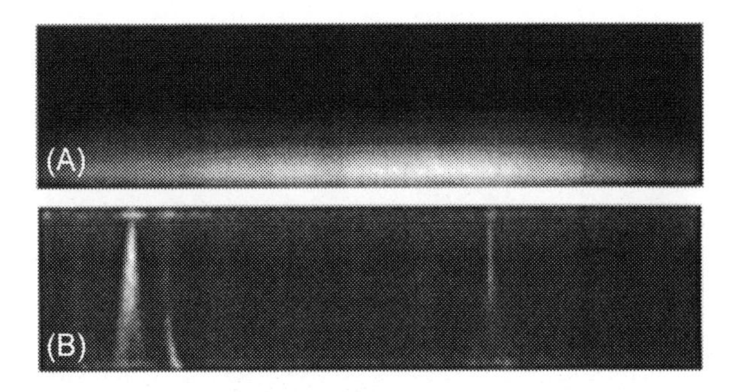

Fig. 9 Typical 10 ns exposure time photographs of a 4 mm gas gap during: (A) an APGD (Vamax = 11 kV) and (B) a filamentary discharge (Vamax = 14 kV). *(From Gherardi, N., Gouda, G., Gat, E., Ricard, A., Massines, F., 2000. Transition from glow silent discharge to micro-discharges in nitrogen gas. Plasma Sources Sci. Technol. 9 (3), 340–346.)*

(OAUGDP), which was also based on DBD. The proposed OAUGDP is driven by radiofrequency (RF) power source and can be operated in air. Roth has proposed the ion trapping mechanism for RF uniform OAUGDP. His group has investigated a number of different industrial applications based on this OAUGDP.

The mechanism for generating a glow-mode DBD is to initiate a Townsend breakdown instead of a streamer breakdown. To form an avalanche under a lower electric field and avoid growing a large amount of positive space charges, sufficient initial seed electrons should exist in the gap before breakdown (Gherardi and Massines, 2001). In DBDs, residual species from the previous half period provide the seed electrons or enhance the initial field for next discharge cycle. This is the so-called memory effect (Kogelschatz, 2002b; Roth, 2001). In helium barrier discharges, residual electrons—trapped in the low-field region of the positive column—are still present when the subsequent discharge is initiated. Residual ions on the dielectric surface of the cathode also play a role for a homogeneous glow. In molecular gases, such as nitrogen, long-life residual metastable species can generate many seed electrons due to the Penning ionization effect at the beginning of the next discharge cycle. The mechanism of formation of Townsend-like DBD and glow-like DBD is still not fully understood. Recent work on the discharge dynamics of glow DBD can be found in (Starostin et al., 2009).

Numerous groups have also reported the realization of atmospheric glow DBD plasmas and their applications. In particular, biological and medical applications using this kind of mild diffuse discharges have been widely studied. Besides the preionization effects by residual species in the gap, Aldea et al. (2009) have also argued another principle for generation of stable atmospheric glow DBDs. It is suggested that an APG DBD can be stabilized by an electronic feedback to fast current variations and by using adequate dielectric surfaces. Special barrier materials can also help to establish the spatially diffuse volume discharge. Laroussi et al. (2002) have introduced a resistive barrier discharge (RGB). By using a high resistivity (several mega ohms/cm) layer instead of a dielectric layer, the spatially diffuse discharge can be generated in atmospheric pressure helium gas. The high resistivity layer serves as a distributed resistor to limit the discharge current, which is same as the function of a lumped resistor in the normal glow discharge. RGB can be driven by power frequency AC or even a DC power supply. When it is driven by DC the discharge current shows a pulse current form with a few microseconds duration and of few tens of kilohertz frequency. This is explained by a voltage feedback effect of the resistive layer.

Although the mechanism is still under investigation, electrical characteristics of the glow DBD driven by an AC voltage is same in most publications as shown in the comparative waveforms of (Fig. 10). The feature of a single-current peak per half period or a single step in the charge measurement is accepted as one of signatures for obtaining a diffuse glow discharge. It is noted that in some papers, the glow-like

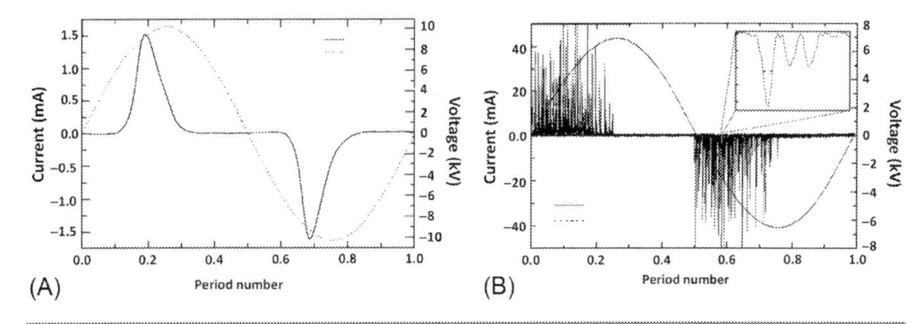

Fig. 10 (A) Discharge current of an APGD between polymers (Vamax = 10 kV, gap = 2 mm). (B) Typical electrical characteristics of a filamentary discharge in nitrogen. (Vamax = 6.4 kV, gap = 1 mm) *(From Gherardi, N., Gouda, G., Gat, E., Ricard, A., Massines, F., 2000. Transition from glow silent discharge to micro-discharges in nitrogen gas. Plasma Sources Sci. Technol. 9 (3), 340–346.)*

DBD are also used to describe the appearance of relatively uniform DBD plasmas. However, these DBDs are not real glow discharge, but a series of overlapping fine streamers. The exact glow DBD should be diagnosed by a fast ICCD camera.

3.3 Applications of DBD APNTP

The DBD nonthermal plasma source has many advantages. The DBD device is simple, stable, reliable, and economical. These advantages have led to a number of applications including industrial ozone generation, surface modification of polymers, and plasma-chemical vapor deposition, as well as pollution control, excitation of CO_2 lasers, excimer lamps, and large-area flat plasma-display panels. In emerging applications such as plasma medicine and food, very "mild" homogeneous DBD plasma is used to treat biological matter.

4 Glow Discharge APNTP

4.1 Low Pressure Glow Discharge

A glow discharge, which is a kind of classical self-sustaining discharge, has been extensively studied for several decades. Glow discharge current is in an order of mA and sustained by the SEE from the cold cathode as a result of positive ion bombardment. Glow discharges are likely the most studied and widely used gas discharges. It is a typical nonthermal equilibrium discharge. It is found in numerous applications ranging from light sources to plasma reactors for materials processing (Raizer, 1991; Lieberman and Lichtenberg, 2005).

A glow discharge is generally operated under low pressure and is well-known for its unique luminous pattern, which is in contrast to the relatively low current Townsend dark discharge. The light emission pattern of the glow has a stratification structure, which is divided into several typical layers. Each layer has a unique name and corresponds to a part of the glow discharge process. Starting from the cathode side, the first layer is called Aston dark space (sometimes covered by cathode glow) where the electron is ejected from the cathode at low energy. The next thin layer is the cathode glow, in which the atoms are excited by electrons. Following the cathode glow is the cathode dark space, where electron multiplication via impact ionization occurs. Aston dark space, cathode glow, cathode dark space, and negative glow compose the cathode layer. The next bright zone is negative glow, which is sharply separated from the dark cathode layer. The negative glow gradually becomes dim toward the anode, becoming the Faraday dark space. The positive column then begins to lighten (but not as bright as the negative glow) with a long uniform pattern. Near the anode, the positive column is first transferred into anode dark space and finally into the narrow zone of the anode glow. Glow color changes with the gas, which reflects the gas spectrum. Pattern of glow also changes with gas pressure and gap distance.

Glow discharge is the further development of Townsend dark discharge. It lies between the Townsend dark discharge and streamer breakdown. In glow discharge, the space charge effect and the SEE process tend to be predominant. A distinctive feature of glow discharge is the existence of the cathode layer, which is vital for sustaining the discharge. In cathode layer, the slow-moving positive ion space charge in the cathode dark space forms a considerable potential drop of 100–400 V between the cathode and negative glow. This potential drop is known as the cathode fall, which is an important parameter in the glow discharge. It determines the SEE process of the cathode and the consequent ignition of the glow discharge. The cathode fall is also a signature for distinguishing subnormal glow discharge, abnormal glow discharge, and normal glow discharge. In normal glow discharge, the conduction current density at the cathode is constant. The glow discharge current only partially covers the cathode surface, and the current value is proportional to the cathode glow area. In this situation the normal cathode fall is a constant, which has been experimentally measured for many combinations of gas composites and cathode materials. It depends on the gas properties and cathode materials. The cathode fall can be derived from the Townsend self-sustaining discharge condition. In each normal glow discharge the cathode surface current density, cathode fall, and cathode layer thickness are fixed for the specific gas composition, pressure, and cathode material. Similar to the minimum pd value on the Paschen curve, in a glow discharge the discharge current density, which corresponds to a fixed pd_c (product of the pressure and cathode fall thickness), implies a minimum of the cathode fall. This is the minimum power principle of the von Engel-Steenbeek theory on the realization of the minimum possible cathode fall in a normal discharge (Raizer, 1991). The higher power results in an abnormal glow discharge, and even the transition into arc discharge. A typical normal glow current density is 100 μA/cm^2 at about 1 torr. The thickness of the normal cathode layer at this pressure is about 0.5 cm. The normal cathode potential drop is about 200 V and does not depend on pressure and temperature.

If the interelectrode separation is sufficiently large, a relatively long and homogeneous positive column can be seen between the Faraday dark space and the anode. Maintaining electron production in the positive column is still by electron impact ionization in the constant field. The positive column is not an essential part of the cathode layer and it only serves as the electrical connection between the cathode layer and the anode. The positive column of a glow discharge is typically a weakly ionized nonthermal equilibrium plasma (sustained by an electric field). The electron density in a diffuse positive column in a glow discharge is in the order of 10^8–10^{12}.

4.2 Atmospheric Pressure Glow Discharge

The glow discharge has unique merits which make it a widespread homogeneous nonthermal plasma source for many applications, such as material treatment and glow discharge lasers. However, glow discharge is generally produced at low pressure. Since about 2000 APGD has attracted more attention and been studied widely. Nevertheless, at elevated pressure or increased discharge volume the homogeneous glow is often unstable and tends to contract. The transition from glow to spark can always occur. Thus, at atmospheric pressure, design considerations for generating glow discharge plasma are important. In fact, more than 80 years ago German researchers von Engel et al. observed glow discharges between metal electrodes in atmospheric pressure gases (Raizer, 1991; Kogelschatz, 2002b). In the following decades the glow discharge in transversely excited atmospheric-pressure lasers has been an extensive research topic. Preionization by UV radiation, electron beam, or electric discharge has been studied as an approach to homogenize the glow phase and avoid instability. Preionization can provide sufficient seed electrons for formation of a Townsend breakdown. In more recent decades, atmospheric-pressure glow discharges (APGD) are increasingly produced to obtain homogeneous treatment of various targets, such in biomedical area. Besides the preionization, several principal approaches have been proposed to generate a stable and homogeneous APGD. For instance, Okazaki et al. (1993) showed that a stable and homogeneous discharge could be obtained under specific conditions: (1) power supply frequency of over 1 kHz, (2) at least one dielectric layer between the two metal electrodes, and (3) helium dilution gas. The fundamental is to prevent the discharge contraction in the formation of current filament and avoid transition from glow into spark. Based on various kinds of APGDs, researchers have suggested that glow discharge at atmospheric pressure has similar properties to normal glow discharge under lower pressure. The current density is still independent of gas pressure and applied voltage. The luminous pattern of APGD also manifests stratification, however the size of the light emission is smaller compared with low-pressure glow. APGD still remains a nonthermal equilibrium state. But due to the high pressure, gas heating can be considerable up to a few thousand kelvin, while most low-pressure glows are close to room temperature.

4.3 MICRODISCHARGES

Similar to low-pressure glow discharge, APGD can be also be generated between metal electrodes without any dielectric barrier layer. To avoid streamer breakdown and prevent glow-to-arc transition, glow discharge is operated in the range of low *pd* values (generally 10 cm torr). Therefore, according to the similarity law of gas discharge at atmospheric pressure the gap distance should be scaled down to a submillimeter range. This micrometer-sized discharge is usually called microdischarge and the corresponding plasma is referred as microplasma. The minimum gap of a microdischarge can be theoretically only of the order of the cathode fall length, which depends on the gas property, pressure, and cathode material. It should be noted that with the reducing gap distance (<10 μm), the breakdown voltage actually decreases instead of increasing as predicted from the Paschen law. The breakdown mechanism deviates from the Paschen law for microscale breakdown (Ono et al., 2000; Slade and Taylor, 2002; Peschot et al., 2014). In the case of a very small spacing (<2 μm) the field emission plays a significant role. In the transition region ($2\ \mu m < d < 10\ \mu m$) between the field emission and Townsend breakdown, the ion-enhanced field emission dominates. David B. Go et al. have recently developed an analytical model that accounts for both Townsend ionization and ion-enhanced field emission mechanism (Go and Pohlman, 2010; Tirumala and Go, 2010). This model provides a modified Paschen's curve which can give a consistent description of the microscale breakdown.

Discharge in such a short distance results in high-power densities, which usually makes the discharge prone to instabilities: in particular, thermal instabilities (Raizer, 1991; Kunhardt, 2000; Schoenbach and Zhu, 2012). Limiting the current density and dissipation of heat are required for the stable operation of micro APGD. Alternatively, a stable micro APGD can also be realized by means of a pulse discharge with pulse widths that are shorter than the characteristic times for the development of the dominant instability. Generally, lower current is better for stable operation of a micro APGD. And atomic gases and lighter elements with a relatively high flow rate can make the discharge more stable. The discharge, of course, also depends on the cathode material (SEE efficiency). Microdischarges can be driven by DC, AC, pulsed, and RF power supply.

DC micro APGDs have been studied in very short gaps using pin-plate electrodes and parallel-plate electrodes with small dimensions. The typical circuit to generate micro APGDs is shown in Fig. 11, which is the same as for low-pressure glow

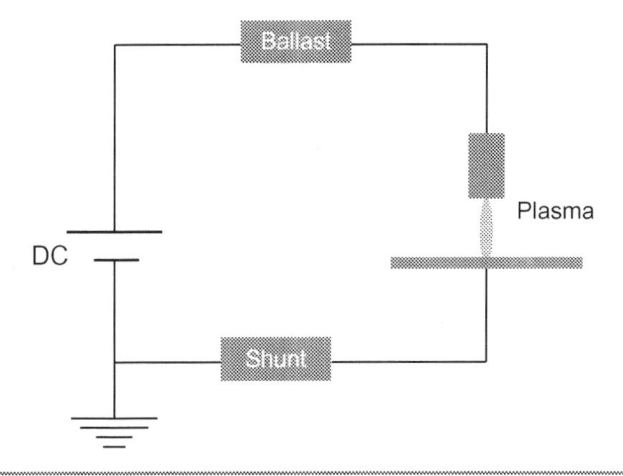

Fig. 11 Circuit diagram of DC glow discharge.

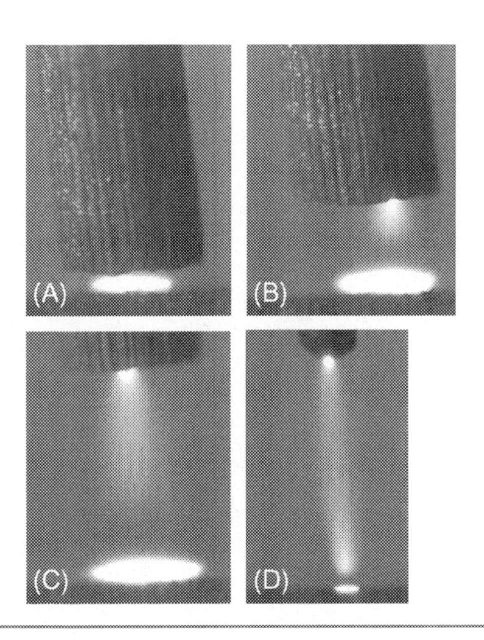

Fig. 12 Images of DC micro APGD in air at (A) 0.1 mm, (B) 0.5 mm, (C) 1 mm, and (D) 3 mm electrode spacings. *(From Staack, D., Farouk, B., Gutsol, A., Fridman, A., 2005. Characterization of a dc atmospheric pressure normal glow discharge. Plasma Sources Sci. Technol. 14 (4), 700–711.)*

discharge. It simply consists of a DC power source, a current-limiting resistor, two metal electrodes separated by a small distance, and a shunt resistor. The shunt resistor is used to measure the discharge current. Fig. 12 shows a group of micro APGD images in air taken by Staack et al. (2005), clearly showing the glow discharge pattern.

4.4 Hollow Cathode Discharge

Hollow cathode discharge (HCD) is a special glow discharge. Low-pressure HCD is widely used as the spectral lamp gas discharge laser. In normal glow discharge with parallel-plane electrodes, the accelerated electron beam moves perpendicular to the plane cathode. In HCD, with a hollow cylinder cathode, the angular electron beams converge and the negative glow regions merge into a brighter negative glow zone, which enhances the electron production and current density. Under the same cathode fall, current density in HCD is much higher than the value of normal glow discharge. With the increasing current density the cathode fall of the HCD doesn't increase, and consequently neither does the cathode temperature.

Conventional HCDs are operated under lower pressure and the macroscale. At atmospheric pressure micro HCD can be used to generate stable glow discharge. Micro HCD is also known as microcavity discharge, which is an efficient approach to stabilize the APGD. It is a basic structure of the microplasma source, which is also known as microcavity plasma. The properties of HCD mainly depend on two scaling laws (Becker et al., 2006). The fist scaling law is Paschen's law, which the product (pd) of the pressure p and the interelectrode separation d obeys the well-known Paschen curve. It determines the required breakdown voltage. The second scaling law, special to HCD, involves the product pD, in which D is the dimension of the cathode microcavity. If pD is in the range of 0.1–10 torr cm, the discharge can develop in different stages, each with its distinctive characteristic. Thus, at atmospheric pressure, microcavity size should be in the micrometer ranges.

A variety of microcavity geometries have been developed, mainly including cylinders, spirals or slits in the cathode, dielectric, and/or anode. Fig. 13 shows several electrodes geometries (Becker et al., 2006). The range of electrode materials employed to date for microplasma devices is equally broad, ranging from refractive metals to semiconductors. Microcavity can be formed by different techniques, including mechanical or laser drilling, ultrasonic milling, and wet and dry chemical etching. For large arrays, silicon bulk micromachining techniques are mostly employed. The arrangement of these microdischarges in arrays is usually important. The implementation of semiconductor microelectronics and MEMS microfabrication techniques has enables the realization of microplasma arrays with a large number of discharge units. Fig. 13 shows various microplasma geometries suitable as single discharges or as an "elementary cell" or pixel in arrays.

4.5 Glow Discharge With Liquid Electrodes

Using water as one electrode is also an approach to generate a relatively large-volume glow discharge at atmospheric pressure. This phenomenon has been studied by many groups and attracted increased attention for the interest of generating large-volume

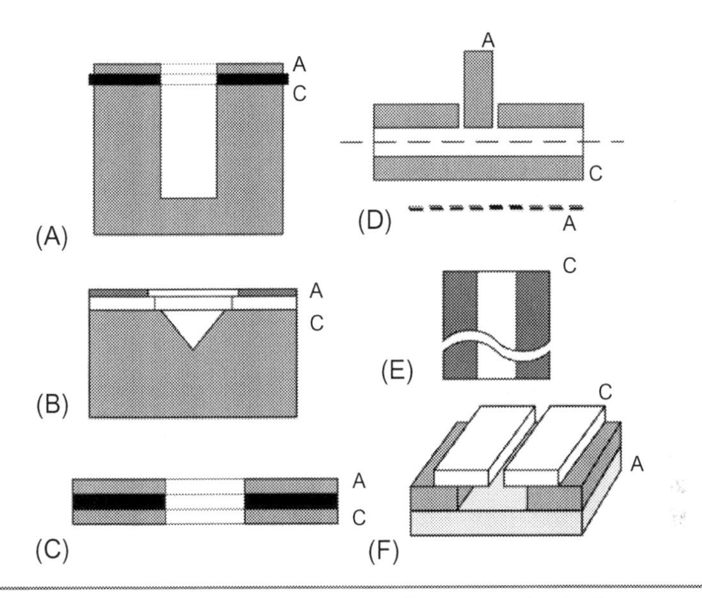

Fig. 13 Various microplasma geometries suitable as single discharges or as an "elementary cell" or pixel. *(From Becker, K.H., Schoenbach, K.H., Eden, J.G., 2006. Microplasmas and applications. J. Phys. D. Appl. Phys. 39 (3), R55.)*

APGD. Laroussi's group reported the generation of atmospheric pressure air glow discharge using the water as an electrode (Laroussi et al., 2003; Lu et al., 2003; Lu and Laroussi, 2003, 2005). The air plasma is generated in the gap between a disc-shaped, water-cooled metal upper electrode, and the surface of water contained in a glass dish. The gap distance can be adjusted from 3 mm to few centimeters, which is much larger than micro-APGD. The discharge is driven by a high voltage AC (60 Hz) power supply. In the positive half cycle of the applied voltage the water electrode is the cathode; in the negative half cycle the water electrode is the anode. The discharge may be characterized by electrical and optical diagnostics. Visible light images clearly show the different glow discharge patterns depending on the water electrode polarity. Ignition phase and steady-state structures are also investigated by high-speed imaging in the case of a 1.3 cm gap. In both the positive and negative half cycle of applied voltage the discharge is initiated from the cathode and anode-directed. When the water electrode is the cathode, high-speed imaging shows that before breakdown, fine ripples arise on the water surface; this enhances the local electric field strength. When the applied voltage increases to breakdown threshold, discharges initiate at each separated point. Consequently, the plasma volume in the gap has several discrete contact points at the water surface cathode. This makes the plasma volume wider. When the water electrode is the anode, most of the voltage drop is across the cathode fall near the metal cathode, and the field strength near the water anode is too low to form the water ripple. The corresponding glow plasma column has only one channel in the gap and is brighter than the plasma with the water cathode. In the case of a 16 kV (RMS) driving voltage, when water is the cathode the gas temperature is between 800 and 900 K, and the peak current is 67 mA; with water as the anode, the gas temperature is in the 1400–1500 K range, and the peak current is 81 mA.

When water is used as one electrode, in the case of smaller gap distances, the water surface deformation influences the initiation process of gas breakdown. When the Coulomb force strength at the water surface is stronger than the gravitational and the surface tension, the water surface forms a Taylor-cone structure, which enhances the local-field strength (Obata et al., 2015; Melcher and Smith, 1969). Bruggeman et al. (2007) have investigated the water surface deformation and its influence on DC electrical breakdown in a metal pin-water electrode system. The diameter of the metal pin electrode is 1 cm, so that the corona effect at small interelectrode distances can be excluded. For gap distances smaller than 7 mm the Taylor-cone instability at the water surface triggers the electrical breakdown. High-speed images show that at breakdown, the water surface has a Taylor cone-like shape. In the case of larger gap discharge (>7 mm) the breakdown voltage threshold is larger than the Taylor-cone instability limit value. Before breakdown the water surface elevation is not significant and the discharge channel is formed between the pin electrode and the water surface at its original position prior to breakdown. In 2008 Bruggeman published a series of systematic research work on the characteristics of DC glow discharge with liquids as one electrode in atmospheric air. The detailed work can be found in (Bruggeman et al., 2008a,b,c,d). Fig. 14 shows a series of discharge appearances with different electrode polarities and different exposure times. A further topical review article on nonthermal plasmas in contact with liquids is recommended (Bruggeman and Leys, 2009). Fig. 15 shows an overview of the different electrode configurations used in gas-phase electrical discharges with liquid electrodes. Various electrical breakdown processes in a metal-to-water electrode system are summarized according to the different values of the ratio r/d (r the electrode radius, and d the interelectrode distance).

5 Atmospheric Pressure Plasma Jets

The plasma jet is a broad concept covering various kinds of configurations which can, in common, realize the operation of gas discharge in a nonsealed ("open") electrode arrangement and the projection of the discharge plasma species into an open environment (Winter et al., 2015). Conventional plasma jet devices are typically the thermal arc jet, which are used for applications of cutting, welding, and other material techniques. In last 10 years the interest in plasma jets have been increasingly shifted to atmospheric pressure nonthermal (cold) plasma jets (APNPJs) (Schutze et al., 1998). The distinctive feature of this jet configuration is its ability to launch the stable plasma species to a separate environment, where the electric field can be

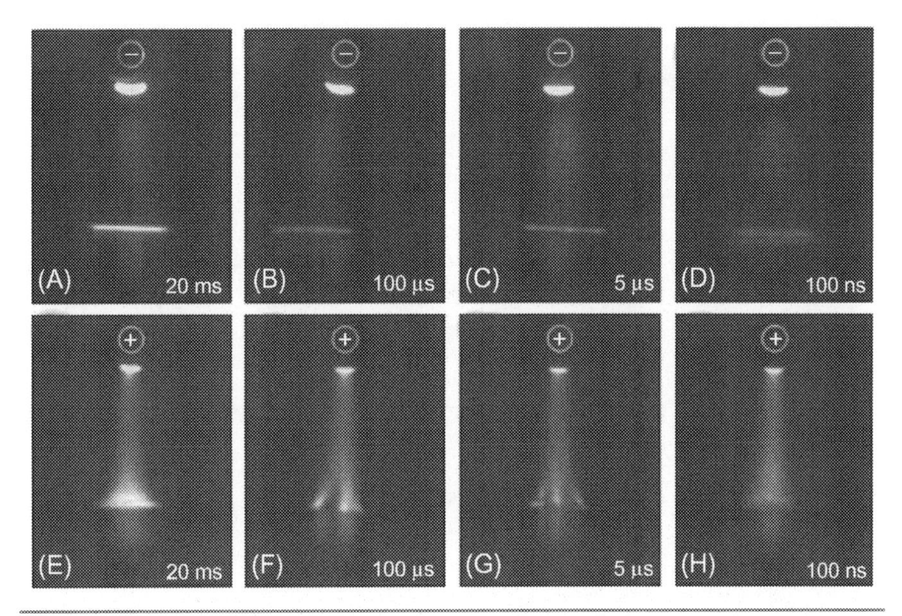

Fig. 14 Spatial optical emission patterns of discharges burning at 30.9 mA for positive and negative polarity of the water electrode on different timescales (20 ms–100 ns). The polarity of the pin electrode is indicated in the image. The inter-electrode distance is 5 mm. The shutter opening time of the CCD is indicated on the right bottom part of the image. The images for negative and positive polarity of the water electrode have the same gain for the same shutter opening times. *(From Bruggeman, P., Liu, J., Degroote, J., Kong, M.G., Vierendeels, J., Leys, C., 2008. DC excited glow discharges in atmospheric pressure air in pin-to-water electrode systems. J. Phys. D. Appl. Phys. 41 (21), 2008.)*

very low (Lu et al., 2012). This allows the treated target to no longer be needed to be confined within the plasma source device (the object is placed between the discharge gap). Thus, plasma jets can be used for direct treatment of various objects without limitation of the treatment size. On the other hand, the spatial separation of the plasma source and the plasma-target interaction regions allows for more freedom in source designs, to vary and control both plasma dynamics and reaction chemistry. APNPJ has been widely applied in the biomedical field, such as in the inactivation of bacteria, wound healing, and cancer treatment (Fridman and Friedman, 2013; Kong et al., 2009; Morfill et al., 2009). It is becoming a universal plasma source for a quick setup of interdisciplinary experiments. In last decade, APNPJs have been widely studied; in particular a series of systematic works on the mechanisms and applications of APNPJs have been published by Laroussi and Lu's group and Kong's group (Lu et al., 2012; Kong et al., 2009; Morfill et al., 2009; Laroussi and Akan, 2007). A most recent review paper by Weltmann's group gives an updated overview of devices and new directions of APNPJs since 2012 (Winter et al., 2015).

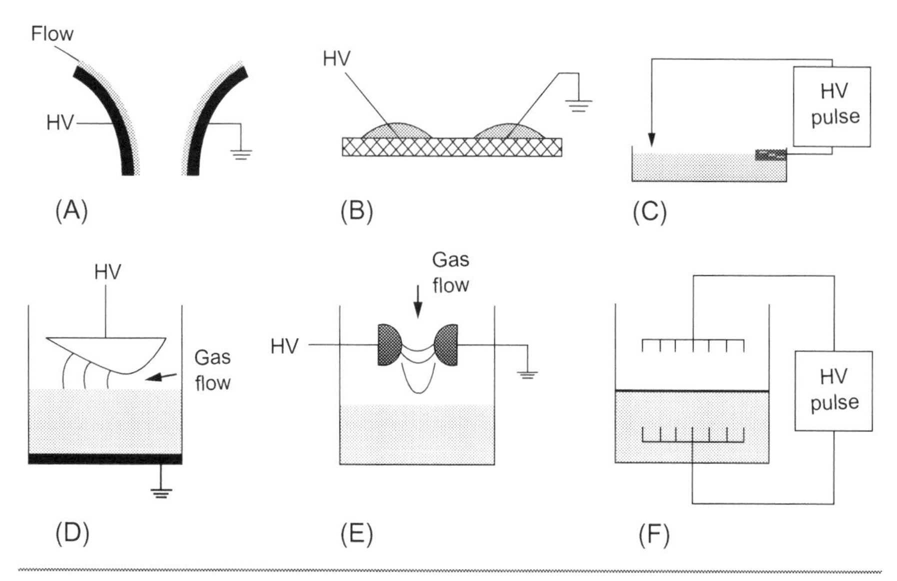

Fig. 15 Overview of the different electrode configurations used in gas phase electrical discharges with liquid electrodes. *(From Bruggeman, P., Leys, C., 2009. Nonthermal plasmas in and in contact with liquids. J. Phys. D. Appl. Phys. 42 (5), 053001.)*

Various electrode configurations have been developed to generate APNPJ. Lu et al. (2012) classify these configurations into four categories: dielectric-free electrode jets, DBD jets, DBD-like jets, and single electrode jets. DBD jets are relatively safe and stable configurations. Fig. 16 shows a schematic of DBD plasma jets. They can operate with a kHz AC power supply, RF, microwave, and pulsed DC supply. In the case of some electrode configurations, APNPJ can be driven by a DC power source. The orientation of the electric field with respect to the gas flow direction are different for various plasma jet configurations. The electric field can be parallel to the gas flow or perpendicular to the gas flow. Using noble gases, it is easy to generate a plasma jet at atmospheric pressure; in particular with helium gas, APNPJ with a length up to tens of centimeters can be generated. However, the noble gas plasma jets are not as reactive as air plasma jets. Therefore, most APNPJs work with noble gas mixed with a small percentage of reactive gases, such as O_2 or air.

The plasma jet is sometimes called a plasma bullet, which suggests its nature (Teschke et al., 2005). The continuous plasma plume in appearance is indeed made up of fast propagating bullet-like structures. Lu and Laroussi have proposed a streamer model to explain the bullet behavior (Lu et al., 2012). The plasma bullet is regarded as a guided ionization wave, which is also observed in the sprite discharge in the upper atmosphere (Lu et al., 2014). In contrast to the normal cathode-directed streamer, which stochastically propagates in free space, in APNPJ the bullet-like streamer repeatedly propagates along the predetermined path guided by the jet

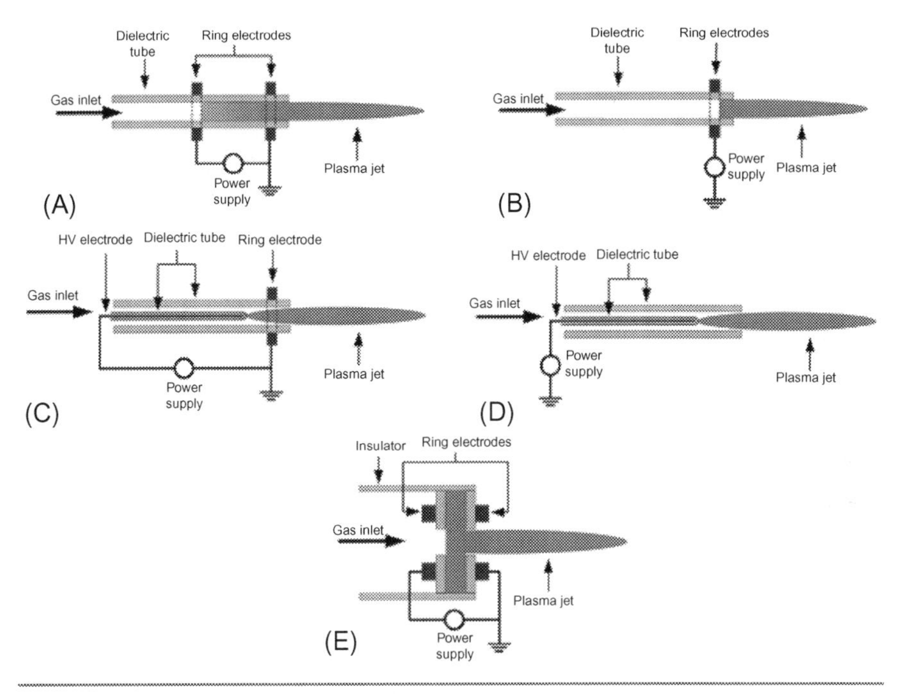

Fig. 16 Schematic of DBD plasma jets. *(From Lu, X., Laroussi, M., Puech, V., 2012. On atmospheric-pressure non-equilibrium plasma jets and plasma bullets. Plasma Sources Sci. Technol. 21 (3), 260.)*

configuration. Electrical and gas parameters determine the plasma bullet dynamics. When the applied voltage reaches a threshold, the plasmas jet exhibits plasma bullet behavior, while beneath the voltage threshold, the jet operates in a chaotic mode (Walsh et al., 2010; Lu et al., 2012). Fig. 17 shows stochastic and bullet modes of APNPJ driven by pulsed DC voltages with low and high amplitudes.

The size of the APNPJ can be very flexible. The radial size of the plasma jet can be reduced to micrometer ranges by reducing the dielectric tube diameter. This enables studies on the interaction of plasma jets with a few biological cells, or even a single cell. For applications which require a large area or bulk treatment, the APNPJ can be moveable along the object surface or operated in an array mode. 1D or 2D operations of plasma jet arrays have been studied by many groups. The stability and uniformity of the parallel operation of multiple jets requires voltage and current balancing in the driving circuit. Fig. 18 shows two examples of 2D plasma jet arrays (Sun et al., 2012). In addition, recent interesting studies on the impact of multiple plasma jets with small diameters onto the same location of the target have been performed (Douat et al., 2012). This approach allows for the merging of multiple plasma jets and consequently increasing of the deposited dose of plasma irradiation onto a fine location. It can also tune the reactive species composition using different feeding gas mixtures to each jet.

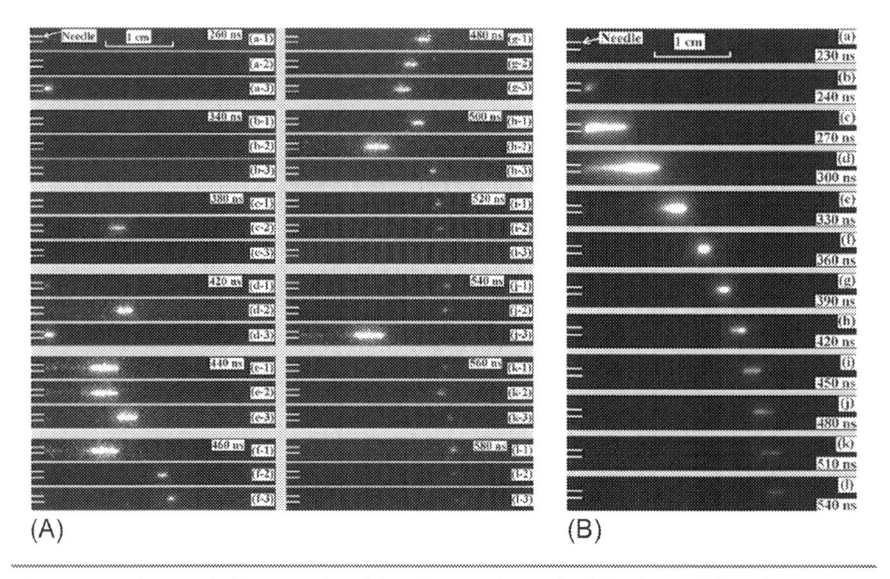

Fig. 17 High-speed photographs of the plasma plume for (A) 8 kV and (B) 9 kV. For (A), due to the randomness of the discharges, three photographs are taken for every delay time. Pulse frequency: 10 kHz, pulse width: 500 ns, working gas: He/O_2 (20%), and total flow rate: 0.4 L/min. *(From Lu, X., Laroussi, M., Puech, V., 2012. On atmospheric-pressure non-equilibrium plasma jets and plasma bullets. Plasma Sources Sci. Technol. 21 (3), 260.)*

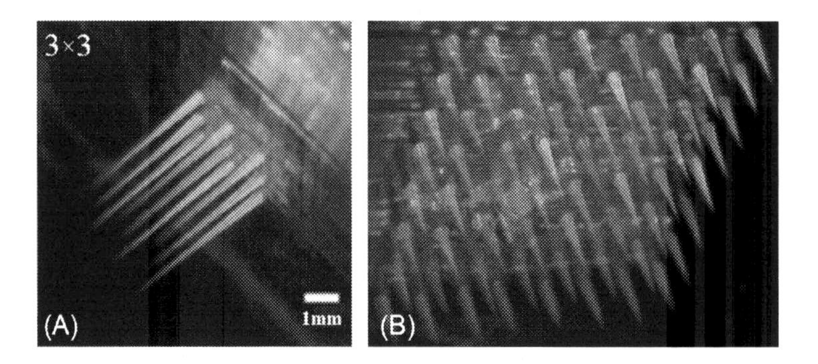

Fig. 18 (A) 3 × 3 array of microjets operating with a He backing pressure of 860 torr and (B) array of 64 plasma jets in an 8 × 8 configuration and operating with a He backing pressure of 780 torr. The amplitude of the 20-kHz voltage driving the array is 0.88 kV rms. *(From Douat, C., Bauville, G., Fleury, M., Laroussi, M., Puech, V., 2012. Dynamics of colliding microplasma jets. Plasma Sources Sci. Technol. 21 (3), 034010.)*

6 High Voltage Pulsed Discharge Produced APNTP

At atmospheric pressure, the streamer discharge is the general discharge form. To avoid the transition of streamer into spark, an external current-limiting resistor is necessary to inhibit a spark or arc. Another approach is to use pulsed power (or pulsed voltage) to drive the atmospheric pressure discharge (Chu and Lu, 2014; Kawai et al., 2010). In the previous sections, the pulsed power driven discharge has been mentioned, in particular with the corona discharge. Based on its advantages and significant potentials in the field APNTP, pulsed discharge-induced APNTP is emphasized here as an independent section. The research on pulsed discharge has a long history and is an important subject in gas discharge physics and high voltage engineering. One early research on spark formation time under overvoltage was a typical pulsed discharge experiment (Kuffel et al., 2000). Pulsed discharge features include pulse rising time and pulse duration (width). In recent decades, nanosecond pulsed discharge has been a focus of attention, not only in fundamental science but also for various applications. The study of nanosecond discharge relies on state-of-the-art pulsed power generators.

Traditional nanosecond pulse power generators use a spark gap switch to control the discharge; the discharge is usually operated under single pulse mode or low frequency. With the development of pulse power technology, solid state high voltage switches, such as the power semiconductor switch and magnetic switch, have been more used as they improve the repetition rate of the pulsed power generator (Bluhm, 2006; Pai and Zhang, 1995). Various ns/ms pulse power generators with shorter pulse rise times and higher repetition rates have been developed based on different circuit topologies, such as magnetic pulse compression, solid state marx generator, linear transformer driver, and even traditional transmission lines with a fast gap switch. These pulsed power generators can deliver flexible pulse parameters for nonthermal plasma generation, such as pulse voltage amplitude, polarity, duration, and repetition rate (Akiyama et al., 2007; Jiang et al., 2014). The advancement of these pulsed power generators has facilitated research of high repetition rate nanosecond pulsed discharges. The typical pulsed voltage parameters are 100–500 ns pulse duration (fwhm), 10–100 pulse rising time, and repetition rates up to tens of kHz. Nanosecond pulsed discharge enables application of higher voltage amplitudes to energize electrons within the nanosecond range, achieving a highly nonequilibrium state. It is worth mentioning that due to the pulse mode of operation with very small pulse energy, the operation of high voltage pulsed discharges can be safer than the conventional DC or low frequency AC high voltage discharges. Most discharge configurations mentioned in previous sections can be driven by nanosecond pulsed voltages and exhibit distinctive discharge characteristics from DC or AC driven discharge.

For DBDs, nanosecond pulsed DBD has been increasingly studied in recent years. Electrical and plasma characteristics have shown some distinct advantages over AC-DBDs. In nanosecond pulsed DBD, both discharge voltage (voltage across gap) and discharge current exhibit the bipolar pulse forms. Two discharge pulses are generated during the pulse rising time and falling time of the applied pulsed voltage, respectively (Walsh and Kong, 2007; Tao et al., 2008). The discharge current can reach hundreds of ampere of peak values; the average current density of a single microdischarge is in the order of $\sim$10 A cm^{-2}, which is much higher than the values of AC-DBDs ($\sim$10–100 mA/cm^2). This can significantly improve the energy deposition, electron density, and electron temperature (Chu and Lu, 2014). On the other hand, high repetition rate nanosecond high voltage pulse enables realization of uniform diffuse discharge in a wider range of conditions (Walsh and Kong, 2007; Tao et al., 2008, 2010). Fig. 19 shows comparative images of diffuse glow-like and filamentary DBDs driven at 1 kH nanosecond HV pulses (Tao et al., 2008). Although it is not a real glow discharge, diffuse "glow-like" mode is much more uniform and homogenous than the irregular filamentary discharge. The plasma jet sources introduced in last section have also been widely driven by nanosecond HV pulses.

Atmospheric pressure discharges between various bare metallic electrodes, driven by nanosecond pulsed high voltages, have also been widely studied. Based on different pulse parameters and electrode configurations, nanosecond pulsed discharges can exhibit not only streamer discharge but also glow discharge. Pai et al. (2009, 2010) have investigated different regimes of nanosecond pulsed discharge at atmospheric pressure air with pin electrodes. By changing the voltage amplitude corona, glow and spark regions have been found and the electrical and thermal characteristics have been studied for each regime. Moreover, high repetition rate nanosecond pulsed discharges can realize the diffuse glow-like discharge in larger areas. Fig. 20 shows the corona, diffuse and spark mode of nanosecond pulsed discharges with different repetition rates (Shao et al., 2013). The gap distance between pin-plate

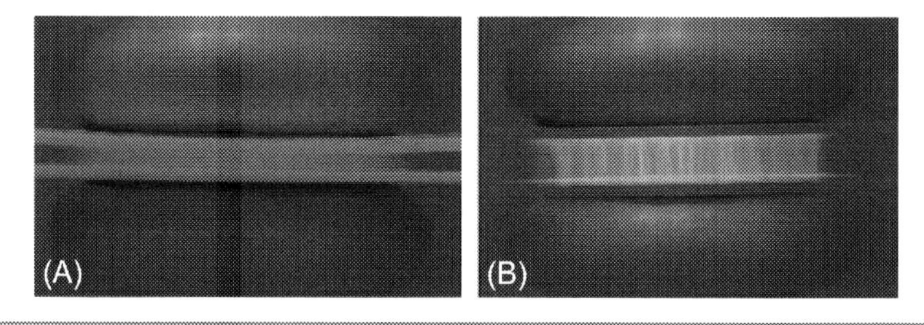

Fig. 19 DBD images of two typical discharge modes (the applied PRF is 1 kHz, (A) the air gap is 2 mm, and the thickness of PTFE is 2 mm and (B) the air gap is 8 mm, and the thickness of glass is 2 mm). *(From Tao, S., Kaihua, L., Cheng, Z., Ping, Y., Shichang, Z., Pan, R., 2008. Experimental study on repetitive unipolar nanosecond-pulse dielectric barrier discharge in air at atmospheric pressure. J. Phys. D. Appl. Phys. 41 (21), 215203.)*

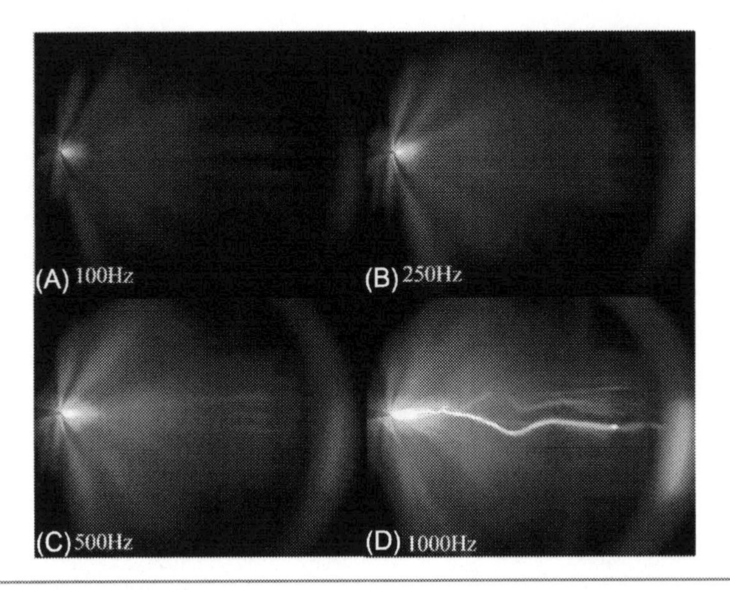

Fig. 20 Discharge images at different PRFs. (A) 100 Hz, (B) 250 Hz, (C) 500 Hz, and (D) 1000 Hz (applied voltage: 120 kV and gap spacing: 8 cm). *(From Shao, T., Tarasenko, V.F., Zhang, C., Baksht, E.K., Zhang, D., Erofeev, M.V., Ren, C., Shutko, Y.V., Yan, P., 2013. Diffuse discharge produced by repetitive nanosecond pulses in open air, nitrogen, and helium. J. Appl. Phys. 113 (9), 093301.)*

electrodes is 8 cm. Fig. 21 shows uniform diffuse discharges between 110 cm-long wire-wire electrodes (Liu et al., 2014). Besides the realization of diverse discharge modes, the nanosecond HV pulsed discharge can enable the well-defined control of the extreme discharge energy, which is hardly realized by other traditional methods. The dedicated control of discharge energy via extreme short pulse duration, flexible pulse repetition rate, and operation duty cycle has induced some emerging novel applications, such as nanomaterial synthesis. Pai and Ostrikov's group has synthesized well-defined MoO3 nanoscale architectures using nanosecond spark discharge and has shown the high energy efficiency in nanoscale synthesis using nanosecond plasmas, compared with other approaches (Pai et al., 2013).

7 Conclusion

APNTP sources is broad technological area. A wide variety of APNTPs exist, which span a range of electron densities and electron energies. Consequently, the application of APNTPs has covered a wide range of areas. Specific technological application requires application of specific plasma sources, that is, specific electric discharges. In this chapter the fundamental APNTP technologies are reviewed, in which APNTPs are mainly generated by the direct application of electric fields

Fig. 21 Images of discharge due to high-voltage repetitive nanosecond pulses in atmospheric air with the different electrode spacing (D). The top electrode connects to the output end of the HRNPG while the bottom one is directly grounded. The length of the line-line electrode is ~110.0 cm and the electrode spacing changes from 3.0 to 7.0 cm. The length of the reference ruler is 30.0 cm. (A) No discharge, (B) $110.0 \times 3.0\ cm^2$, (C) $110.0 \times 4.0\ cm^2$, (D) $110.0 \times 5.0\ cm^2$, (E) $110.0 \times 6.0\ cm^2$, and (F) $110.0 \times 7.0\ cm^2$. *(From Liu, Y.-L., Lee, L., Yu, B., Ge, Y.-F., Yang, L., Hu, W., Ma, N., Lin, F.-C., 2014. Large-scale nonthermal plasma generated by repetitive nanosecond pulses and barrier-free wire electrodes in atmospheric pressure air. IEEE Trans. Plasma Sci. 42 (10), 2946–2950.)*

across the electrodes. Electromagnetic wave is also a general energy form to ignite a breakdown and produce a plasma. This approach is called wave-heated plasma. However, a wave-heated APNTP is not introduced. APNTP has been progressing continuously not only in terms of plasma source technology, but also the scientific mechanism. Novel plasma sources are emerging and the corresponding mechanisms behind them is attracting more interest in plasma physics and chemistry. The rapid development of APNTP has also opened many new areas of application.

References

Akiyama, H., Sakugawa, T., Namihira, T., Takaki, K., Minamitani, Y., Shimomura, N., 2007. Industrial applications of pulsed power technology. IEEE Trans. Dielectr. Electr. Insul. 14 (5), 1051–1064.

Aldea, E., Peeters, P., De Vries, H., Van De Sanden, M.C.M., 2009. Atmospheric glow stabilization. Do we need pre-ionization? Surf. Coat. Technol. 200, 46–50.

Babaeva, N.Y., Kushner, M.J., 2014. Self-organization of single filaments and diffusive plasmas during a single pulse in dielectric-barrier discharges. Plasma Sources Sci. Technol. 23 (6), 065047.

Becker, K.H., Schoenbach, K.H., Eden, J.G., 2006. Microplasmas and applications. J. Phys. D. Appl. Phys. 39 (3), R55.

Bluhm, H., 2006. Pulsed Power Systems Principles and Applications. Springer, New York, ISBN: 3-540-26137-0.

Briels, T.M.P., Kos, J., Van Veldhuizen, E.M., Ebert, U., 2006. Circuit dependence of the diameter of pulsed positive streamers in air. J. Phys. D. Appl. Phys. 39 (24), 5201–5210.

Briels, T.M.P., Kos, J., Winands, G.J.J., Van Veldhuizen, E.M., Eber, U., 2008. Positive and negative streamers in ambient air: measuring diameter, velocity and dissipated energy. J. Phys. D. Appl. Phys. 41 (23), 234004.

Bruggeman, P., Leys, C., 2009. Nonthermal plasmas in and in contact with liquids. J. Phys. D. Appl. Phys. 42 (5), 053001.

Bruggeman, P., Graham, L., Degroote, J., Vierendeels, J., Leys, C., 2007. Water surface deformation in strong electrical fields and its influence on electrical breakdown in a metal pin water electrode system. J. Phys. D. Appl. Phys. 40 (16), 4779–4786.

Bruggeman, P., Ribežl, E., Maslani, A., Degroote, J., Malesevic, A., Rego, R., Vierendeels, J., Leys, C., 2008a. Characteristics of atmospheric pressure air discharges with a liquid cathode and a metal anode. Plasma Sources Sci. Technol. 17 (2), 025012.

Bruggeman, P., Guns, P., Degroote, J., Vierendeels, J., Leys, C., 2008b. Influence of the water surface on the glow-to-spark transition in a metal-pin-to-water electrode system. Plasma Sources Sci. Technol. 17 (4), 887.

Bruggeman, P., Van Slycken, J., Degroote, J., Vierendeels, J., Verleysen, P., Leys, C., 2008c. DC electrical breakdown in a metal pin-water electrode system. IEEE Trans. Plasma Sci. 36 (4), 1138–1139.

Bruggeman, P., Liu, J., Degroote, J., Kong, M.G., Vierendeels, J., Leys, C., 2008d. Dc excited glow discharges in atmospheric pressure air in pin-to-water electrode systems. J. Phys. D. Appl. Phys. 41 (21), 215201.

Chang, J.-S., Lawless, P.A., Yamamoto, T., 1991. Corona discharge processes. IEEE Trans. Plasma Sci. 11 (6), 1152–1166.

Chu, P.K., Lu, X.P., 2014. Low Temperature Plasma Technology Methods and Applications. CRC Press, London, ISBN: 978-1-4665-0991-7.

Connolly, J., Valdramidis, V.P., Byrne, E., Karatzas, K.A., Cullen, P.J., Keener, K.M., Mosnier, J.P., 2013. Characterization and antimicrobial efficacy against E. coli of a helium/air plasma at atmospheric pressure created in a plastic package. J. Phys. D. Appl. Phys. 46 (3), 035401.

Dong, L., Fan, W., He, Y., Liu, F., 2008. Self-organized gas-discharge patterns in a dielectric-barrier discharge system. IEEE Trans. Plasma Sci. 36 (4), 1356–1357.

Douat, C., Bauville, G., Fleury, M., Laroussi, M., Puech, V., 2012. Dynamics of colliding microplasma jets. Plasma Sources Sci. Technol. 21 (3), 034010.

Fridman, A., Friedman, G., 2013. Plasma Medicine. Wiley, New York, ISBN: 978-0-470-68969-1,

Gherardi, N., Massines, F., 2001. Mechanisms controlling the transition from glow silent discharge to streamer discharge in nitrogen. IEEE Trans. Plasma Sci. 29 (3), 536–544.

Gherardi, N., Gouda, G., Gat, E., Ricard, A., Massines, F., 2000. Transition from glow silent discharge to micro-discharges in nitrogen gas. Plasma Sources Sci. Technol. 9 (3), 340–346.

Go, D.B., Pohlman, D.A., 2010. A mathematical model of the modified Paschen's curve for breakdown in microscale gaps. J. Appl. Phys. 107 (10), 103303.

Hackam, R., Akiyama, H., 2000. Air pollution control by electrical discharges. IEEE Trans. Dielectr. Electr. Insul. 7 (5), 654–683.

Jiang, W., Sugiyama, H., Tokuchi, A., 2014. Pulsed power generation by solid-state LTD. IEEE Trans. Plasma Sci. 42 (11), 3603–3608.

Katsuki, S., Akiyama, H., Abou-Ghazala, A., Schoenbach, K.H., 2002. Parallel streamer discharges between wire and plane electrodes in water. IEEE Trans. Dielectr. Electr. Insul. 9 (4), 498–506.

Kawai, Y., Ikegami, H., Sato, N., Matsuda, A., Uchino, K., Kuzuya, M., Mizuno, A., 2010. Industrial Plasma Technology: Applications from Environmental to Energy Technologies. Wiley-VCH, Weinheim, ISBN: 9783527325443.

Kogelschatz, U., 2002a. Dielectric-barrier discharges: their history, discharge physics, and industrial applications. Plasma Chem. Plasma Process. 23 (1), 1–46.

Kogelschatz, U., 2002b. Filamentary, patterned, and diffuse barrier discharges. IEEE Trans. Plasma Sci. 30 (4), 1400–1408.

Kong, M.G., Kroesen, G., Morfill, G., Nosenko, T., Shimizu, T., Van Dijk, J., Zimmermann, J.L., 2009. Plasma medicine: an introductory review. New J. Phys. 11.

Kuffel, E., Zaengl, W.S., Kuffel, J., 2000. High Voltage Engineering Fundamentals, second ed. Newnes, Oxford, ISBN: 978-0-7506-3634-6.

Kunhardt, E.E., 2000. Generation of large-volume, atmospheric-pressure, nonequilibrium plasmas. IEEE Trans. Plasma Sci. 28 (1), 189–200.

Laroussi, M., Akan, T., 2007. Arc-free atmospheric pressure cold plasma jets: a review. Plasma Process. Polym. 4 (9), 777–788.

Laroussi, M., Alexeff, I., Richardson, J.P., Dyer, F.F., 2002. The resistive barrier discharge. IEEE Trans. Plasma Sci. 30 (1), 158–159.

Laroussi, M., Lu, X., Malott, C.M., 2003. A non-equilibrium diffuse discharge in atmospheric pressure air. Plasma Sources Sci. Technol. 12 (1), 53–56.

Lieberman, M.A., Lichtenberg, A.J., 2005. Principles of Plasma Discharges and Material Processing. Wiley-Interscience, Hoboken, ISBN: 0-471-72001-1.

Liu, Y.-L., Lee, L., Yu, B., Ge, Y.-F., Yang, L., Hu, W., Ma, N., Lin, F.-C., 2014. Large-scale nonthermal plasma generated by repetitive nanosecond pulses and barrier-free wire electrodes in atmospheric pressure air. IEEE Trans. Plasma Sci. 42 (10), 2946–2950.

Lu, X.P., Laroussi, M., 2003. Ignition phase and steady-state structures of a nonthermal air plasma. J. Phys. D. Appl. Phys. 36 (6), 661–665.

Lu, X.P., Laroussi, M., 2005. Atmospheric pressure glow discharge in air using a water electrode. IEEE Trans. Plasma Sci. 33 (2), 272–273.

Lu, X.P., Leipold, F., Laroussi, M., 2003. Optical and electrical diagnostics of a non-equilibrium air plasma. J. Phys. D. Appl. Phys. 36 (21), 2662–2666.

Lu, X., Laroussi, M., Puech, V., 2012. On atmospheric-pressure non-equilibrium plasma jets and plasma bullets. Plasma Sources Sci. Technol. 21 (3), 260.

Lu, X., Naidis, G.V., Laroussi, M., Ostrikov, K., 2014. Guided ionization waves: theory and experiments. Phys. Rep. 540 (3), 123–166.

Massines, F., Rabehi, A., Decomps, P., Gadri, R.B., Se´gur, P., Mayouxb, C., 1997. Experimental and theoretical study of a glow discharge at atmospheric pressure controlled by dielectric barrier. J. Appl. Phys. 83 (6), 2950.

Massinesa, F., Se´gurb, P., Gherardia, N., Khamphanb, C., Ricardb, A., 2003. Physics and chemistry in a glow dielectric barrier discharge at atmospheric pressure: diagnostics and modelling. Surf. Coat. Technol. 174–175, 8–14.

Matsumoto, T., Wang, D., Namihira, T., Akiyama, H., 2011. Discharge appearances of 2- and 5-ns pulsed power. IEEE Trans. Plasma Sci. 39, 2262–2263.

Meek, J.M., Graggs, J.D., 1953. Electrical Breakdown of Gases. Clarendon Press, Oxford.

Melcher, J.R., Smith, C.V., 1969. Water surface deformation in strong electrical fields and its influence on electrical breakdown in a metal pin water electrode system. Phys. Fluids 12 (4), 4779.

Morfill, G.E., Kong, M.G., Zimmermann, J.L., 2009. Focus on plasma medicine. New J. Phys. 11.

Namihira, T., Tsukamoto, S., Wang, D., Katsuki, S., Hackam, R., Akiyama, H., Uchida, Y., Koike, M., 2000. Improvement of NO_X removal efficiency using short-width pulsed power. IEEE Trans. Plasma Sci. 28 (2), 434–442.

Nijdam, S., Moerman, J.S., Briels, T.M.P., Van Veldhuizen, E.M., Ebert, U., 2008. Stereo-photography of streamers in air. Appl. Phys. Lett. 92 (10), 101502.

Nijdam, S., Geurts, C.G.C., Van Veldhuizen, E.M., Ebert, U., 2009. Reconnection and merging of positive streamers in air. J. Phys. D. Appl. Phys. 42, 045201.

Obata, D., Tasaka, H., Katsuki, S., Akiyama, H., 2015. Formation of liquid cone jet dependent on rise time of driving voltage. J. Electrost. 76.

Okazaki, S., Kogoma, M., Uehara, M., Kimura, Y., 1993. Appearance of stable glow discharge in air, argon, oxygen and nitrogen at atmospheric pressure using a 50 Hz source. J. Phys. D. Appl. Phys. 26 (5), 491.

Ono, R., Oda, T., 2003. Formation and structure of primary and secondary streamers in positive pulsed corona discharge—effect of oxygen concentration and applied voltage. J. Phys. D. Appl. Phys. 36 (16), 1952–1958.

Ono, T., Sim, D.Y., Esachi, M., 2000. Micro-discharge and electric breakdown in a micro-gap. J. Micromech. Microeng. 10, 445–451.

Ono, R., Teramoto, Y., Oda, T., 2010. Gas density in pulsed positive streamer measured by laser shadowgraph. J. Phys. D. Appl. Phys. 43, 345203.

Ono, R., Nakagawa, Y., Oda, T., 2011. Effect of pulse width on the production of radicals and excited species in a pulsed positive corona discharge. J. Phys. D. Appl. Phys. 44 (48), 485201.

Pai, S.T., Zhang, Q., 1995. Introduction to High Power Pulse Technology. World Scientific Publishing, Singapore, ISBN: 9810217145.

Pai, D.Z., Stancu, G.D., Lacoste, D.A., Laux, C.O., 2009. Nanosecond repetitively pulsed discharges in air at atmospheric pressure—the glow regime. Plasma Sources Sci. Technol. 18 (4), 045030.

Pai, D.Z., Lacoste, D.A., Laux, C.O., 2010. Nanosecond repetitively pulsed discharges in air at atmospheric pressure—the spark regime. Plasma Sources Sci. Technol. 19 (6), 065015.

Pai, D.Z., Ostrikov, K., Kumar, S., Lacoste, D.A., Levchenko, I., Laux, C.O., 2013. Energy Efficiency in Nanoscale Synthesis Using Nanosecond Plasmas. Nature Publishing Group, London. Scientific Reports 3.

Peschot, A., Bonifaci, N., Lesaint, O., Valadares, C., Poulain, C., 2014. Deviations from the Paschen's law at short gap distances from 100 nm to 10 μm in air and nitrogen. Appl. Phys. Lett. 105 (12), 123109.

Purwins, H.-G., 2011. Self-organized patterns in planar low-temperature ac gas discharge. Plasma Sources Sci. Technol. 39 (11), 2112–2113.

Raizer, Y.P., 1991. Gas Dishcarge Physics. Springer, Berlin, ISBN: 3-540-19462-2.

Roth, J.R., 2001. Industrial Plasma Engineering: Applications to Nonthermal Plasma Processing, vol. 2. IOP Publishing, Bristol, ISBN: 0-7503-0544-4.

Samaranayake, W.J.M., Miyahara, Y., Namihira, T., Katsuki, S., Sakugawa, T., Hackad, R., Akiyama, H., 2000. Pulsed streamer discharge characteristics of ozone production in dry air. IEEE Trans. Dielectr. Electr. Insul. 7 (2), 254–260.

Schoenbach, K.H., Zhu, W., 2012. High-pressure microdischarges: sources of ultraviolet radiation. IEEE J. Quantum Electron. 48 (6), 768–782.

Schutze, A., Jeong, J.Y., Babayan, S.E., Park, J., Selwyn, G.S., Hicks, R.F., 1998. The atmospheric-pressure plasma jet: a review and comparison to other plasma sources. IEEE Trans. Plasma Sci. 26 (6), 1685–1694.

Shao, T., Tarasenko, V.F., Zhang, C., Baksht, E.K., Zhang, D., Erofeev, M.V., Ren, C., Shutko, Y.V., Yan, P., 2013. Diffuse discharge produced by repetitive nanosecond pulses in open air, nitrogen, and helium. J. Appl. Phys. 113 (9), 093301.

Slade, P.G., Taylor, E.D., 2002. Electrical breakdown in atmospheric air between closely spaced (0.2 μm–40μm) electrical contacts. IEEE Trans. Plasma Sci. 25 (3), 390–396.

Smulders, E.H.W.M., Van Heesch, B.E.J.M., Van Paasen, S.S.V.B., 1998. Pulsed power corona discharges for air pollution control. IEEE Trans. Plasma Sci. 26 (5), 1476–1484.

Staack, D., Farouk, B., Gutsol, A., Fridman, A., 2005. Characterization of a dc atmospheric pressure normal glow discharge. Plasma Sources Sci. Technol. 14 (4), 700–711.

Starostin, S.A., Premkumar, P.A., Creatore, M., Van Veldhuizen, E.M., De Vries, H., Paffen, R.M.J., Van De Sanden, M.C.M., 2009. On the formation mechanisms of the diffuse atmospheric pressure dielectric barrier discharge in CVD processes of thin silica-like films. Plasma Sources Sci. Technol. 18 (4), 045021.

Stollenwerk, L., 2010. Interaction of current filaments in a dielectric barrier discharge system. Plasma Phys. Controlled Fusion 52 (12), 124017.

Sun, P.P., Cho, J.H., Park, C.-H., Park, S.-J., Eden, J.G., 2012. Close-packed arrays of plasma jets. IEEE Trans. Plasma Sci. 40 (11), 2946–2950.

Tao, S., Kaihua, L., Cheng, Z., Ping, Y., Shichang, Z., Pan, R., 2008. Experimental study on repetitive unipolar nanosecond-pulse dielectric barrier discharge in air at atmospheric pressure. J. Phys. D. Appl. Phys. 41 (21), 215203.

Tao, S., Yu, Y., Cheng, Z., Dongdong, Z., Zheng, N., Wang, J., Ping, Y., Yuanxiang, Z., 2010. Excitation of atmospheric pressure uniform dielectric barrier discharge using repetitive unipolar nanosecond-pulse generator. IEEE Trans. Dielectr. Electr. Insul. 17 (6), 1830–1837.

Teramoto, Y., Ono, R., Oda, T., 2012. Production mechanism of atomic nitrogen in atmospheric pressure pulsed corona discharge using two-photon absorption laser-induced fluorescence. J. Appl. Phys. 111, 113302.

Teschke, M., Kedzierski, J., Finantu-Dinu, E.G., Korzec, D., Engemann, J., 2005. High-speed photographs of a dielectric barrier atmospheric pressure plasma jet. IEEE Trans. Plasma Sci. 33 (2), 310–311.

Tirumala, R., Go, D.B., 2010. An analytical formulation for the modified Paschen's curve. Appl. Phys. Lett. 97 (15), 151502.

Van Heesch, E.J.M., Winands, G.J.J., Pemen, A.J.M., 2008. Evaluation of pulsed streamer corona experiments to determine the O* radical yield. J. Phys. D. Appl. Phys. 41, 234015.

Van Veldhuizen, E.M., 2000. Electrical Discharges for Environmental Purposes: Fundamentals and Applications. Nova Science Publishers, New York.

Walsh, J.L., Kong, M.G., 2007. 10 ns pulsed atmospheric air plasma for uniform treatment of polymeric surfaces. Appl. Phys. Lett. 91 (25), 1–3.

Walsh, J.L., Iza, F., Law, V.J., Janson, N.B., Kong, M.G., 2010. Three distinct modes in a cold atmospheric pressure plasma jet. J. Phys. D. Appl. Phys. 43 (7), 7520.

Wang, D., Jikuya, M., Yoshida, S., Namihira, T., Katsuki, S., Akiyama, H., 2007. Positive- and negative-pulsed streamer discharges generated by a 100-ns pulsed-power in atmospheric air. IEEE Trans. Plasma Sci. 35 (4), 1098–1103.

Winands, G.J.J., Liu, Z., Pemen, A.J.M., Van Heesch, E.J.M., Yan, K., Van Veldhuizen, E.M., 2006. Temporal development and chemical efficiency of positive streamers in a large scale wire-plate reactor as a function of voltage waveform parameters. J. Phys. D. Appl. Phys. 39, 3010–3017.

Winter, J., Brandenburg, R., Weltmann, K.-D., 2015. Atmospheric pressure plasma jets: an overview of devices and new directions. Plasma Sources Sci. Technol. 24 (6), 064001.

Yan, K., Van Heesch, E.J.M., Pemen, A.J.M., Huijbrechts, P.A.H.J., 2001a. From chemical kinetics to streamer corona reactor and voltage pulse generator. Plasma Chem. Plasma Process. 21 (1), 107–137.

Yan, K., Van Heesch, E.J.M., Pemen, A.J.M., Huijbrechts, P.A.H.J., 2001b. Elements of pulsed corona induced nonthermal plasmas for pollution control and sustainable development. J. Electrost. 51–52.

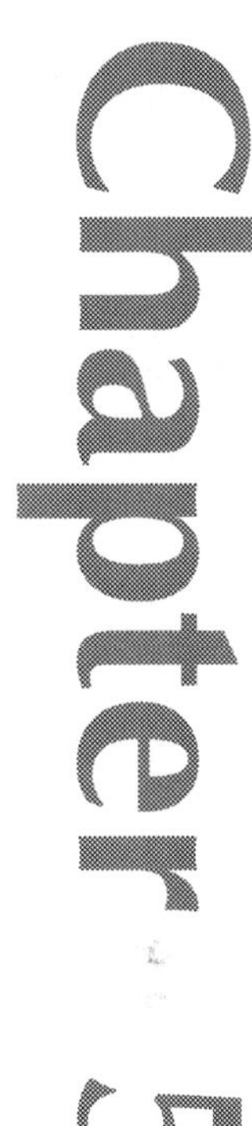

Plasma Diagnostics

K. Ishikawa
Nagoya University, Nagoya, Japan

1 Introduction

When considering nonequilibrium plasmas for various industrial applications, it is important to be able to characterize the plasma source for process analysis and control. The complexity of plasma processes arises from the involvement of a myriad of species including: electrons, ions, radicals, and photons (hereafter referred to as "plasma components"). Measurement and analysis of the plasma parameters and chemistry allow relating the plasma processes to observed changes in the matrix, say a given food, thus opening the scope for fine-tuning the process. Next, in-line characterization of the source also allows to control the behavior of plasma through established toolboxes available from control engineers. The characterization of the plasma sources involves the use of many sophisticated diagnostic tools and this field is commonly referred to as "plasma diagnostics."

Over the years, various diagnostic methods to characterize plasmas have been introduced with, as mentioned previously, the aim of better understanding and controlling plasmas. Plasma sources are categorized by their geometry, source gas, electrical power source, and other characteristics. Usually, users control plasma

Cold Plasma in Food and Agriculture. http://dx.doi.org/10.1016/B978-0-12-801365-6.00005-6

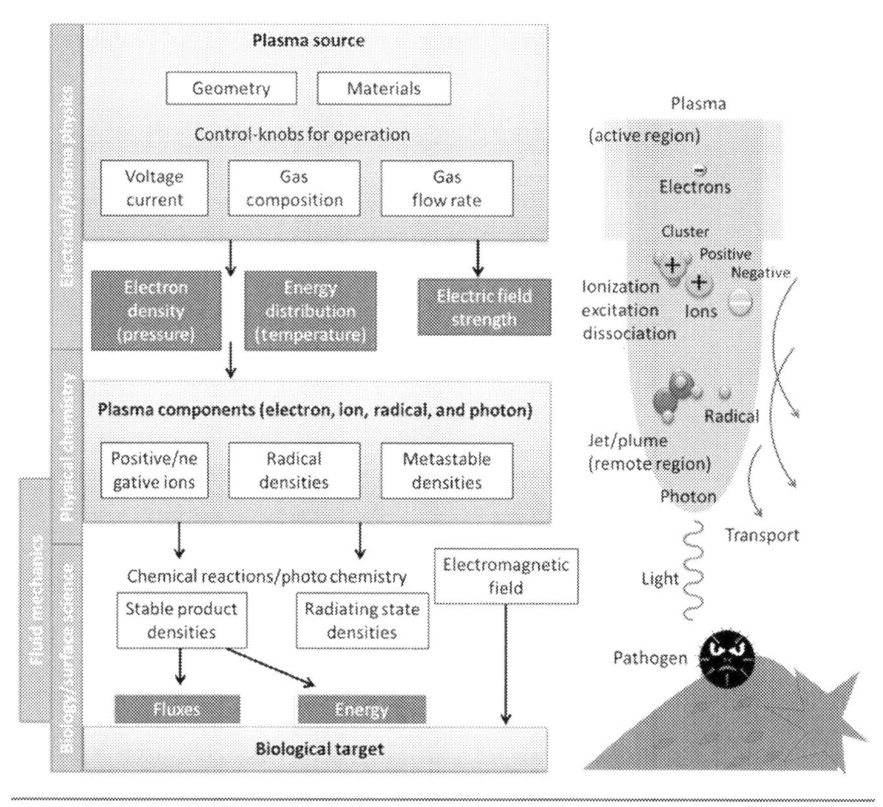

Fig. 1 Relationship between device and operation parameters, plasma physics and chemistry, and applications.

characteristics, such as gas flow rate and electric power, by operating instrument control knobs (Fig. 1). These parameters act to modify the physicochemical characteristics of plasma, such as their electron density, electron energy distribution function (EEDF), and concentration of reactive species. For most practical purposes, however, physicochemical parameters, such as the composition and concentration of radicals, should be controlled directly instead of the instrumental parameters (control knobs). To determine the best practices for this process, plasma diagnostic techniques focus on clarifying the relationship between the control parameters and the plasma parameters.

Once electric power is applied to the plasma source, a discharge initiates, which is further sustained by means of the electromagnetic power. In a microscopic view of the inside of the plasma, electrically charged particles are generated by the discharge. Subsequently, several of these particles disappear from the plasma, eg, through recombination, and these diffuse outward from the discharge region. A balance between the generation and loss in plasma is established when a stationary phase

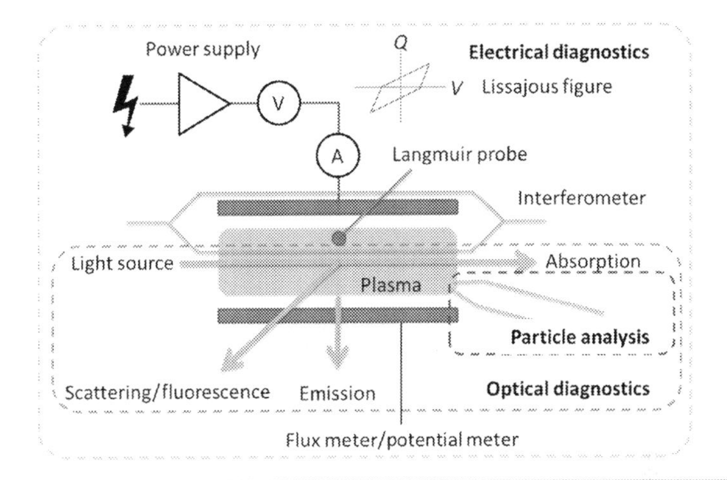

Fig. 2 Schematic summary of plasma diagnostic techniques.

is reached. In particular, it is worthwhile noting that plasma demonstrates both dynamic and static behaviors. Considering this important feature of "dynamic equilibrium" in cold plasmas, diagnostics must preferably be conducted in situ at real time. The diagnostic tools available for probing plasmas could broadly be classified into electrical and optical methods. Fig. 2 shows a schematic summary of electrical and optical diagnostics, and particle analysis.

2 Electrical Diagnostics of Plasma

Plasma can behave as a conductor or an insulator. An oscillating electric field E with angular wavenumber ω, $E(t) = E(\omega)e^{-i\omega t}$, is assumed to propagate through the plasma, which is a medium with conductivity $\hat{\sigma}(\omega)$ and dielectric function $\hat{\varepsilon}(\omega)$, where i is the imaginary unit, $i^2 = -1$, and t is time. The electric displacement (polarization) is defined as $p(\omega) = \hat{\varepsilon}(\omega)E(\omega)$; the induced polarization is a response due to electron motion as schematically shown in Fig. 3. Hence, we can introduce the effective plasma dielectric function as given by the Drude model (Chabert and Braithwaite, 2011), $\hat{\varepsilon}(\omega) = 1 - \frac{{\omega_p}^2}{\omega^2 - i\omega_\tau\omega}$, where ω_p is the electron plasma wavenumber $\sqrt{\frac{n_e e^2}{m_e \varepsilon_0}}$, ω_τ is the damping wavenumber, ε_0 is the permittivity, n_e is the electron density, e is the elemental charge, and m_e is the mass of an electron. Similarly, the electric current and the plasma conductivity are given by $j(\omega) = \hat{\sigma}(\omega)E(\omega)$ and

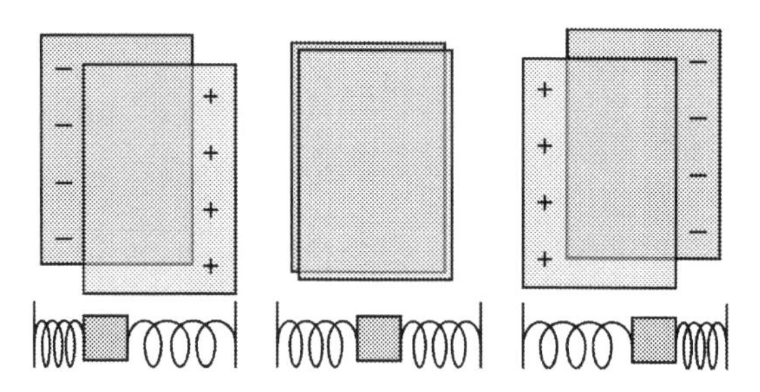

Fig. 3 Schematic representation of plasma response under an applied oscillating electromagnetic field. Negatively charged electrons and positively charged ions in each sheet respond as an oscillator.

$\hat{\sigma}(\omega) = \frac{{\omega_p}^2}{i\omega - \omega_\tau}$, respectively. The relationship between the refractive index n and extinction coefficient k is defined as $(n+ik)^2 = \hat{\varepsilon}$, and $nk = \sigma$. When the plasma wavenumber satisfies $\omega > \omega_p$, the electromagnetic waves propagate in the plasma (Lieberman and Lichtenberg, 2005).

Electrically charged particles move in plasma. Positive ions attract electrons, which are negative particles. For the sake of simplicity, positive ions are regarded as immobile. Each positive ion is then shielded by the surrounding electrons. As a result, an electrostatic potential abruptly forms to shield the positive ions in the plasma; this phenomenon is called Debye shielding. The characteristic length of the shield depends on the number density and velocity of the electrons, as well as the Debye length $\lambda_d = \sqrt{\varepsilon_0 k T_e / e^2 n_e} \approx 740\sqrt{T_e[V]/n_e\left[\mathrm{cm}^{-3}\right]}\,[\mathrm{cm}]$, where k is the Boltzmann constant. Here, T_e, which has units of volts, is used instead of kT_e. Namely, an observer cannot detect one charge in a bulk plasma region shielded with a length of λ_d. Some fast-moving electrons and most stationary ions present in the plasma exist in front of an object surface (wall) placed in the plasma, because the kinetic velocity of the electrons is faster than that of the ions. This snapshot reveals the form of the ion sheath surrounding the object, as shown in Fig. 4 (Chabert and Braithwaite, 2011).

2.1 Langmuir Probe

By immersing a small electrode, which is used as a probe, into the plasma, the current-voltage (*I-V*) characteristic curve is measured. The device used in this method is called a Langmuir or electrostatic probe (Chen, 1965). In general, this method requires that the collisionless sheath, which is defined as the diameter r_p of the probe, is shorter than the mean free path λ_e of electrons (ie, $r_p \ll \lambda_e$) and that

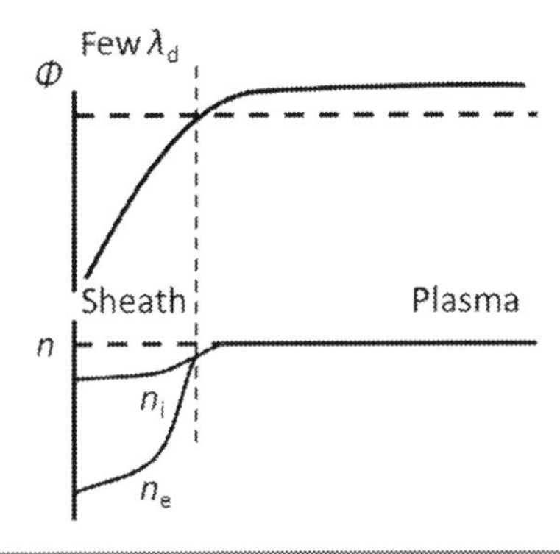

Fig. 4 Potential profile of space charge sheath. The number densities of the ions and electrons decrease near the wall.

the Debye length satisfies $\lambda_d \ll \lambda_e$, called the Laplace limit. If these requirements are not satisfied, the plasma parameters, such as the electron density, space (floating) potential, and EEDF, cannot be estimated because of the disturbances caused by the sheath structure and collisions in the sheath.

Fig. 5 shows a typical *I-V* curve that satisfies the given requirements. The curve is divided into three regions: (1) the ion saturation region, (2) the electron repulsion (transition) region, and (3) the electron current saturation region. Electric currents are defined as the sum of the (negative) electron current $-I_e$ and the (positive) ion current I_{ion}. If the EEDF is assumed to have a Boltzmann distribution, the electron current is given by

$$I_e \approx \frac{en_e S}{4}\sqrt{\frac{8kT_e}{\pi m}}\exp\left(\frac{-e(V_s - V_p)}{kT_e}\right). \tag{1}$$

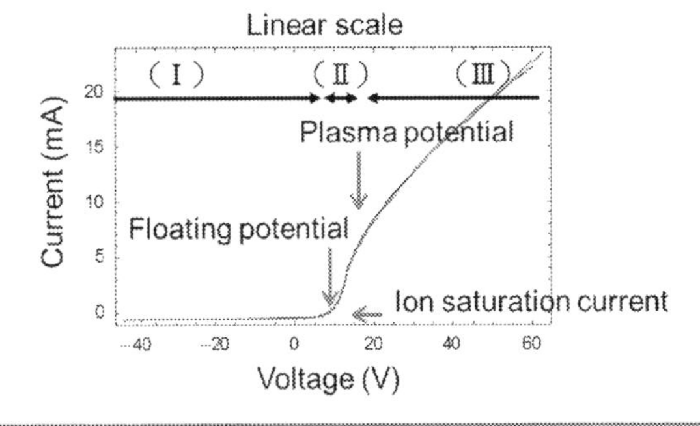

Fig. 5 Typical current-voltage characteristic curve of plasma.

The ion current is given by

$$i_{\text{ion}} = ken_e s\sqrt{\frac{kT_e}{m_i}}, \tag{2}$$

where k is the ratio of the temperatures of the electrons and ions and is approximately $1/\sqrt{e} \approx 0.61$, as $T_i \ll T_e$. From Eqs. (1) (2), we can obtain the plasma parameters.

In contrast to the usual collisionless case, atmospheric-pressure plasmas satisfy the conditions $r_p \gg \lambda_d$ and $\lambda_d \gg \lambda_e$ and strongly reduce the electric current. In this continuous medium (defined as a medium in which the conditions $\lambda_e \ll \lambda_d \ll r_p$ are fulfilled), the thermal motion of electrons is impeded, and the electron density in front of the probe surface decreases, which causes a reduction in the electric current. The transport of both the ions and electrons is governed by mobility and diffusion. This is called the blocking effect. For a cylindrical probe, the reduction in current is given by

$$I_{\text{em}} = I_e\left[1 + \frac{3r_p}{4\lambda_e}\ln\left(\frac{L}{r_p}\right)\right]^{-1}, \tag{3}$$

where I_{em} is the measured electron current reduced by the blocking effect and L is the probe length (Zakharowa et al., 1960; Sakai et al., 2005).

Hence we get

$$I_e \propto \left(\frac{eV_p}{kT_e}\right)^{\alpha} \exp\left(-\frac{eV_p}{kT_e}\right), \tag{4}$$

where the term $\frac{eV_p}{kT_e}$ is regarded as a potential and α is explicitly determined an exponent parameter in the range of 1–5/3 and is defined as a function of r_p/λ_d. Depending on the situation, appropriate corrections of the measured data are necessary. In contrast to the collisionless scenario, the thick sheath is taken into account in the potential solved by the Laplace equation (Laplace limit, $r_p/\lambda_d \ll 1$). Various models are suggested in the probe theories for describing the *I-V* curves; however, one should take into account appropriate prerequisites of each model. These regimes are illustrated in Fig. 6.

If there is no well-defined ground electrode in the plasma, the floating double-probe method is generally used. In this method, two probes are used to measure the imposing potential. The currents to probes 1 and 2 are defined as i_1 and i_2, and thus the total current is $i = \frac{i_2 - i_1}{2} + \frac{i_1 + i_2}{2}\tanh\left(\frac{eV}{2kT_e} - \frac{1}{2}\ln\frac{s_2}{s_1}\right)$, where the probe surface areas of the probes are s_1 and s_2. The *I-V* curve of the double-probe device demonstrates point symmetry, as shown in Fig. 7. The zero crossing corresponds to the floating potential and depends on the second term in the argument

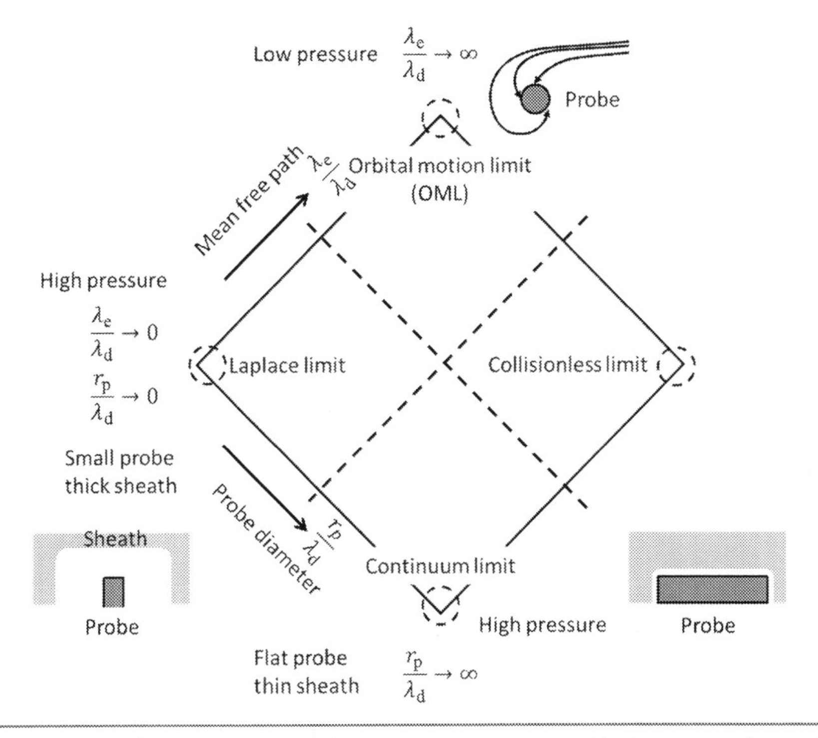

Fig. 6 Probe theory regimes of the ion current region with respect to the probe characteristics.

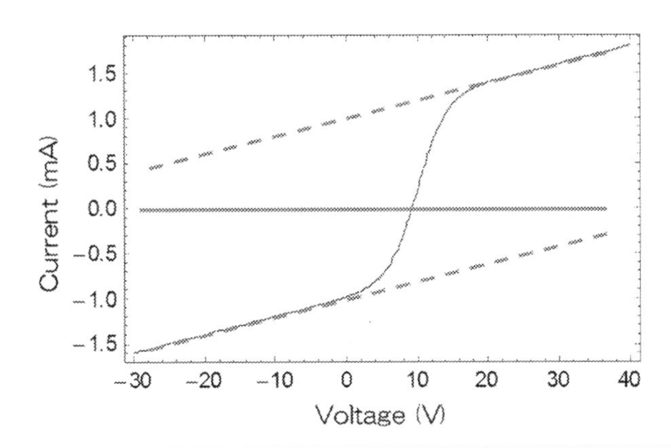

Fig. 7 Typical current-voltage characteristic curve of the double-probe measurement.

of the hyperbolic tangent function, that is, the ratio of the surface areas s_1 and s_2. The slope at the floating potential can be used to determine T_e.

2.2 Equivalent Circuit Model

A high-voltage power supply for the generation of atmospheric-pressure plasma provides high-frequency continuous and pulsed waveforms with frequencies ranging from direct current to gigahertz. The rise and fall rates are within times on the order of nanoseconds, while voltages are on the order of tens of kilovolts, and currents are on the order of milliamperes. The temporal behavior of the voltage $V(t)$ is measured by a high-voltage probe, which records the temporal variation in the voltage with an analog-digital converter, that is, a digital oscilloscope. For high voltages above 500 V, a voltage divider with certain characteristics (eg, a peak voltage of 20 kV and a maximum frequency of 75 MHz is typically used).

The temporal behavior of an electric current $I(t)$ is measured by voltage detection at the terminals of elements with known impedances. Another method is to measure pulsed and alternative currents using current transformers and Rogowski coils, as shown in Fig. 8.

A typical equivalent circuit for an atmospheric-pressure plasma source is shown in Fig. 9. The total electric current $I(t)$ is the sum of the discharge current of the gap capacitor and the dissipated power of the plasma generation. A constant voltage V is maintained and is defined as the sum of voltages at the terminals $V_{\text{dielectic}}(t)$ and the gap $V_{\text{discharge}}(t)$. Furthermore, $I(t) = I_{\text{discharge}}(t) + C_{\text{gap}}\frac{dV_{\text{discharge}}}{dt}$, $V = V_{\text{dielectic}}(t) + V_{\text{discharge}}(t) = \text{const.}$, and $V_{\text{dielectic}}(t) = R\left[I_{\text{discharge}}(t) - \left(C_{\text{gap}} + C_{\text{source}}\right)\frac{dV_{\text{dielectric}}}{dt}\right]$, where $I_{\text{discharge}}$ is the discharge current and C_{source} is the series capacitance. For example, the actual pulsed current is given by $I_{\text{discharge}}(t) = I_0 \exp\left(-\frac{t}{\tau}\right)$, which yields $V_{\text{dielectric}}(t) = \frac{I_0 R}{1 - RC/\tau}\left(\exp\left(-\frac{t}{\tau}\right) - \exp\left(-\frac{t}{RC}\right)\right)$. If the product RC is negligibly small, the last term is negligible. The product RC of a system should be smaller than τ to avoid slow response (Naudé et al., 2005).

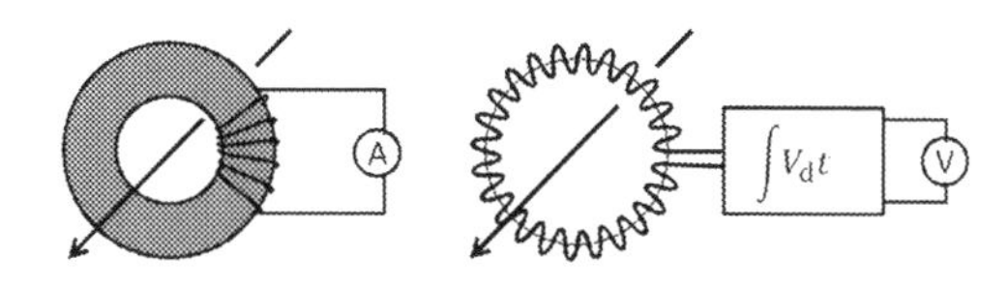

Fig. 8 Current transformer *(left)* and Rogowski coil *(right)*.

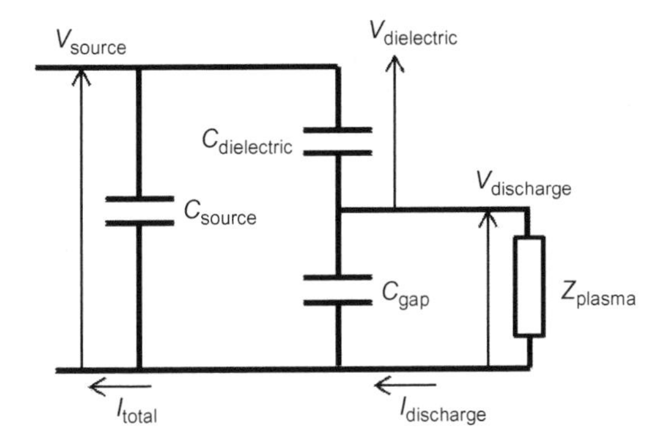

Fig. 9 Simplified electrical equivalent circuit of the gap and the solid electrode.

The power of the discharge is given by $P = \frac{1}{T}\int_{0}^{T} V(t)I(t)dt$. If the load is a capacitor and the applied voltage is a sinusoidal wave, the applied voltage vs. the current (Lissajous curve) is ellipsoidal. If the load is a resistance, the relationship is linear. A typical Lissajous curve is shown in Fig. 10. Multiplying the frequency $1/T$ and the capacitance C, the area surrounding the Lissajous curve corresponds with the power, as defined previously.

2.3 Interferometry

The refractive index of a plasma can be characterized and measured by interferometric methods. In atmospheric-pressure plasmas, a change in the gas number density greatly affects the refractive index. Thus, separating the changes in the electron and gas number densities is an important step.

In such methods, two beam paths are provided for only detection of frequency shift. The heterodyne interferometry probing laser beam, used to measure the plasma refractive index, is separated into two paths. One of the paths is shifted to frequency;

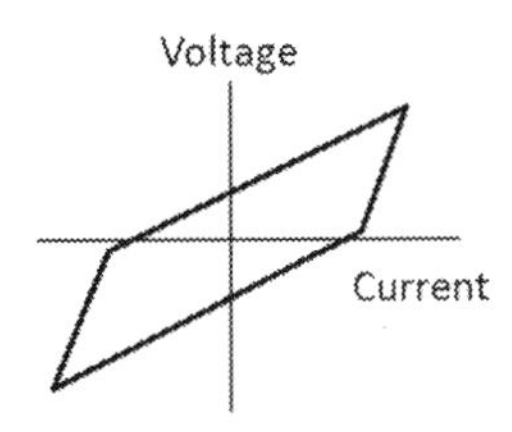

Fig. 10 Lissajous curve.

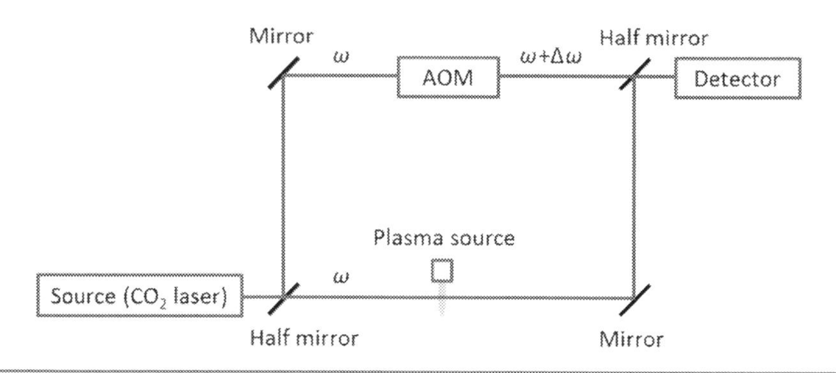

Fig. 11 Schematic of CO_2-laser heterodyne interferometry using a Mach-Zehnder interferometer equipped with AOM.

this shift is related to the change in the neutral and electron density in the discharge. A phase shift appears due to a change in the refractive index of the medium. After the two beams merge at the interferometer, the phase difference of the two beams can be measured using heterodyne techniques.

In CO_2-laser heterodyne interferometry at a wavelength of 10.6 μm, the oscillator frequency is provided by an acousto-optic modulator (AOM), and the original laser beam is separated into two paths (see Fig. 11). After the two beams are merged, the detected signal is transmitted through a low-pass filter to filter out unwanted high-frequency components, including ω and its harmonics. The beat frequency component $\Delta\omega$ is included as a phase shift signal. The AOM modulation is given by $E \cos \omega t$, the electric field of the light is $U_1 = E_1 \cos(\omega + \Delta\omega t)$, and the other electric field passing through the plasma with a phase shift Φ is $U_2 = E_2 \cos(\omega t + \phi)$. Therefore, the light intensity I is

$$I = |U_1 + U_2|^2 = E_1{}^2 + E_2{}^2 + E_1 E_2 \cos(\Delta\omega t - \phi). \tag{5}$$

Thus, the phase shift can be measured by a lock-in amplifier at $\Delta\omega$ (Choi et al., 2009). If the plasma can be assumed spatially homogenous, plasma density can be calculated by the relationship of laser path length along the plasma source and the refractive index.

3 Optical Diagnostics of Nonthermal Plasma

3.1 Instrumentations

Plasma emits light, consisting of electric and magnetic fields, which comprise electromagnetic waves; different types of electromagnetic radiation include X-rays, ultraviolet (UV) radiation, visible light, infrared radiation, and microwaves. Light is a form of energy, and each wavelength or frequency is associated with a certain amount of energy. The relationship between energy E and frequency ν is given by $E = h\nu$, where h is the Planck constant.

Spectroscopy is the study of the interaction between matter and electromagnetic radiation. In our discussion, instrumentation for the detection of light is first introduced. Instruments for spectroscopic measurements are typically constructed by the use of four components, as shown in Fig. 12: (1) a light source, (2) a light guide, (3) a spectrometer, and (4) a detector. For example, in plasma optical emission spectroscopy, plasma is used as a light source. Another method, called absorption spectroscopy, calls for the insertion of plasma as an absorber in the path of the light guide.

A representative example of a spectrometer is the diffraction grating spectrometer. A typical grating has blazed grooves with a given spacing d. Fig. 13 shows

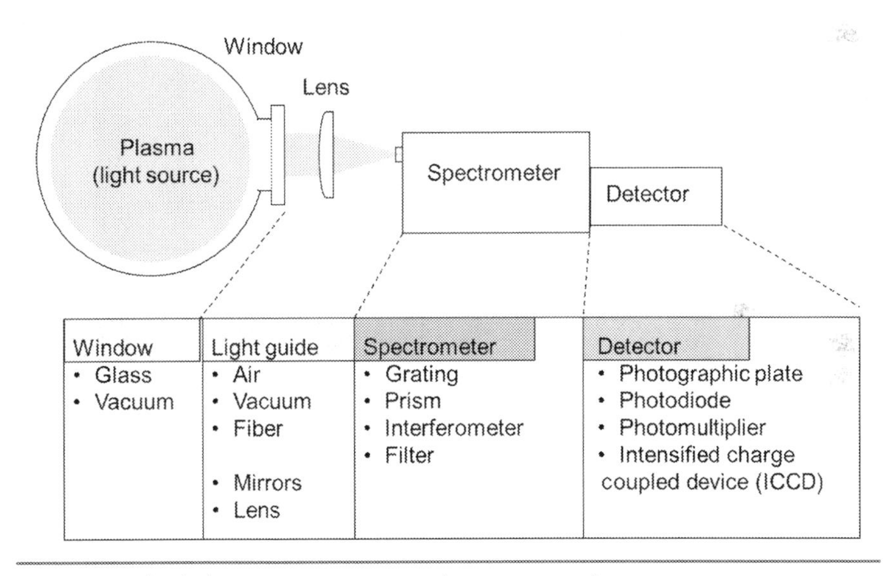

Fig. 12 Method of creating an apparatus for spectroscopic measurements.

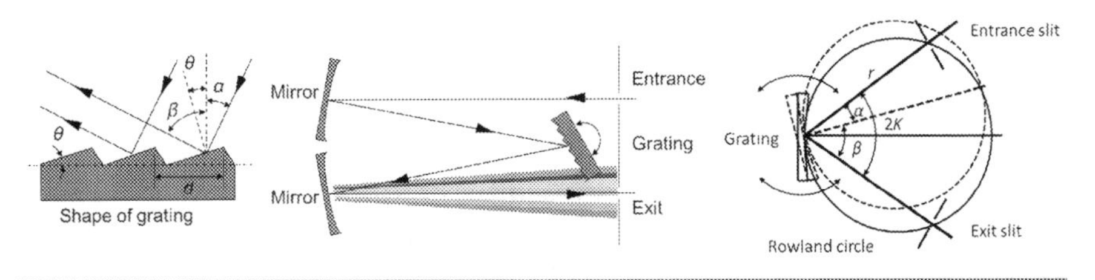

Fig. 13 Grating structure *(left)*, Czerny-Turner configuration *(center)*, and Seya-Namioka configuration *(right)*.

grating structures and diffraction conditions. The blaze wavelength λ and the diffraction angles α and β must satisfy the condition $d(\sin\alpha \pm \cos\beta) = m\lambda$, where m is the diffraction order. The spectral resolution is determined by $d\lambda/dl = \cos\beta/gf$, where l is the distance to the focal plane, g is the number of grooves, and f is the focal length. In practice, a condition with a focal length of 250 mm and a grating of 1200 lines/mm can obtain a maximum resolution of 0.15 nm. The Czerny-Turner design is commonly used for spectrometers. As the grating is rotated with monochromatic light directed through it, the light shining through the exit slit changes.

In the vacuum ultraviolet (VUV) range, grating monochromators generally use concave gratings, which reduce the required number of mirrors for low-transmission or refraction optics. Such a grating on a spherical mirror combines both dispersion and focusing in one element. For maximum resolution, two slits and a grating are mounted on the Rowland circle, which has a diameter equal to the principle radius of the spherical surface. The Seya-Namioka monochromator was designed for a VUV spherical grating without using the Rowland circle. The dimensions and locations of the entrance and exit slits are fixed, and the wavelength is scanned by rotating the grating around its axis tangent to the zeroth groove at the grating center. Here, the diffraction condition is satisfied by $m\lambda = 2d\cos K\sin\theta$. Thus r, and r', and $2K = \alpha - \beta$ are constant, $\alpha = K + \theta$, and $\beta = \theta - K$, where θ is the rotation angle of grating as defined in Fig. 13. If the focal length is 200 mm, $2K$ is approximately 70° and $R/r = 1.222$, where R is the radius of curvature of the spherical grating.

3.2 Optical Emission Spectroscopy

The emission and absorption processes are transitions between quantum states; that is, the behavior of the particles is controlled by the rules of quantum mechanics. Radiative processes related to electron transitions are divided into (1) transitions between upper- and lower-bound states (bound-bound transitions), (2) transitions between bound and free-electron states (free-bound transitions), and (3) free electrons gaining energy that is lost by positively charged particles emitting photons (free-free transitions, called Bremsstrahlung radiation). Usually, the first bound-bound transitions are important.

As discharges are sustained by introducing electromagnetic power, ionization and excitation processes result in spontaneous or diffusive losses. Ionized and excited states decay spontaneously by emitting a photon. Furthermore, all generated upper-bound electrons spontaneously lose energy by emission and reach an equilibrium state.

Spectral line width is inversely proportional to radiation lifetime. Mathematical processes, such as Fourier transformation, convert the time-domain functions into the frequency-domain function, thereby imparting clarity and better resolution. Infinite oscillation in the time domain converts it into a delta function without spectral

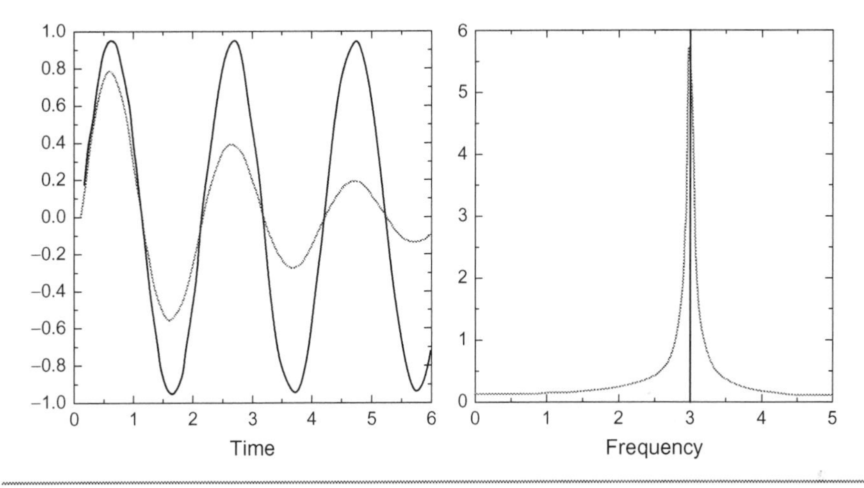

Fig. 14 Fourier transformation takes a time-domain function *(left)* and converts it into a frequency-domain function visualized as the spectrum *(right)*.

width in the frequency domain; this is shown in Fig. 14. Namely, a sinusoidal wave for optical emissions terminates after a time τ. If the sinusoidal waveform decays as $\exp(-t/\tau)$, the spectral profile is represented by a Lorentzian shape of $I(\nu) = \dfrac{I_0}{1 + (2\pi\tau\Delta\nu)^2}$.

The spectral broadening represents various kinds of information about the light emitters in the plasma. The spontaneous transition from the aforementioned upper state is determined by the narrowest spectral broadening width, called the Natural broadening.

3.3 Spectral Profile (Voigt)

The emitters in plasma take thermal motions, usually assumed to be the Maxwellian velocity distribution. This is called Doppler broadening, and full width at half maxima (fwhm) is analytically expressed by $\Delta\lambda = 7.16 \times 10^{-7}\lambda_0\sqrt{T/m}$, where λ_0 is the wavelength of emission line (in nm) under consideration, T is the gas temperature (in K), and m is the relative atomic mass (in Da). This essentially means that the measured $\Delta\lambda$ can be used for determination of gas temperature. The spectral profile is represented by a Voigtian shape.

3.4 Plasma Density (Stark Broadening)

Stark broadening occurs when an electric field is applied to the atom, polarizing the surrounding electron of the positively charged nuclei. An electric dipole moment is induced in proportion to the electric field. As a result, the energy shifts linearly with respect to the product of the electric field and the dipole moment. This leads to the energy shift of the atom, which obeys the square of the electric field; this is called the second Stark effect. For atomic hydrogen, the first Stark effect with a larger shift is observed. For atoms emitting photons, the shift is dependent on $\Delta\nu = kE^m$, where m is the order of the Stark effect. Stark broadening depends on the electron density. In the Balmer series of atomic hydrogen, broadening lines of H_α, H_β, H_γ, H_δ, and H_ε are observed as electron density increases. For the Balmer H_β line, a relationship between the plasma density and the spectral full width at half maxima $\Delta\lambda$ has been published in the existing literature: $n_e = C(n_e, T_e)[\Delta\lambda]^{3/2}$, where C is coefficient (Fig. 15) (Becker et al., 2004). If no hydrogen is contained in the targeted plasma, a trace amount of hydrogen can be applied without disturbing the plasma.

3.5 Optical Absorption Spectroscopy

Electromagnetic waves are transmitted through plasma or reflect at the plasma boundary. Regarding particles inside the plasma as absorbers, the transmittance of the light depends on the incident light intensity, satisfying the Beer-Lambert law. From the absorption coefficient k and the absorption length L, the absorbance $G(kL)$, is given as

$$G(kL) = 1 - \frac{I_a}{I_0} = \frac{\int f_a(\nu)[1 - \exp(-k_a f_l(\nu)L)]d\nu}{\int f_l(\nu)d\nu}, \tag{6}$$

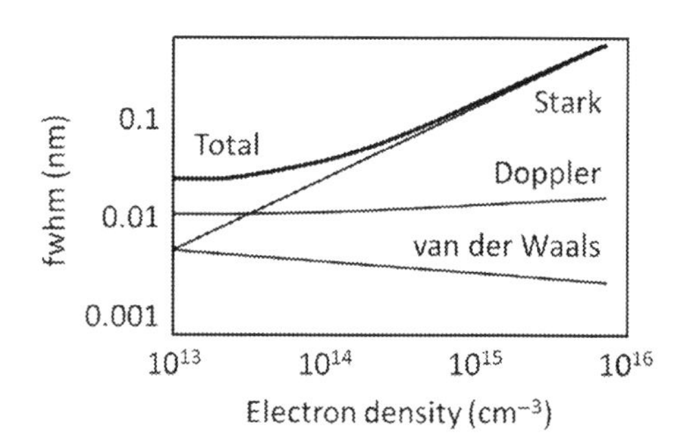

Fig. 15 Typical relationships between full-width at half maxima (fwhm) vs. electron density. If the electron density is larger than 10^{14} cm^{-3}, the Stark broadening determines the measured spectral width; however, high-resolution observations are still needed. *(From Becker, K.H., Kogelschatz, U., Schoenbach, K.H., Barker, R.J., 2004. Non-Equilibrium Air Plasmas at Atmospheric Pressure. CRC Press, Institute of Physics, London.)*

where $f(\nu)$ is the spectral profile function and subscripts "a" and "l" indicate the light source and absorber. The number N of atoms is related to the absorption of the transition of one atom as

$$N = \frac{8\pi v_s{}^2}{c^2}\frac{g_l}{g_u}\frac{1}{A}k_a\int f_a(\nu)d\nu. \tag{7}$$

The energy of the atomic lines remains within the VUV range. VUV light cannot be transmitted through ambient atmospheric air, therefore a vacuum light guide is necessary. Additionally, the spectral width of the atomic line is very narrow. Each atomic line is difficult to discriminate, even when a high-resolution spectrometer is used. Thus, when using a continuous light source, other dispersive methods, such as Fourier transform interferometry, are preferable. For simplification, this method uses resonant spectroscopy with the line emission of an identical atom. At a given temperature, the partition function of each level must obey the Boltzmann relation. As a result, if the line profiles of a light source and an absorber are identical, measurements can be taken independently with the spectral resolution of the detected system. Even if the main peak absorption is saturated, as shown in Fig. 16, the shoulder part of the main peak can be transmitted through in the case of different peak shape. This indicates that the spectral line shapes for both light source and resonant absorber are required to be identical. Additionally, the optical absorption length should be optimized.

To exemplify the significance of absorption spectroscopy, consider an oxygen atom, for which the absorptions ($^3S_1^o \leftarrow {}^3P_j J = 1,2,3$) arise from the transition between a triplet of the lower state and the degenerative upper state, as shown in Fig. 17. In accordance with the spin degeneracy, the ratio of the statistical weights is 5:3:1. Using a reference light source and a spectrometer (without spectral resolution of each lines), optical absorption measurements can be conducted.

3.6 LASER-INDUCED FLUORESCENCE

In laser-induced fluorescence (LIF), plasma species are studied by excitation with a laser. After the resonant absorption of the laser light, spontaneous emission from an excited state to a lower state occurs nondirectionally. This process is regarded as a type of scattering. In LIF, a narrow spectral width laser provides selectively electronic excitation. Because the fluorescence wavelength is different from the excitation wavelength, highly sensitive measurements are simple to perform.

In principle, a transition from a lower state X to a higher state A occurs, and the subsequent transition to another lower state B causes fluorescence (see Fig. 18). Because transition probabilities are determined by Einstein coefficients, the excitations of $A \leftarrow X$ are sufficiently frequent to saturate the population of state A, and the

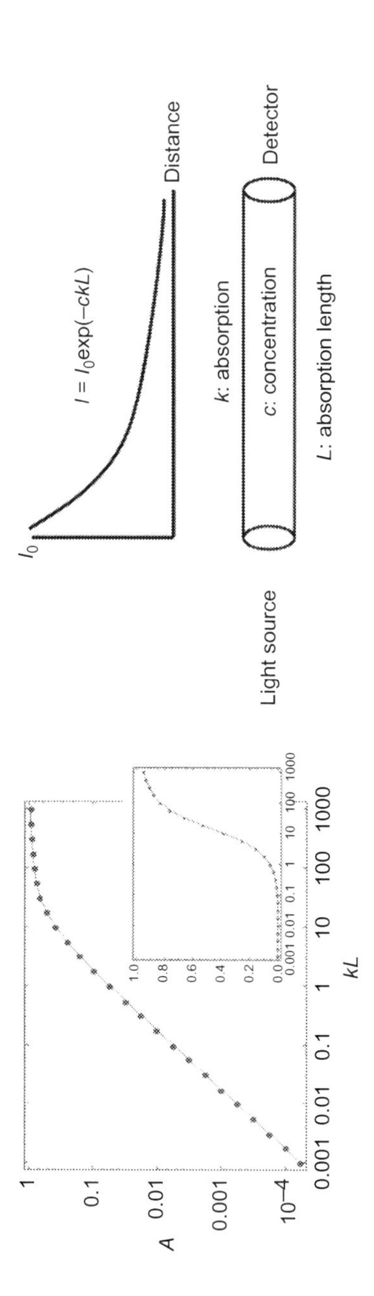

Fig. 16 Absorption coefficient A as a function of optical thickness kL.

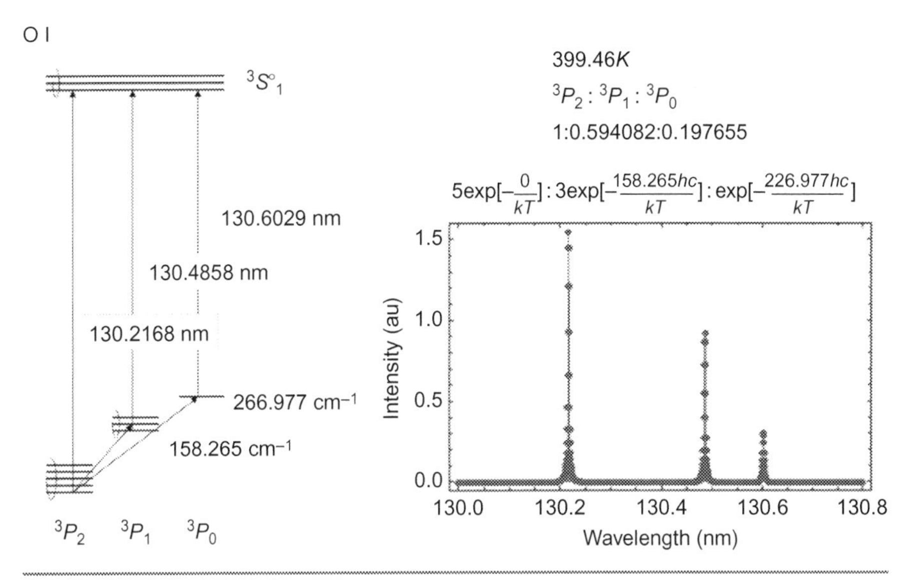

Fig. 17 Grotrian (term) diagram and high-resolution VUV spectrum for oxygen atom.

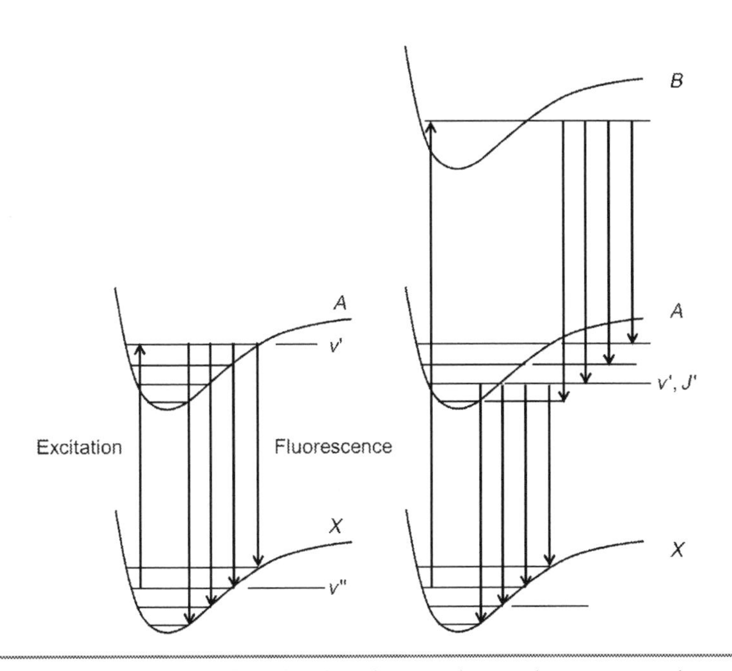

Fig. 18 Two-level *(left)* and three-level *(right)* LIF scheme after transitions between the grounded state *(X)* and the excited state (A and B).

fluorescent emission of $A \rightarrow X$ remains constant. In this situation, the population of each state is determined only by the statistical weight g. This indicates that the fluorescence intensities reflect the number of targeted atoms or molecules; otherwise, nonradiative quenching occurs. Using a sheet of laser light, the number density of the particles can be spatially mapped if quenching has not occurred. If quenching has occurred, the mapping can be estimated quantitatively.

In practice, atomic oxygen is excited by VUV light with a wavelength of 130 nm. Also, when a 225 nm-wavelength laser light (at half of the energy) irradiates, two photons are simultaneously absorbed; this causes a transition and a fluorescence at 844 nm to be detected. This is called two-photon absorption LIF (TALIF).

LIF easily provides the spatial distribution of fluorescence, which indicates the number density. The optical absorption spectroscopy can obtain a line integral intensity of the number density of absorbers, however a spatial distribution cannot be observed. Furthermore, LIF allows the two-dimensional mapping of the density contour, employing a sheet of laser beams. In TALIF, sheet beams are not commonly used to produce strong laser intensity, because a sufficient amount is gained through the nonlinear effect.

3.7 Laser Scattering

Light scattering arises from the electromagnetic radiation caused by the movement of scatterer electrons that vibrate incident to the electromagnetic field. Atoms, molecules, and small particles induce dipole oscillations called Rayleigh scattering. Vibrations are modulated and shifted as a result of molecular rovibrations; this phenomenon is called Raman scattering. Particles with diameters larger than incident wavelengths, but no larger than a dipole, induce Mie scattering.

Thomson scattering is the elastic scattering of electrons and is caused by free electrons. The difference between the velocities of electrons induces the Doppler effect and the spectral widths are widened by the Doppler shift $\Delta\lambda$ in proportion with the electron velocity v_e, estimated to be the EEDF. Parameters for the scattering depend on the excitation wavelength λ. If λ is shorter than the Debye length, the electron scatters incident light; otherwise, the Doppler shift of the ions causes scattering, which is governed by the group motion of the electrons. The scattering intensity depends linearly on the density of scatterer electrons. Low electron density is difficult to observe because of the strong overlapping of Rayleigh and Raman scattering signals. The Rayleigh signal should always be eliminated, and this is commonly performed using a triple grating spectrometer with a spatial filter as shown in Fig. 19. The total cross-section of Thomson scattering is $\sigma_T = \frac{8\pi r^2}{3}$, where the classical radius of an electron is $r = \frac{e^2}{4\pi\varepsilon_0 m_e c^2} = 2.88 \times 10^{15} [m]$.

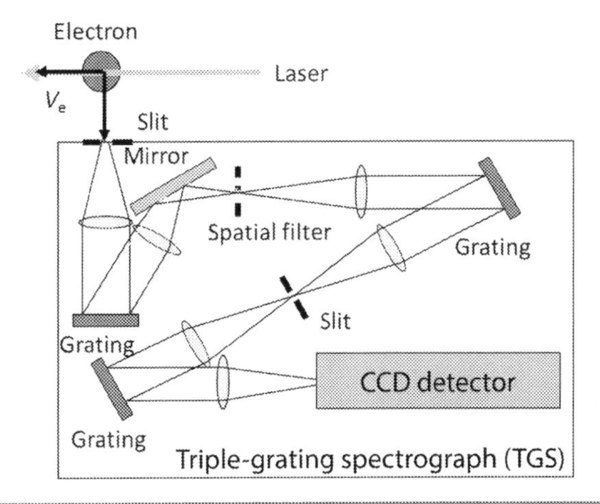

Fig. 19 Triple grating spectrometer for detection of Laser Thomson scattering.

In practice, the measured spectrum after obtaining the spatially filtered Rayleigh signal is shown in Fig. 20. Raman scattering is inelastic scattering, induced by the rovibrational motions of molecules. Raman signals are composed of Stokes (S-branch) and anti-Stokes (O-branch) scattering. Constituents of air, nitrogen, and oxygen yield strong Raman signals. In high-resolution measurements, rotational Raman spectra can be subtracted from simulated spectra as shown in Fig. 20.

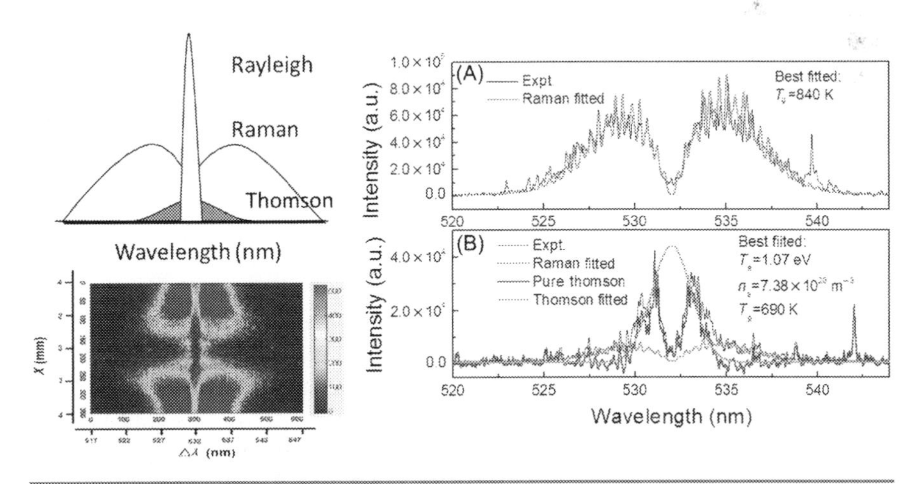

Fig. 20 Schematic illustration of atmospheric pressure plasma analysis *(right)*. Thomson and Raman scattering *(left)*. *(From Jia, F., Sumi, N., Ishikawa, K., Kano, H., Inui, H., Kularatne, J., Takeda, K., Kondo, H., Sekine, M., Kono A., Hori, M., 2011. Laser scattering diagnosis of a 60-Hz non-equilibrium atmospheric pressure plasma jet. Appl. Phys. Expr., 4, 026101.)*

This process yields discriminated Thomson scattering signals. Depending on the slit width of the spatial filter, an electron density of 10^{13} cm^{-3} and an electron temperature of 0.7 eV are the minimum requirements. From the Raman scattering spectra, the rovibrational temperature of the targeted molecules can be estimated.

In general, when scattering probabilities are not sufficiently high, the systems are not suitable for the measurement of deficient particles. Phase-matching nonlinear optical methods, such as coherent anti-Stokes Raman scattering, significantly enhance the detection efficiency of Raman signals for the diagnosis of radicals inside plasmas.

3.8 Infrared Spectroscopy

In the infrared region, transitions between rovibrational states and molecular rotation and vibration are observed instead of electronic transitions. This method is superior to detect molecular species. Fourier transform spectrometry is advantageous for the detection, with a high spectral resolution and high signal-to-noise ratio. The major types of noise associated with solid-state electronic devices are thermal, shot, and flicker noise. Thus, interferometers create time-domain interferograms by moving a mirror of optical paths. The time-domain interferogram is converted into a frequency-domain spectrum by Fourier transformation.

Tunable quantum cascade lasers were developed as high-brightness sources. The central wavelength of the laser depends on the operating temperature. Additionally, the electric current of the laser excitation allows wavelengths to be scanned for absorption. Homodyne detection, using a lock-in amplifier, allows high-throughput infrared spectroscopy.

4 Electron Spin Resonance

To detect radical species generated in plasma, electron paramagnetic spectroscopy can be used to directly measure unpaired electrons (dangling bonds, radicals, transition metal complexes, etc.). If there is no magnetic field, the quantum states of dangling bonds are degenerative. However, when an external magnetic field is applied, the degenerative spin states split into their different spin states; that is, electrons behave like small magnets. This phenomenon is called Zeeman splitting. Thus, absorption with respect to transition between these levels can be performed to detect dangling bonds.

When detecting the resonant absorption, it is considered that energy splitting is proportional to the magnitude of the external magnetic field. For a magnetic field of 300 mT, the energy splitting of dangling bonds usually corresponds with the energy of a microwave of 9 GHz, which is called the *X*-band. Depending on the magnetic field strength, microwaves with frequencies between 1 and 500 GHz are used.

Atoms and radicals with an odd number of electrons are paramagnetic and contain a permanent magnetic dipole. The hydrogen atom contains a single electron, and the magnetic dipole of the atom is caused by the motion of the electron around the nucleus. For an enclosed loop of the motion of this electron, the current is determined by the electronic charge $-q$, speed v, and circular orbit area $A = \pi r^2$. The magnetic moment is $\mu = -evr/2 = -eL/2m_e$, where L is the orbital angular momentum of the electron. The angular momentum is quantized as $L = h/2\pi\sqrt{l(l+1)}$ where l is the orbital quantum number (0, 1, 2, ...). The energy of the magnetic dipole under a magnetic field H directed along the z-axis is given by $E = -\mu H = \mu_b H m_l$, where μ_b is the Bohr magnetron, $\mu_b = -\dfrac{eh}{4\pi m_e} = 9.274\ 10^{-24 J/T}$, and m_l is $\pm l$, $\pm(l-1), \ldots, \pm 1, 0$. Thus, the energy splits into $(2\ l+1)$ levels. An s orbital ($l=0$) contains two electrons with spins of $m_s = \pm 1/2$. Three p orbitals ($l=1$) exist, and each can contain two electrons with spins of $m_s = \pm l$, 0. Additionally, fulfilled $3d$ shells ($l=2$) pose spin angular momentum.

Free radicals contain an unpaired electron. Two unpaired electrons with parallel spins have a total spin quantum number of $S=1$, which represents a triplet state. The spin angular momentum S of the electron gives rise to a magnetic moment, similar to that from the orbit. S is quantized as $S = h/2\pi\sqrt{s(s+1)}$, where s is 1/2 for an electron. The energy is defined as $E = g\mu_b H m_s$, where g is the g-factor. A transition between the two Zeeman levels resonantly takes place in the absorption of the electromagnetic radiation. For a hydrogen atom, $g = 2.00228$. Thus, the energy splits into $(2\ s+1) = 2$ Zeeman levels (Fig. 21).

In the classical Larmor precession picture, the resonance of a magnetic moment under static magnetic field, H, with an angular frequency of $\omega = e/2m_e H$, is expressed by the Bloch equation. The z-comportment of magnetization (M) spin is given by

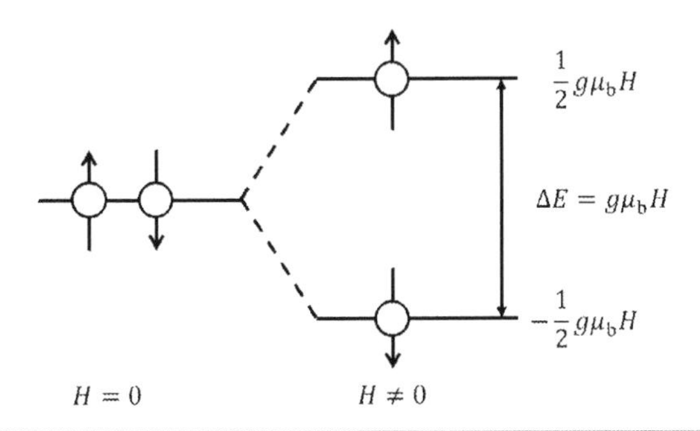

Fig. 21 Schematic representation of Zeeman splitting and resonant condition of electron spin resonance.

$$\frac{dM_z}{dt} = -\frac{M_z - M_0}{T_1}\ \frac{dM_x}{dt} = -\frac{M_x}{T_2}\ \frac{dM_y}{dt} = -\frac{M_y}{T_2}.$$

The rotating magnetic field follows the magnetic moment, providing spin torque. When the spin state is in resonance, the energy of higher state relaxes to the surrounding lattice. The rate of relaxation is expressed by the relaxation time, spin-lattice T_1 and spin-spin T_2 relaxation. If the relaxation rate is too slow to dissipate the absorbed energy required for the transition between the spin states, saturation occurs. The absorption does not increase in proportion to $\sqrt{P}$, where P is the microwave power.

Nuclei cause spin angular momentum. The magnetic moment along the z-axis of the magnetic field is $g\mu_N Hm_I$. The energy has a hyperfine structure, given by $E = g\mu_b Hm_s + Am_I m_s + g\mu_N Hm_I$, where μ_N is the nuclear magneton, and A is the hyperfine coupling constant. Each m_s level of the hydrogen atom splits into two levels, corresponding to $m_I = +1/2$ and $-1/2$ for an intensity of $I = 1/2$ of the proton nucleus. The selection rules are $\Delta m_s = 1$, and $\Delta m_I = 0$. The two transition lines are separated by the hyperfine constant, which is 50.5 mT for hydrogen atom. In a polyatomic paramagnetic species the magnetic moment of unpaired electrons interacts with several magnetic nuclei. These give rise to the level splitting in the degree of degeneracy. The relative intensity of lines coinciding with the degeneracy can be obtained from the Pascal triangle. For a total spin of $S = 1$ or higher, a fine structure or zero-field splitting appears. In the case of two radicals separated by a fixed distance, the spectrum is in the doublet state. The fine structure constant D depends on the distance between the radicals of the pair.

Electron nuclear double resonance (ENDOR) is a combination of electron and nuclear magnetic resonance. The hyperfine coupling A is obtained from the difference between the transition energies. Nuclear magnetic resonance (NMR) transitions are allowed according to the selection rule $\Delta Iz = 0$. If the electron spin resonance (ESR) transitions become saturated, nuclear spin flips are driven by radio frequency power, which is equal to the NMR transition. In ENDOR methods, additional frequencies of up to 150 MHz are applied (see Fig. 22).

In pulsed ESR, samples are exposed to a series of short intense microwave pulses. The traditional pulse sequence is two pulses. Free induction decay (FID) is performed for liquid samples, and electron spin echo modulation is performed for solid samples. Two pulses separated by a time t are applied. The second pulse is twice as long as the first. At a time t after the second pulse, a transient response appears from the sample; this is called the spin echo. By monitoring the echo amplitude as a function of time, a spin echo envelope can be recorded. The pulses are shorter than the spin relaxation times, and they excite spins. The resonator will stretch the pulses to make them longer, acting as a low-pass filter. The x- and y-components of M induce

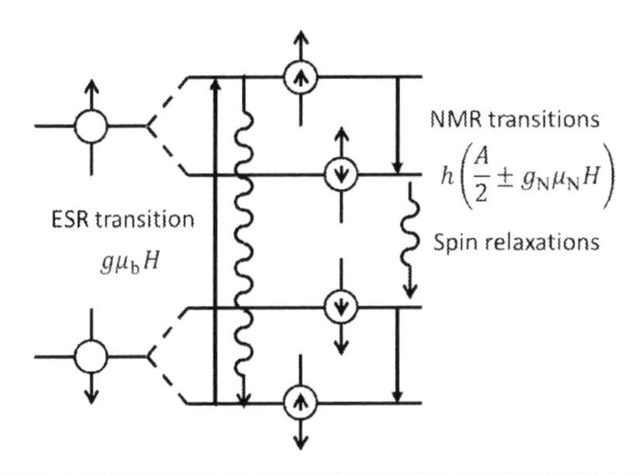

Fig. 22 Principle of ENDOR and electrochemical impedance spectroscopy.

precession (FID). FID occurs at precisely the same resonant frequency and decay as the T_2 spin relaxation time. The second pulse has the T_1 spin-lattice relaxation time of the z-component of M leftover by the first pulse. After a time of $2t$, the spin echo is induced.

Narrow lines are characterized by using the high-field ESR. Splitting due to different g-factors is proportional to the microwave frequency. Using both higher magnetic field and higher microwave frequency, high spectral resolution can be obtained.

In continuous microwave measurements, a microwave-resonant cavity is used. When the magnetic field is swept and microwave absorptions are recorded, electron resonant absorption is observed.

For short-lived radicals, such as reactive oxygen species which include hydroxyl radicals and superoxide, an approach known as spin trapping is required. Typically, the free radical adds to the double bond of a nitrone (Fig. 23) to form a radical adduct with a half-life of several minutes, which is long enough to allow detection.

5 Mass Spectrometry of Plasma

Electrically charged particles are analyzed by applying an electric or magnetic field via energy or mass separation. In the detection of electrical neutrals, the neutrals

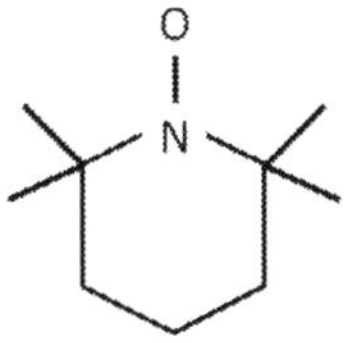

Fig. 23 Nitroxyl radical.

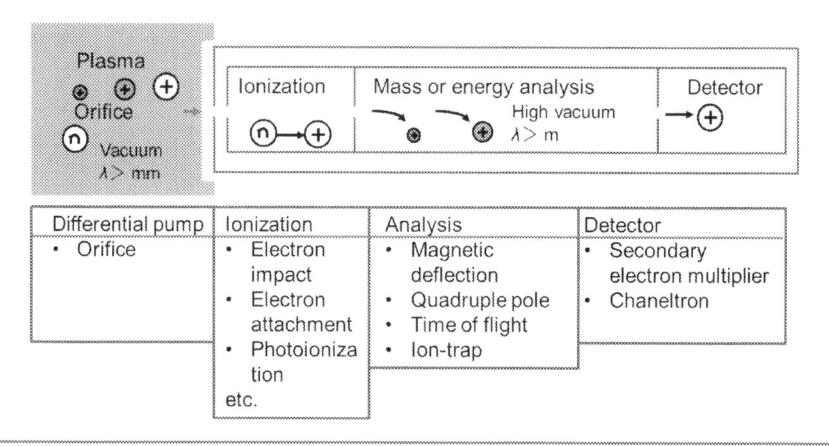

Fig. 24 Schematic representation of components in the mass spectrometer.

should be ionized before analysis. Compared with optical diagnostics, the sensitivity of detection is relatively high because electrically charged particles can sense electric currents. However, the transport of particles extracted from plasmas is disadvantageous. To conquer even these invasive, noncontact methods, it is informative if similar particles are detected at sampled surface locations, because it is important to know what species exist on the surface.

The apparatus is designed by the construction of a differential pumping system, ionization system, mass or energy separation system, and detector (Fig. 24).

For mass spectroscopic analysis, gaseous species are ionized, since only ions can be detected. Electron-induced ionization is commonly used. The electron energy is set to approximately 70 eV to maximize the ionization efficiency. Furthermore, the molecule is fragmented into small molecules. In most molecules, the ionization threshold energy appears around several tens of electron volts. Thus, an energy scan of the detected ions can analyze the molecule by their threshold energy. If the electron impact energy is set to a value above the appearance energy, and below the ionization threshold energy, then the radical can be selectively detected without the fragmental disturbance of the ionization of the parent molecule.

6 Concluding Remarks

Plasma diagnostics play an important role in explaining the plasma chemistry and dynamics for various technological applications. The study of plasma in general, and cold plasma in particular, is a complicated subject. There is a growing interest to understand and unravel the fundamental basis of its nonlinearity, the nonequilibrium nature, and the associated many-body problem, among other areas. The responses and fluctuations of plasmas have been extensively studied to clarify these phenomena. The development of real-time in situ diagnostics is also accelerating, which is likely to thrust forward the understanding of plasma dynamics in the near future.

Acknowledgments

The author would like to acknowledge Professors Masaru Hori, Makoto Sekine, Toshio Hayashi, and Keigo Takeda of Nagoya University; Professor Fendong Jia of Chinese Academy of Science; Professors Hitoshi Ito, Osamu Oda, Hiroki Kondo, Hiromasa Tanaka, and Hiroshi Hashizume of Nagoya University; Professors Takayuki Ohta and Masafumi Ito of Meijo University; and Dr N. N. Misra of Dublin Institute of Technology.

References

Becker, K.H., Kogelschatz, U., Schoenbach, K.H., Barker, R.J., 2004. Non-Equilibrium Air Plasmas at Atmospheric Pressure. Institute of Physics, CRC Press, London.

Chabert, P., Braithwaite, N., 2011. Physics of Radio-Frequency Plasmas. Cambridge University Press, Cambridge.

Chen, F.F., 1965. Electric probes. In: Huddlestone, R.H., Leonard, S.L. (Eds.), Plasma Diagnostic Techniques. Academic Press, New York, pp. 113–200.

Choi, J.Y., Takano, N., Urabe, K., Tachibana, K., 2009. Measurement of electron density in atmospheric pressure small-scale plasmas using CO_2-laser heterodyne interferometry. Plasma Sources Sci. Technol. 18 (3), 035013.

Jia, F., Sumi, N., Ishikawa, K., Kano, H., Inui, H., Kularatne, J., Takeda, K., Kondo, H., Sekine, M., Kono, A., Hori, M., 2011. Laser scattering diagnosis of a 60-Hz non-equilibrium atmospheric pressure plasma jet. Appl. Phys. Express 4, 026101.

Lieberman, M.A., Lichtenberg, A.J., 2005. Principles of plasma discharges and materials processing, second ed. Wiley, Hoboken, NJ.

Naudé, N., Cambronne, J.P., Gherardi, N., Massines, F., 2005. Electrical model and analysis of the transition from an atmospheric pressure Townsend discharge to a filamentary discharge. J. Phys. D. Appl. Phys. 38, 530–538.

Sakai, O., Kishimoto, Y., Tachibana, K., 2005. Integrated coaxial-hollow micro dielectric-barrier-discharges for a large-area plasma source operating at around atmospheric pressure. J. Phys. D. Appl. Phys. 38, 431–441.

Schmidt, M., 2013. Plasma diagnostics. In: Meichsner, J., Schmidt, M., Schneider, R., Wagner, H. (Eds.), Nonthermal Plasma Chemistry and Physics. CRC Press, Boca Raton, pp. 312–387 (Chapter 6).

Zakharowa, V.M., Kagan, Y.M., Mustafin, K., Perel, V.I., 1960. Probe measurements at medium pressures. Soviet Phys.-Tech. Phys. 5, 411.

Chapter 6

Principles of Nonthermal Plasma Decontamination

S. Patil*, P. Bourke*, P.J. Cullen*,†
*Dublin Institute of Technology, Dublin, Ireland, †University of New South Wales, Sydney, NSW, Australia

1 Introduction

1.1 Plasma as a Tool for Biodecontamination

A common process used for microbial inactivation is sterilization, which is a physical and/or chemical process (Moisan et al., 2002). Generally, a physical process such as heat, and chemical process like ethylene oxide and gamma radiation are used for microbial inactivation. Heat-sensitive materials cannot be sterilized by application of heat because it can cause irreversible damage to the product. Ethylene oxide is highly inflammable and toxic, thus it becomes important to ensure safety of operating personnel. Radiation processes have the unique requirements of isolated site and safety of operating personnel that makes its application undesirable. A number of such disadvantages drive research for a novel, highly efficient sterilization process that would be nontoxic, operate at low temperature, and is thus compatible for a wide range of materials (Sureshkumar et al., 2010).

Cold Plasma in Food and Agriculture. http://dx.doi.org/10.1016/B978-0-12-801365-6.00006-8

Among the microbial inactivation strategies, atmospheric pressure nonthermal plasma has achieved increased attention among advanced nonthermal technologies. It presents numerous potential advantages over conventional methods, such as its nontoxic nature, low process operational costs, short treatment time at low temperatures, significant reduction of water consumption throughout disinfection processes, and its applicability to a wide variety of goods (Chiang et al., 2010; Korachi et al., 2010; Song et al., 2009). Nonthermal plasmas constitute charged particles like positive and negative ions, electrons, quanta of electromagnetic radiation, and excited and non-excited molecules (Misra et al., 2011). The antimicrobial agents generated attack multiple cellular targets, making nonthermal plasma highly effective for inactivation. However, it is thought that there would be a limited possibility of emergence of a resistance mechanism, due to the range of active agents, mechanisms of action, and target sites. Literature reports a number of studies highlighting cold plasma antimicrobial efficiency and its potential application in food, industrial, clinical, biomedical, and healthcare areas. Generally, diverse modes of electrical discharges such as corona discharge, micro hollow cathode discharge, atmospheric pressure plasma (APP) jet, gliding arc discharge, one atmospheric uniform glow discharge (OAUGD), dielectric barrier discharge (DBD), and plasma needle may be used for plasma generation (Nehra et al., 2008).

Thus we review findings on the role of different plasma species in microbial inactivation and their effects on multiple cellular targets in addition to the microbial inactivation kinetics of nonthermal plasma.

2 Role of Plasma Species in Microbial Inactivation

Most nonthermal plasmas operate at low temperatures; therefore, no substantial thermal effects are expected on bacterial cells (Laroussi and Leipold, 2004). Thus, heat is not a major contributing factor in plasma inactivation efficiency. The major plasma sterilizing factors are decided by the plasma source type and plasma characteristics, which further dictates different sterilization results due to different plasma parameters (Deng et al., 2006). Different plasma parameters such as applied voltage, mode of plasma exposure (direct or indirect/remote), gas type, treatment time, and relative humidity have influence on the generation of reactive species. In general, optical emission spectroscopy (OES) is used as a diagnostic tool for reactive species measurement. An emission spectrum detected by OES can provide a quick overview of the excited species present (Stoffels et al., 2008).

2.1 Reactive Oxygen Species (ROS) and Reactive Nitrogen Species (RNS)

Reactive species generated differ depending on system design and key process characteristics, including inducer gas composition and applied voltage. Laroussi and

Leipold (2004) ruled out the role of heat and UV radiation in *Bacillus* spores using DBD-APP in two different gas compositions, mainly pure helium (He), and 97% He and 3% oxygen (O_2), but attributed increased inactivation efficacy to the presence of chemically reactive species such as NO, NO_2, O, ozone (O_3), hydroxyl radicals (OH) and other reactive species. Lu et al. (2008) recorded improved inactivation efficiency using nonthermal atmospheric plasma jet operated in He/O_2 gas mixture, and it was attributed to the ROS like oxygen atoms, O_3 and metastable state O_2^*. Boxhammer et al. (2012) demonstrated that only higher concentrations of nitrates and H_2O_2 generated by cold atmospheric plasma (CAP) resulted in enhanced bacterial reductions.

Heise et al. (2004) reported different DBD atmospheric plasma efficiency in different gases for spore inactivation. Sureshkumar et al. (2010) studied the effect of RF plasma on *Staphylococcus aureus* in N_2 and N_2-O_2 gases. Improved sterilization was obtained by 2% O_2 addition into the N_2 plasma, than using pure N_2 plasma alone. Based on OES, the authors concluded the highly reactive species like atomic nitrogen, oxygen radicals, and UV photons were possible major species responsible for enhanced antimicrobial effect. Synergistic action of argon (Ar) plasma generated reactive species, such as ionized Ar gas molecules, O_x, NO_x, and UV light for bactericidal effects, was also reported (Shimizu et al., 2008). Deng et al. (2010a) demonstrated argon plasma generated reactive species such as $^{\cdot}OH$, N_2, N_2^+, O radicals, and O_3 produced in Ar+O_2 plasma jet played a significant role in *Bacillus subtilis* inactivation. Deng et al. (2010b) indicated important contribution of $^{\cdot}OH$ radicals in bacterial inactivation. The $^{\cdot}OH$ intensities were increased in Ar+H_2O_2 gas plasma jet with enhanced inactivation effect. Deng et al. (2006) concluded that reactive species, such as oxygen atoms, metastable oxygen molecules, OH, and RNS produced using CAP either in pure He or He-O_2, were responsible for microbial inactivation. Atmospheric He plasma was more effective than He-O_2 plasma.

Eto et al. (2008) concluded O_3 was one of the main sterilizing factors, besides UV emissions that were generated by DBD plasma in dry air for inactivation of spores. While in moist air, it was due to the action of synergistic OH radicals. Patil et al. (2014) reported that relative humidity was a critical factor in bacterial spore inactivation by high-voltage atmospheric cold plasma (HVACP), where a major role of plasma-generated species other than ozone was noted. Kuzmichev et al. (2000) concluded DBD discharge in moistened air produced reactive species O_3 and OH and NO and NO_x species in air that enhanced the lethality of the killing process. Akitsu et al. (2005) introduced water vapor in He atmospheric plasma glow discharge in order to generate OH radicals. It was claimed that addition of O_2 to He generated O-atoms that react with microbial cell walls and destroy DNA. Schwabedissen et al. (2007) showed that surface DBD plasma generated O_3 as a reactive species in ambient air. However, addition of water vapor generated a number of reactive species like H_2O_2, nitric acid (HNO_3), nitrous acid (HNO_2), N_2O_4, and N_2O_5. Rodriguez-Mendez et al. (2013) investigated pulsed DBD plasma for bacterial inactivation in water and determined discharges in water generated chemical species OH, O, H, O_2^+ radicals.

Ferrell et al. (2013) found atomic oxygen, ROS, RNS, and in particular NO, were produced by pulsed-based nonthermal plasma discharge. Hernández-Arias et al. (2011) concluded reactive species, mainly atomic oxygen and O_3, tuned out to be the major species responsible for germicidal effects in water by pulsed DBD plasma.

Rapid bacterial inactivation by direct plasma treatment was ascribed to the synergistic action of short-living plasma agents and O_3 by Vaze et al. (2010). The direct impact of reactive species such as O_3, NO, hydrogen peroxide (H_2O_2), O_2, and OH on the outermost membranes and cell walls of microorganisms was reported (Laroussi et al., 2000; Laroussi, 2005). The ROS such as O_3, atomic oxygen, singlet oxygen, superoxide, peroxide, OH radicals were reported to be involved in bacterial inactivation (Joshi et al., 2011). Abramzon et al. (2006) reported a radio frequency (RF) high-pressure plasma jet generated mainly NO and OH reactive species with a strong impact on cell wall and cell membrane. Deng et al. (2007a) demonstrated protein destruction by He-O_2 atmospheric pressure glow discharge (APGD) plasma, and showed excited atomic oxygen and excited nitride oxide (NO) were main protein-destructing reactive agents.

The role of APP jet-generated, long-lived O_2 and O_3 molecules in inactivation was reported (Herrmann et al., 1999). Efremov et al. (2000) concluded O_3 generated by atmospheric pressure DC glow discharge in dry air was responsible for microbial inactivation. Gaunt et al. (2006) showed ROS like O, H_2O_2, superoxide, OH radicals were involved in spore inactivation. In addition, strong emission lines of He species, as well as UV and OH radicals, were noted. Jung et al. (2010) observed a significantly enhanced OH and excited oxygen atomic emission (OI) spectra emitted by APP with catalyst TiO_2. The reactive oxygen radical was suggested as a dominant factor for spore inactivation. De Geyter and Morent (2012) reviewed the role of plasma generated reactive species in the sterilization process. At high pressures, various reactive species like O, O_2^*, O_3, $OH^\cdot$, NO, NO_2 were suggested to play a significant role in plasma sterilization process (Laroussi, 2002, 2005; Fridman, 2008). The inactivation of *S. aureus* was attributed to hydroxyl radicals generated in water by He-O_2, APP (Bai et al., 2011). Bayliss et al. (2012) recorded that increasing the applied plasma voltage led to greater production of oxygen atoms.

Uhm et al. (2007) demonstrated that the Ar/O_2 plasma jet was highly effective for *Bacillus atrophaeus* inactivation and oxygen radical was suggested to play a dominant role. The major role of DBD plasma-generated OH radicals in microbial inactivation was reported by Hähnel et al. (2010) and Helmke et al. (2011). Hong et al. (2009) reported oxygen radicals generated by RF atmospheric plasma in He-0.2%O_2 mixture were effective for bacterial cells and endospore destruction. Colagar et al. (2010) investigated sterilization of *Streptococcus pyogenes* by afterglow DBD in O_2 and CO_2 gas. DBD discharge in CO_2 gas generated reactive agents such as CO_2^+, CO^+ and CO_3^-, and discharge in O_2 gas generated reactive species like O, O^+, O^-, and O_2^+. The results showed that DBD in O_2 gas was efficient for bacterial inactivation in liquid and on surface-cultured media, and it was attributed to active chemical agents: mainly O_3 and excited single-state oxygen O_2 ($a^1 \Delta_g$). Huang et al. (2007) studied bacterial deactivation effects of low-temperature argon atmospheric plasma brush with oxygen addition. Based on the results, authors suggested that the interaction of oxygen-based plasma species like O and O_3 with biomolecules of bacteria promotes the plasma efficiency with O_2 addition in Ar plasma. Superoxide anion radicals ($O^-{}_2{}^{\cdot}$) were reported to be one of the key elements for bacterial inactivation by plasma (Ikawa et al., 2010). Additionally, authors discussed the formation of peroxynitrite anion ($ONOO^-$) that can be formed by reaction between plasma-generated nitric oxide ($NO^{\cdot}$) and $O^-{}_2$ radicals in the presence of nitrogen gas, which might contribute to bacterial inactivation.

The emission spectra of thin-layer DBD plasma treatment, for the microbial inactivation of fresh pork and beef samples, was evaluated recently. The nitrogen and oxygen molecular spectra such as NO, N_2, $N_2{}^+$ were observed in the emission spectrum, where ambient air was used in the plasma system. O_3 was also generated by thin-layer DBD plasma (Jayasena et al., 2015). Surface microdischarge (SMD) plasma generated O_3 was attributed as the major agent for bacterial inactivation (Jeon et al., 2014). RF atmospheric plasma with H_2O_2 entrained in the feedstock gas showed effective oral bacterial inactivation and it was dependent on $^{\cdot}OH$ concentration (Kang et al., 2011). Kim et al. (2009a) recorded the primary role of ROS such as O, NO, and OH species in bacterial inactivation by RF atmospheric pressure microplasma.

Li et al. (2013) identified that direct current atmospheric pressure oxygen plasma generated neutral reactive oxygen species, such as O_3 and O_2 ($a^1 \Delta_g$), as dominant sources for oxidative stress on *Escherichia coli* cells. Lim et al. (2007) reported atomic oxygen radicals played a significant role in spore inactivation by atmospheric pressure Ar-O_2 plasma. Marchal et al. (2012) suggested that low-temperature plasma jet treatment of biofilms resulted in OH radicals production from water contained in biofilm or close to the biofilm.

Pavlovich et al. (2013a) measured the chemistry of gas-phase indirect air DBD treatment and reported O_3 as an important active agent in bacterial inactivation, although H_2O_2, nitrate and nitrite were also present. Schnabel et al. (2014) reported inactivation of microorganisms due to major active agents such as $NO^{\cdot}$, $NO^{\cdot}{}_2$, whereas other components like CO_2, H_2O, NO_2, NO_3 might play a secondary role in microwave plasma treatment. Sohbatzadeh et al. (2010) reported the main antimicrobial agents generated by atmospheric pressure DBD plasma were O_3, monoatomic oxygen and OH radicals. Takai et al. (2013) studied in vitro solution sterilization using low-temperature APP jet. The results of that study indicated that the bacterial inactivation was proportional to the concentration of hydroperoxy radicals ($HOO^{\cdot}$) in the solution. The high permeation of these radicals into the cell membrane was reported to play a key role for efficient bactericidal inactivation, using plasma and reduced pH method. Thiyagarajan et al. (2013) revealed primary reactive nitrogen species such as NO and NO_2 were effective for *S. aureus* and *E. coli* cells inactivation by atmospheric pressure resistive barrier discharge air plasma jet. Van Gils et al. (2013) used RF atmospheric pressure argon plasma jet for bacterial inactivation in solutions. The importance of plasma-induced chemistry at the gas-liquid interface was discussed in detail. It was shown that bactericidal effects were solely ascribed to the plasma-induced liquid chemistry, which produced stable and transient chemical species such as HNO_2, peroxynitrite $ONOO^-$, and H_2O_2. Under the tested experimental conditions, RNS were suggested to play a major role in bacterial inactivation rather than ROS. Wattieaux et al. (2013) revealed that microwave argon plasma jet at atmospheric pressure produced the potential biocide active species as atomic oxygen, O_3 and UV radiations. The interaction of important plasma species (OH, H_2O_2, O, O_3, O_2, and H_2O) with bacterial cell wall component, peptidoglycan, by reactive molecular dynamics simulations was investigated by Yusupov et al. (2013). It was reported that reactive species such as OH, O, O_3, and H_2O_2 could break structurally important bonds of peptidoglycan, that is, C—O, C—N, or C—N, that could consequently lead to bacterial cell wall destruction.

Ali et al. (2014) identified a dominant presence of OH radicals, Ar atoms and RNS/ROS in nonthermal annular plasma jet (NAPJ) and nonthermal soft plasma jet (NSPJ) emission spectrum. Inactivation of *Propionibacterium acnes* by NSPJ in shorter exposure time was attributed to various reactive species, such as atomic oxygen (O), $O_2{}^+$, OH and atomic nitrogen (N), which collectively affected the cells. Chang and Chen (2014) showed the presence of reactive species like OH radicals, neutral, and atomic species of nitrogen produced using low-temperature APP that played important role in microbial inactivation. Dolezalova and Lukes (2015) demonstrated the generation of H_2O_2, nitrite ($NO_2{}^-$), and nitrate ($NO_3{}^-$) ions in atmospheric pressure argon plasma jet treated water. The concentration of generated ROS and RNS generally increased with increasing plasma exposure time, and increased lipid peroxidation in bacteria was observed. Edelblute et al. (2015) suggested that reactive species O_3 and NO_2 were responsible for bacterial inactivation by atmospheric plasma.

2.2 UV Photons

The role of UV photons in nonthermal APP is debatable. Few studies reported their role in microbial inactivation by APP. In plasma sterilization, direct destruction of the microbial genetic material by UV radiation was suggested by Moisan et al. (2001). The major role of UV photons in microbial inactivation at reduced gas pressure and in oxygen-containing gas mixtures, in addition to the significant contribution of oxygen atoms as an erosion agent, was reported by Moisan et al. (2002). The presence of UV radiation by direct exposure of filamentary mode DBD plasma as one of the possible sporicidal mechanism was suggested by Trompeter et al. (2002). A relatively high level of UV emission generated by microwave argon plasma with a strong etching action was reported to be responsible for *E. coli* and methicillin resistant *S. aureus* inactivation (Lee et al., 2005). Sharma et al. (2005) and Rahul et al. (2005) stated microbial inactivation due to UV photons was produced by RF sustained plasma discharge and also by the action of highly reactive radicals present in the plasma afterglow. Plasma sterilization at reduced gas pressure in O_2 containing mixtures evaluating the respective roles of UV photons and oxygen atoms was investigated by Philip et al. (2002). Authors reported predominant role of UV radiation over erosion by oxygen atoms for spore inactivation. The major role of UV radiation in inactivation of *E. coli* using microwave-driven plasma discharge in Ar gas was also reported (Sato et al., 2006; Shimizu et al., 2008).

Boudam et al. (2006) conducted a set of experiments where *B. subtilis* spores were subjected to APP treatment in N_2-N_2O gas mixture. Experiments were performed under two different conditions; in one case microorganisms were exposed to plasma where UV radiation was strong, whereas in the other case, there was absence of UV radiation. Two situations were obtained by simply varying the percentage of the oxygen-containing molecule N_2O added to N_2. Whenever the UV intensity was high enough (at 0, 40, and 220 ppm of N_2O), the UV photons dominated the spore inactivation process, although oxygenated species were acting on the spore surface through erosion. Whereas in the absence of UV radiations (1000 ppm), the process was controlled by the action of reactive species.

The major sterilization effect of RF plasma in Ar gas on *S. aureus* due to the synergistic effect of UV radiation and argon-free radicals was reported (Sureshkumar, 2009). Eto et al. (2008) performed DBD plasma sterilization experiments using *Geobacillus stearothermophilus* spores with various ratios of N_2/O_2 mixtures. It was demonstrated that the UV emission intensities were dominant and increased with an increase in N_2 gas addition. When the N_2/O_2 gas ratio was equal, a weak peak in the UV wavelength range was observed. In N_2 atmosphere, mainly the UV radiation generated by DBD plasma in dry air accounted for sterilization. Sureshkumar et al. (2010) ascribed a significant role of UV radiation in effective bacterial inactivation using RF plasma in N_2 gas.

Chang and Chen (2014) attributed the inactivation of oral bacteria, *Enterobacter faecalis*, to the excited species, charged particles, and UV radiation generated by low-temperature APP. Guo et al. (2015) reviewed the results on the influence of UV radiation generated in the nonthermal plasma on the inactivation of microorganisms. The gas composition used for plasma generation was reported to have an influence on the role of UV radiation in microbial inactivation processes.

A minor role of the UV radiation in *B. subtilis* inactivation by APP jet was reported by Deng et al. (2010a). This finding was consistent with previously reported results by Hong et al. (2009). In that study, authors examined the atmospheric plasma inactivation effect of UV and temperature, where *B. subtilis* endospores were treated under a thin fused quartz plate, which allowed 90% of the UV (190–270 nm) to pass through. Results showed that the UV from the atmospheric plasma only slightly affected the viability of bacterial spores. Bacterial population of *Pseudomonas aeruginosa* was completely eliminated in 10 min using nonthermal plasma treatment with a very low UV output (Laroussi, 1996).

In contrast, no major contribution of UV photons in the inactivation of microorganisms by APP has been published in the majority of research studies (Choi et al., 2006; Deng et al., 2006; Helmke et al., 2011; Herrmann et al., 1999; Ikawa et al., 2010; Kelly-Wintenberg et al., 1998; Kim et al., 2009b; Laroussi, 1996, 2002, 2005; Laroussi and Leipold, 2004; Lu et al., 2008; Pointu et al., 2005). In APP, no role of UV radiation was assigned, as most of the UV radiation generated was absorbed by air molecules (Xu et al., 2009; Vrajova et al., 2009).

3 Effect of Nonthermal Plasma on Microbial Cells

Nonthermal plasma effects have been evaluated against a number of microorganisms. Plasma inactivation is a multifaceted process involving a varied set of effects on bacterial structure and its components. It is valuable when establishing a nonthermal plasma application to understand the biological target structural and biochemical characteristics to inform the selection of system- and process-related factors, which dictate the reactive species and thus the desired bactericidal effects.

3.1 Effect on Cell Morphology

Researchers assessed nonthermal plasma effectiveness by observing morphological changes in bacteria using microscopy techniques, such as scanning electron microscopy (SEM) and atomic force microscopy (Kuo et al., 2006; Pompl et al., 2009). Sureshkumar et al. (2010) recorded severely damaged *S. aureus* cells following RF plasma treatment in N_2-O_2 gas mixture. Recently, Ali et al. (2014) evaluated the inactivation efficacy of NAPJ and NSPJ devices against *P. acnes* and *P. acnes* biofilms. SEM demonstrated the morphological changes and damage in sessile bacteria in biofilm. The severity of damage was dependent on plasma treatment time. Significant damage in biofilm, with cell bursting as well as cell elimination including membrane damage with extensive leakage, was demonstrated after 1800 s of NAPJ treatment, whereas NSPJ treatment of only 300 and 600 s showed similar bacterial biofilm responses. Rossi et al. (2006) performed low-pressure plasma sterilization of *G. stearothermophilus* spores in different N_2/O_2 mixtures including pure nitrogen and pure oxygen. SEM demonstrated significant changes in dimensions of plasma-treated spores in the presence of high amounts of atomic oxygen.

Conversely, discharge in low atomic oxygen concentration with high UV radiation intensity showed no such effects for similar plasma treatment time. Bai et al. (2011) reported severely deformed *S. aureus* cell surfaces after 10 min of He/O_2 APP micro-jet treatment in water. Bermudez-Aguirre et al. (2013) investigated bacterial inactivation on fresh produce like lettuce, carrots, and tomatoes using atmospheric pressure cold plasma in argon. Authors observed important differences in physical structure of *E. coli* cells compared to the untreated cells. Bacterial cell structure showed extensive structural changes, including a high degree of electroporation, loss and disruption of membrane, and deformation after treatment. Similarly, extensive breakdown of the *E. coli* K-12 cell membrane due to unionized He plasma treatment was reported by Yu et al. (2006). Helium plasma discharge caused slight shrinkage, leakage of cytoplasm content, and breakage of the *B. subtilis* spore membrane due to plasma-generated ROS in the presence of heat, UV photons, electric field and other charged particles (Deng et al., 2006). Cullen et al. (2014) reported obvious shrinkage of *Listeria monocytogenes* using in-package DBD-ACP treatment. The same group reported cell damage after indirect exposure with plasma for *S. aureus* ATCC 25923 and *E. coli* NCTC 12900 (Fig. 1).

Yang et al. (2011) examined the morphological and structural changes in oral bacteria *Streptococcus mutans* and *Lactobacillus acidophilus* using low-temperature argon plasma treatment. Significant alteration in *S. mutans* cell size and morphological changes, including large amount of cell debris formation due to 15 s of plasma treatment, were observed. In case of *L. acidophilus* cells, longer plasma treatments of 60 s showed damages on cell walls. Additional increase in plasma exposure time to 300 s led to more cell damage and fragments of *L. acidophilus* cells were found. More rapid inactivation of *S. mutans* cells was observed due to pronounced cell damages caused by plasma treatment. Authors concluded that the plasma bacterial disinfection efficiency was dependent on the input power of plasma, the cell supporting media, and also on the type of bacteria. Boudam et al. (2006) reported less damage of *B. subtilis* spore structure with a low degree of spore erosion by exposure to APP in N_2-N_2O gas mixtures. Low-temperature APP germicidal effectiveness was evaluated against oral bacteria *Enterococcus faecalis* (Chang and Chen, 2014). A ruptured bacterial cell surface, including an outer cell wall with lost integrity, was noted after plasma exposure. Two min of plasma treatment effectively impaired and destructed the integration of bacteria.

Deng et al. (2010a) reported morphological changes such as direct physical impact on *B. subtilis* cells with leakage of inner substances after DBD-APP jet treatment in either Ar gas type or Ar+3.59% O_2. Hong et al. (2009) observed morphological changes in *E. coli* cells after 2 min of atmospheric plasma treatment at 75 W in He and O_2 mixture using transmission electron microscopy (TEM). Plasma treatment resulted in severe bacterial cytoplasmic deformations with leakage of bacterial DNA. Huang et al. (2007) examined deactivation effects of low-temperature

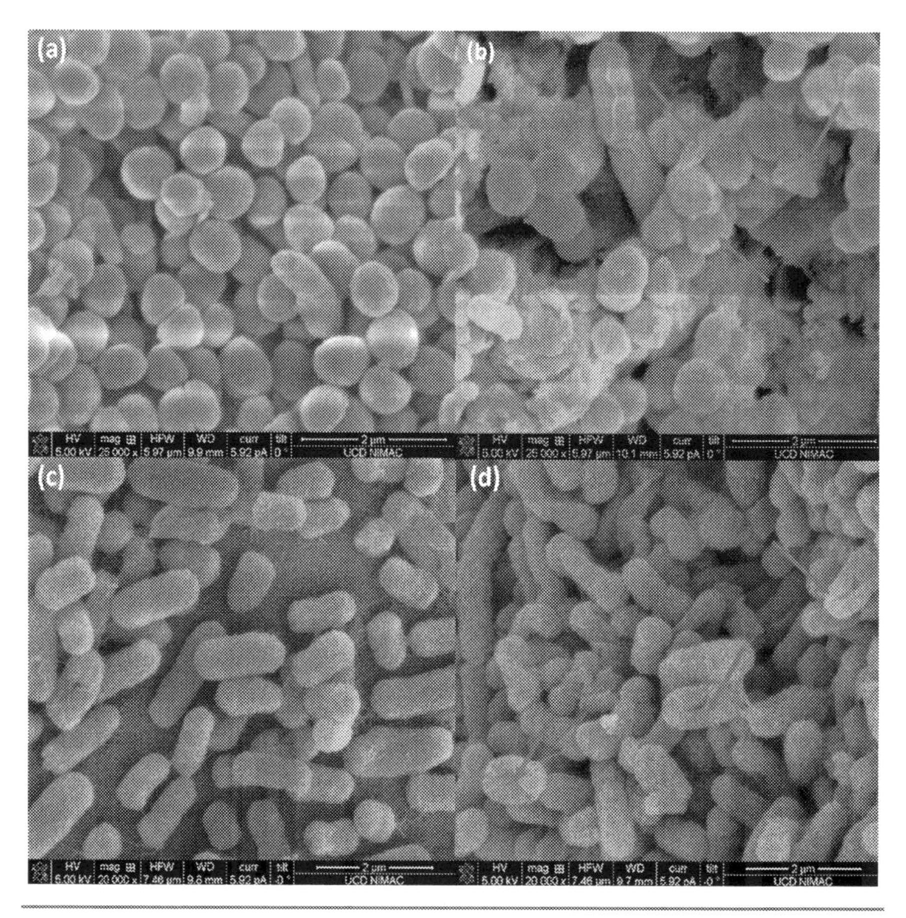

Fig. 1 SEM images of control and treated cells after indirect exposure with plasma at 80 kVRMS for 1 min following 24 h of posttreatment storage. (A) Untreated *S. aureus* ATCC 25923; (B) treated *S. aureus* ATCC 25923; (C) untreated *E. coli* NCTC 12900; (D) treated *E. coli* NCTC 12900. Arrows indicate the significantly different damaging patterns on cell envelopes of *E. coli* and *S. aureus*. *(Permission from Han, L., Patil, S., Boehm, D., Milosavljevic, V., Cullen, P.J., Bourke, P., 2016. Mechanism of inactivation by high voltage atmospheric cold plasma differs between* Escherichia coli *and* Staphylococcus aureus. *Appl. Environ. Microbiol. 82, 450–458. http://dx.doi.org/10.1128/AEM.02660-15. American Society of Microbiology.)*

atmospheric plasma against *E. coli* and *Micrococcus luteus*. SEM demonstrated significant alteration in cell size and transformed morphologies of *E. coli* cells. Distinct cell structure damage was observed after 5 min of Ar plasma treatment, but much smaller amounts of cell debris within 1 min of Ar plasma with an oxygen addition were reported. Authors suggested that the reduced size of cell debris of *E. coli* with 1 min plasma exposure was due to fast etching or erosion of cells by oxygen-enriched plasma species. Similarly, fragments of *M. luteus* were found after 5 min of Ar plasma exposure and the addition of oxygen to Ar plasma also resulted in damaged cell structure, with fine debris observed around each cell, although the cell size did not change much with shorter exposure time. Joaquin et al. (2009) examined sequential structural changes in Gram-negative *Chromobacterium violaceum* biofilm-forming cells after plasma treatment with different exposure times. After 60 min of plasma treatment, broken or amorphous cell structures were observed that were barely recognizable as bacterial remnants.

APP treatment of *S. aureus* cells for 20 min exhibited a complete change in the cell morphology and a clustering behavior was observed (Korachi and Aslan, 2011). Kostov et al. (2010) observed some structural changes in the bacterial surface leading to lysis of *S. aureus* cells and membrane erosion of Gram-negative *E. coli* cells after DBD plasma treatment in air. Microwave-induced argon plasma treatment of *E. coli* and methicillin-resistant *S. aureus* cells for only few seconds ($\geq$5 s) showed significant reduction in cell size, and transformed and amorphous cell morphologies with microscopic debris formation, which demonstrated a strong etching effect responsible for bacterial sterilization by plasma (Lee et al., 2005). Atmospheric positive corona discharge (APCD) induced changes in interior morphology and composition of fungal aerosols of *Penicillum expansum* spores were examined by Liang et al. (2012a). APCD caused noticeable morphological changes like damage to the cell wall and cell contraction with interior cytoplasmic changes that indicated diffusion of reactive species into the cell and their reaction with cytoplasm.

Disruption of the microbial cell walls and leakage of cytoplasmic contents as a result of short exposure of low-temperature plasma has been previously reported (Laroussi et al., 1999; Ma et al., 2008; Montie et al., 2000). Cooper et al. (2010) observed cellular elongation of *Bacillus stratosphericus* cells that was associated with a viable but nonculturable state due to DBD plasma treatment under a wet environment. Hu and Guo (2012) investigated DBD air plasma sterilization of *E. coli* on the surface of medical polyethylene terepthalate film. TEM demonstrated that longer exposure (5 min) caused bacterial cell envelopes to be severely convoluted and the overall density of nuclear material was low; some of the disrupted and fragmented lysed cells were also observed. Authors suggested that the loss of structural integrity and leakage of intracellular content was due to plasma treatment. Gliding arc discharge application for destruction of phytopathogen *Erwinia carotovora* subsp. atroseptica showed that it significantly affected overall structure of the microorganisms. Irregular surfaces with lost elongated shape were observed after 2.5 min of APP discharge in ambient air. Deep surface cracks with a complete decay of organized microbial structure were observed following prolonged plasma discharge exposure to 5 min (Moreau et al., 2007). Severe morphological alterations of the *E. coli* cell wall due to barrier discharge plasma treatment were also reported (Navabsafa et al., 2013). Microwave-induced argon plasma treatment caused severe morphological changes in *B. subtilis* spores and fungal spores *Penicillum citrinum* (Park et al., 2003). Damaged and ruptured cell membranes were observed only after 10 s of plasma treatment. Smaller cell structures with microscopic debris were observed after 20 s. Rapid damage of fungal spores, where transformed amorphous morphology with a significant reduction in cell size and holes on the cell wall, was observed. Unlike bacterial spores, plasma treatment of fungal spores for only 10 s caused microscopic debris formation.

Microwave-induced argon plasma treatment of *E. coli* resulted in deformed cell shape (Sato et al., 2006). *S. aureus* bacterial size reduction, and biofilm architectural damage shortly after exposure to plasma discharge, was observed by Traba and Liang (2011). Van Bokhorst-Van De Veen et al. (2015) evaluated nitrogen CAP-induced morphological changes in different spore formers such as *Bacillus cereus*, *B. atrophaeus*, and *G. stearothermophilus*. SEM showed alterations in CAP-treated spore surface morphology and loss of spore integrity. Severe physical damage, including etching and irregular surfaces on the spores, was observed. Authors recorded distinctive morphological changes in plasma-exposed spores, in comparison with other tested disinfection methods, including mainly UV, heat, and hypochlorite.

TEM demonstrated that nonthermal plasma exposure of 10 min caused alterations in the *S. aureus* cell surface, whereas SEM analysis showed no apparent alterations in bacterial spore structures treated under similar conditions (Venezia et al., 2008). Zhang et al. (2013) used Ar/O_2 (2%) cold plasma microjet to create plasma-activated water (PAW) for disinfection of *S. aureus*. TEM demonstrated damaged cell wall and internal structures of bacteria. Severe cell structural damage of *E. coli* and *M. luteus* cells due to argon plasma treatment was reported by Yu et al. (2006). Similarly, Liang et al. (2012b) observed atmospheric pressure nonthermal plasma-induced severe cell membrane ruptures of *B. subtilis* and *Pseudomonas fluorescens*.

3.2 Effect on Cell Membrane Function

Multiple approaches are generally used to detect cells with damaged cell membranes. A LIVE/DEAD BacLight bacterial viability staining kit is one of the approach used to detect cells with damaged cell membranes. It consists of two dyes, green-fluorescent SYTO-9 that stains live cells and red fluorescent propidium iodide which only penetrates cells with damaged membranes. Alkawareek et al. (2012) examined the effects of atmospheric pressure nonthermal plasma in He/O_2 mixture on *P. aeruginosa* biofilms. Confocal laser scanning microscopy demonstrated that plasma treatment of 240 s caused the majority of biofilm cells to stain red, indicating damaged cell membrane. Alkawareek et al. (2014) assessed atmospheric pressure nonthermal plasma effects on membrane integrity/permeability of *E. coli* cells. Compromised cell membrane integrity was examined by measuring extracellular

ATP concentration using an ATP Bioluminescent kit. Early plasma exposure time caused a slower rate of ATP leakage, which later followed with a much higher rate. Authors further suggested that the presence of threshold membrane damage, and possible pore size, has to be achieved before significant leakage can occur. Similarly, Joshi et al. (2011) demonstrated compromised *E. coli* cell membrane integrity after the exposure to nonthermal DBD plasma that was detected by examining lipid peroxidation. It is generally indicated by malondialdehyde (MDA) formation, one of the products of lipid peroxidation. Bai et al. (2011) evaluated the viability of He/O_2 APP-treated *S. aureus* cells by monitoring the cell membrane integrity using fluorescence microscopy. A mixture of fluorescent dyes, SYTO-9 and propidium iodide were used to stain untreated and plasma-treated cells. Plasma treatment of *S. aureus* (10^8 CFU/ml) for either 8 or 16 min greatly compromised cell membrane integrity, indicated by stained-red dead cells. Han et al. (2016) proposed different mechanisms of action for Gram-positive and Gram-negative microorganisms using in-package, high-voltage DBD-ACP as shown in Fig. 2. Examples A-C show the proposed inactivation mechanism of Gram-negative bacteria: "A" shows the structure of Gram-negative bacteria before treatment; the cell envelope consists of a thin layer of peptidoglycan and lipopolysaccharide; "B" shows ACP-generated ROS attacking both the cell envelope and intracellular components, where the cell envelope is the major target; and in "C" the inactivation is mainly caused by cell leakage, with some DNA damage possible.

Examples D-F show the proposed inactivation mechanism of Gram-positive bacteria: "D" shows the structure of Gram-positive bacteria before treatment; the cell envelope consists a thick rigid layer of peptidoglycan; "E" shows ACP-generated ROS attacking both the cell envelope and intracellular components, where intracellular materials are the major targets; and "F" shows the inactivation mainly caused by intracellular damage (eg, DNA breakage), but not leakage (Fig. 2).

Another approach used for examination of changes in cell membrane integrity is by determination of the release of intracellular material, absorbing at 260 and 280 nm, using optical absorption spectroscopy (Virto et al., 2005). Yang et al. (2011) demonstrated oral bacterial low-temperature atmospheric argon plasma deactivation effects by examining the leakage of intracellular components using a UV-visible spectrophotometer at wavelengths of 260 and 280 nm for DNA and protein absorbance, respectively. A few seconds of plasma exposure led to a large amount of protein and/or nucleic acid leakage, which indicated compromised cell membrane integrity. Deng et al. (2010a) investigated *B. subtilis* devitalization mechanism by atmospheric plasma jet in pure argon or Ar/O_2 mixture. The amount of protein leakage was determined by measuring absorbance at 595 nm by colorimetry, and it was increased with increasing plasma treatment time. The protein leakage quantity was less in Ar+O_2 mixture than in pure Ar, which was due to the action of ROS that could react with protein and consume a small part of the protein. In addition, lipid peroxidation was detected and interestingly, larger MDA production was measured in

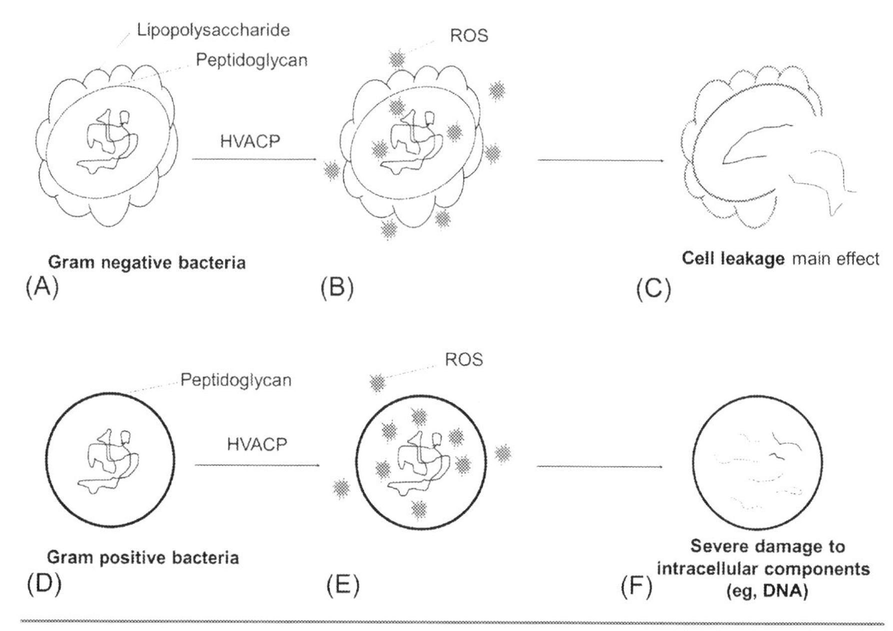

Fig. 2 Proposed mechanism of action of HVACP with Gram-negative and positive bacteria. *(Permission from Han, L., Patil, S., Boehm, D., Milosavljevic, V., Cullen, P.J., Bourke, P., 2016. Mechanism of inactivation by high voltage atmospheric cold plasma differs between* Escherichia coli *and* Staphylococcus aureus*. Appl. Environ. Microbiol. 82, 450–458. http://dx.doi.org/10.1128/AEM.02660-15. American Society of Microbiology.)*

Ar/O_2 plasma, rather than only Ar plasma. Similarly, Deng et al. (2010b) showed significant increase in protein leakage of Ar plasma treated *B. subtilis* cells due to the action of charged particles. However, with plasma exposure in $Ar+H_2O_2$ vapor mixture, the measured protein leakage was smaller, due to leaked protein consumption by OH radicals. Synergistic action of charged particles and chemical reactive species damaged the cell membrane.

Loss of *E. coli* cell membrane integrity due to the atmospheric pressure argon plasma jet was also reported by Dolezalova and Lukes (2015). Increased uptake of propidium iodide by permeabilized *E. coli* cells was observed and membrane permeability was increased with increasing plasma treatment time. Helmke et al. (2011) investigated the DBD plasma impact on the structural and functional level of microorganisms. Authors concluded membrane damage, in addition to DNA degradation, as the main mechanism for bactericidal effects of plasma; those effects were aggregated by milieu acidification. The cell membrane integrity analysis was performed using a LIVE/DEAD BacLight bacterial kit. Fluorescence microscopy revealed severe lethal damage to bacterial cell membranes after 1 s of plasma treatment, that is, 84% of the bacteria were with compromised membrane functions. An additional increase in treatment time to 6 s caused 90% of the bacteria to show damaged cell membranes. OH radicals were significantly involved in bactericidal effects, which could induce cell membrane damage. Deng et al. (2006) reported an interaction of reactive species produced by glow discharge plasma jet in a He-O_2 mixture, with the cell membrane of *B. subtilis*, caused damage and loss of cell membrane integrity. Liu et al. (2008) investigated germicidal efficiency of long-distance oxygen plasma at different treatment conditions against *E. coli* on the surface of medical PTFE film. Plasma exposure induced lipid peroxidation, and it was further suggested that MDA action can destroy structural and functional properties of the bacterial cell membrane. Ma et al. (2008) proposed that ROS kills bacteria by oxidizing bacterial cell membrane with further damage to nucleic acids and proteins within the cell. The concentrations of K^+, protein and nucleic acids leaked from cells were measured. The increasing concentration of leaked K^+ in *S. aureus* and *E. coli* was measured with increasing DBD plasma exposure from 0 to 10 s. Similarly, a sharp increase in leaked protein and nucleic acid concentrations of plasma-treated cells with exposure time >4 s was noted. Later the concentration was decreased with increasing exposure time. The rupture of cell membrane was indicated by the release of these intracellular components, which was consistent with the quick cell death that was observed within first several seconds.

Machala et al. (2009) demonstrated the major role of ROS in microbial cell membrane lipid peroxidation by plasma treatment. Damage of Gram-negative microorganisms' cell wall and outer membrane, which led to leakage of nucleic acid and protein from cells as a result of APP treatment, was reported by Kvam et al. (2012). Marchal et al. (2012) observed low-temperature plasma jet exposure of biofilm cells caused permeabilized or damaged cell membranes due to the action of plasma-generated reactive species. DBD air plasma sterilization of *E. coli* was studied by Hu and Guo (2012). Results indicated the primary mechanism of plasma sterilization as an etching action of electrons, ions, radicals on cell membranes, leading to the effusion of intracellular contents like DNA and protein and bacterial death. Navabsafa et al. (2013) reported barrier corona discharge plasma induced *E. coli* cell membrane rupture, determined by ATP assay. Sato et al. (2006) demonstrated the cytoplasmic and outer membrane destruction of *E. coli* due to plasma exposure. Sureshkumar et al. (2010) showed that N_2-O_2 plasma treatment of *S. aureus* cells led to membrane damage with a relatively higher leakage of intracellular content. Takai et al. (2013) reported the key species hydroperoxy radicals were responsible for inactivation of bacterial cells by nonthermal plasma treatment, and the high permeation of those radicals into the cell membrane was reported to be responsible for efficient inactivation. Traba and Liang (2011) used a LIVE/DEAD staining kit to evaluate the antibiofilm activity of plasma discharge gases, such as nitrogen, oxygen, and argon. Results indicated damaged cell membranes of plasma-treated *S. aureus* cells in biofilm form. Tseng et al. (2012) demonstrated the antimicrobial efficacy of He ACP jet for inactivation of the spores of *Bacillus* and *Clostridium*. Authors showed damage to the spore envelope, and loss of integrity of the spore inner membrane and spore coats, which was evident by the leakage of spore content, dipicolinic acid (DPA) and the inactivation of spore germination. Plasma inactivation of spores was primarily suggested due to spore coat/inner membrane damage detected by the subsequent leakage of DPA.

PAW was created using Ar/O_2 (2%) cold plasma microjet and its effectiveness was tested against *S. aureus* (Zhang et al., 2013). The effects on cell membrane integrity were studied by monitoring release of intracellular contents like K^+ ions, nucleic acids, and protein. Results of that study showed the damaged bacterial cell membrane integrity with greater leakage of cellular components.

3.3 Effect on Nucleic Acids

Alkawareek et al. (2014) investigated the plasma-exposure effect on the integrity of bacterial DNA to explicate the potential mechanism of plasma-mediated cell death. To obtain accurate quantitative analysis, authors used plasmid DNA as a representative, considering both the genomic and plasmid DNA will experience analogous interactions with plasma-produced oxidative reactive species. The intact plasmid is present in supercoiled (SC) form, that transforms into an open circular form when a single strand break (SSB) occurs on either of the SC plasmid DNA strands, whereas it reverts to a linear (LIN) form if there is a double-strand break (DSB) or a further adjacent SSB occurs (Yan et al., 2009). Alkawareek et al. (2014) showed rapid damage of intact plasmid DNA with a complete loss of plasmid DNA conformation after 90 s of plasma exposure. DNA damage as a potentially important aspect of plasma-mediated antibacterial efficacy was concluded. Joshi et al. (2011) investigated plasma induced oxidative effects on bacterial DNA. The effects were demonstrated by two methods involving agarose gel electrophoresis for detection of physical disintegration of DNA, and enzyme-linked immunosorbent assay (ELISA) to demonstrate oxidative DNA damage marker 8-hydroxydeoxyguanosine (8-OHdG). Nonthermal DBD plasma treatment of few seconds resulted in damaged DNA, and increasing appearance of 8-OHdG with increasing plasma treatment time. Similarly, Cooper et al. (2010) demonstrated extensive DNA damage of *B. stratosphericus* after 60 s of DBD plasma exposure, and indicated such short exposure was sufficient to disintegrate DNA in possible minute fragments. *E. coli* DNA degradation in liquid media as a result of plasma exposure was found by Yasuda et al. (2008). Dolezalova and Lukes (2015) reported bacterial DNA damage could be caused by plasma treatment and/or reactive oxygen and nitrogen species. DBD plasma treatment of only 12 s was reported to cause bacterial DNA strand breaks and its degradation (Helmke et al., 2011). Korachi and Aslan (2011) investigated the effect of APP corona discharge on bacterial DNA. Gel electrophoresis results revealed no

DNA bands post plasma treatment. Further effects on DNA were tested by performing amplification of 16S rRNA genes of *E. coli*. Appearance of a very faint DNA band was observed following plasma treatment, in comparison with control. Based on the observation, it was suggested that DNA was affected by the plasma treatment as a result of DNA fragmentation or degradation. Similarly, before and after plasma treatment, effects on other bacterial DNA such as *P. aeruginosa* and *S. aureus* were investigated. In comparison with control, fainter bands for all bacteria post plasma treatment were observed. In addition, *E. coli* DNA damage examined over different plasma treatment time revealed DNA degradation after 15 min. Lackmann and Bandow (2014) highlighted a number of studies that assessed the impact of plasma treatment on DNA in vitro. In vitro plasmid DNA treatment resulted in SSB and DSB, as a result of plasma jet exposure in helium-oxygen gas mixture or helium fed plasma plume, and was dependent on treatment time (O'Connell et al., 2011; Ptasińska et al., 2010). Lackmann et al. (2012, 2013) reported thymine dimer formation induced by plasma-emitted photons, and in vitro loss of DNA integrity due to modifications of nucleobases. Sousa et al. (2012) demonstrated nucleotides oxidation due to reactive species generated by plasma discharge. Li et al. (2013) investigated the effect of APP jet on inactivation of three kinds of *E. coli* substrains with their mutants. Authors concluded an important bacterial destruction mechanism was DNA damage caused by oxidation. Mai-Prochnow et al. (2014) reviewed some interesting studies with particular emphasis on mechanisms of interactions of bacterial biofilms with plasma. Colagar et al. (2013) reported complete degradation of *E. coli* genomic DNA after plasma exposure to >180 s. The conformational changes, like thymine dimer formation and single- and double-strand DNA breakdown, due to action of active plasma species were recorded. Kim and Kim (2006) investigated the damaging effects of helium/oxygen ACP on polypepetides, enzyme proteins, and DNA in the cell and reported that plasma treatment of >20 s led to complete degradation of M13mp18 DNA. Moreau et al. (2007) investigated the influence of glidarc APP discharge on the genomic DNA of a potato pathogen by electrophoresis and direct assay. Gel electrophoresis revealed a marked increase in the spot intensity of genomic DNA, extracted from the pathogen, and treated with glidarc discharge for either 1, 2.5, or 5 min; whereas spectrophotometric DNA assay showed increased concentration of extractible genomic DNA from phytopathogenic cultures after exposure to the glidarc discharge. Yan et al. (2009) investigated APP plume effects on the conformational changes of plasmid DNA pAHC25. Agarrose gel elctrophoresis showed that the percentage of the SC plasmid DNA decreased, whereas the open circular and linearized form of plasmid DNA increased. Kurita et al. (2011) exposed DNA solution to argon plasma jet and observed a marked change in the measured length of DNA molecule. Reactive species generated by gas discharge plasma were reported to induce severe DNA damage (Laroussi, 1996, 2002; Lerouge et al., 2001; Kelly-Wintenberg et al., 1998). Conversely Traba and Liang (2011) observed no bacterial DNA fragmentation after exposure of bacterial biofilms with argon plasma discharge for 10 min. Authors suggested plasma-induced bacterial cell death was mainly from cell membrane damage. Similarly, no detectable DNA fragmentation after 20 min of atmospheric plasma treatment of the vegetative cells of *B. subtilis* and their spore forms was observed, even though the D value for the cells was less than 30 s (Tseng et al., 2012). On the contrary, only 5 min of plasma treatments of naked DNA extracted from cells and spores showed noticeable DNA degradation. Longer treatment of 20 min led to severe DNA fragmentation with no sign of intact chromosomal DNA. Authors suggested the primary mechanism of plasma inactivation probably related to the damage of spore envelopes (Tseng et al., 2012). Venezia et al. (2008) demonstrated no alterations or changes to bacterial DNA after nonthermal plasma treatment. Han et al. (2014) found that 5 s of HVACP treatment compromised membrane integrity with a strong increase in leakage, but no significant impact on DNA damage was noted; this suggests that repair is possible when the microbiological target is subject to very short treatment times, even at high voltage. PCR results revealed multisite DNA strand breakage, but cell viability could be maintained with the low level DNA damage observed with the very short plasma treatment time of 5 s (Fig. 3).

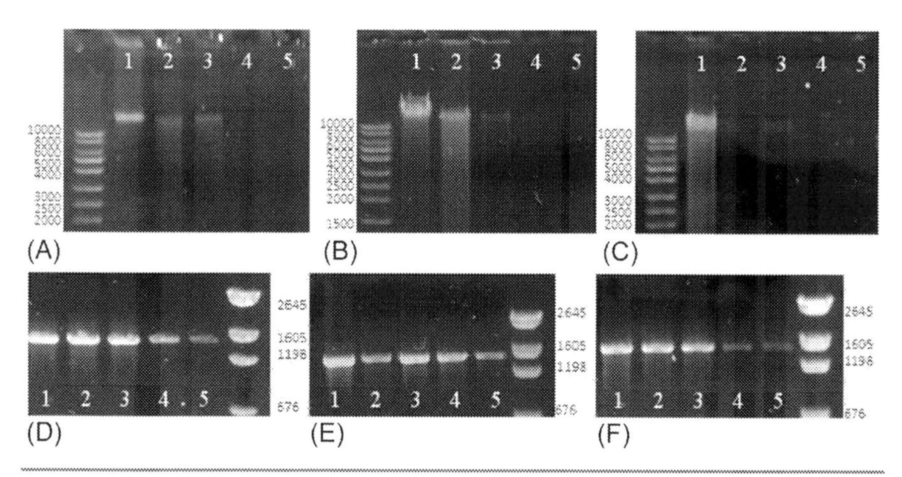

Fig. 3 Agarose gel electrophoresis showing genomic DNA and PCR amplified products of untreated and ACP treated samples. Voltage: 70 kVRMS; treatment time: 0–30 s; posttreatment storage time: 24 h; gas mix: air genomic DNA damage of (A) *E. coli* ATCC 25922; (B) *E. coli* NCTC 12900; (C) *L. monocytogenes* NCTC 11994 16S rRNA PCR results of (D) *E. coli* ATCC 25922; (E) *E. coli* NCTC 12900; (F) *L. monocytogenes* NCTC 11994. *Lane 1*: nonplasma treatment control; 2: 5 s directly treated samples; 3: 5 s indirectly treated samples; 4: 30 s directly treated samples; 5: 30 s indirectly treated samples. *(Permission from Han, L., Patil, S., Cullen, P., Keener, K., Bourke, P., 2014. Bacterial inactivation by atmospheric cold plasma: influence of process parameters and effects on cell leakage and DNA. J. Appl. Microbiol, 116 (4), 784–794. John Wiley and Sons.)*

Roth et al. (2010) investigated the effect of low-pressure plasma on *B. subtilis* spore DNA by monitoring the destruction of particular DNA fragment and revealed DNA to be the primary target for spore inactivation by UV radiation emitted by the plasma. Han et al. (2014) demonstrated severe genomic DNA damage of *E. coli* and *L. monocytogenes* cells after 30 s of HVACP treatment. Authors further performed PCR to assess DNA fragmentation by high-voltage plasma, which revealed that the extent of DNA damage was dependent on the type of bacteria and plasma treatment time.

3.4 Effect on Proteins and Enzyme Activity

Alkawareek et al. (2014) investigated the effect of atmospheric pressure nonthermal plasma on model protein enzyme, serine protease, and proteinase K. Increasing plasma exposure resulted in a reduction of the catalytic activity of proteinase K, but the inactivation rate was significantly lower than the plasmid DNA damage rate. Authors suggested that to severely alter the native enzyme structure and damage the active site, it might require the accumulation of several events, such as oxidation of amino acid side chains or disruption of hydrogen bonding in the tertiary structure of the protein. Lackmann and Bandow (2014) reviewed inactivation of microbes and biological macromolecules by APP jets. Authors reported few studies that analyzed the impact of CAP on the different macromolecules. Elmoualij et al. (2012) employed nitrogen-oxygen plasma treatment for both noninfectious and infectious forms of prion protein inactivation, and reported that plasma jet treatment can inactivate prion proteins; however, an afterglow plasma treatment of 40 min failed to achieve complete prion inactivation as residual infectious particles still remained. Lackmann et al. (2013) recorded glyceraldehyde 3-phosphate dehydrogenase (GapDH) inactivation by helium-oxygen plasma jet treatment in vitro and in vivo. Enhanced GapDH inactivation was observed after in vivo enzyme treatment was done by exposing *E. coli* cells, indicating a liquid environment favorable for protein inactivation. Mass spectrometry analysis revealed the overoxidation of the catalytic cysteine in the enzyme's catalytic center. Deng et al. (2007b) studied atmospheric DBD plasma jet use for the removal of proteinaceous matters from stainless steel surfaces. Bovine serum albumin (BSA) was used as a model protein. Electrophoresis analysis revealed a progressively thinner and weaker protein band after plasma treatment of 4, 60, 180, or 300 s in comparison to the prominent strong band of the control samples. Based on this, authors concluded there was significant damage of the protein that was removed from the surface, whereas if it remained on the surface its integrity was compromised; thus it likely to have a reduced infectivity, if any. Also, Takai et al. (2012) investigated the chemical effects of low-temperature APP on protein in an aqueous solution, using lysozyme as a model. Plasma treatment resulted in decreased enzymatic activity and unfolded a secondary structure due to the increased molecular weight of lysozyme with chemical modification. Plasma-generated reactive species that affect lysozyme and neither UV light nor heat from plasma was suggested to be responsible for this effect. Mogul et al. (2003) assessed the effects of low-temperature O_2 plasma at differing powers on the destruction and removal of specific biomolecules, such as model proteins including BSA and soyabean lipoxygenase (SLO). It was observed that the increasing powers of plasma exposure (50–250 W) caused a nonlinear decrease in the BSA amount and catalytic activity rate of SLO. The resulting loss of sample and enzyme kinetic function was attributed to oxidation, fragmentation, and removal of protein due to the O_2 plasma chemistry. Venezia et al. (2008) examined the lethal effects of the nonthermal plasma-generated sterilants on microorganisms and investigated potential damage to biological macromolecules, including proteins. Electrophoresis analysis showed that the changes in exposed bacterial proteins did not appear altered when proteins were within the bacterial cellular membranes. However, when proteins were directly exposed to plasma treatment, the alterations in the electrophoretic patterns of isolated bacterial proteins did occur and it showed degradation on SDS-PAGE. Mai-Prochnow et al. (2014) reviewed proteomic changes induced in bacterial cells upon exposure to APP. Colagar et al. (2013) assessed the effects of nonthermal atmospheric plasma jet on *E. coli* cells and their macromolecules, exposed for different time durations. The changes in banding patterns of total protein were observed following gel electrophoresis analysis. Plasma treatment led to protein degradation, with degradation of larger proteins (50–90 kDa) first, often causing an observed increase in smaller protein fragments ($\leq$25 kDa) in samples. Kim and Kim (2006) assessed He/O_2 atmospheric pressure cold plasma induced damaging effects on polypeptides using BSA as a model structural protein and alkaline phosphatase as a model enzyme protein. The quantity of intact BSA was decreased with increasing plasma exposure durations. Also, enzyme activity was shown to dramatically decrease upon plasma exposure. In addition, the effects of plasma exposure on intracellular enzyme protein was measured by determining the plasma effects on intracellular catalase activity. Intracellular catalase activity was abrogated in plasma exposure in a time dependent manner. Hou et al. (2008) investigated the effects of DBD atmospheric plasma on membrane-bound proteins and intracellular proteins in *Klebsiella pneumoniae*. It was observed that the concentration of membrane-bound proteins decreased remarkably, whereas the protein concentration in cell supernate increased. The distribution of membrane-bound proteins analyzed by SDS-PAGE indicated that proteins could be decomposed into peptides with low molecular weights by DBD plasma. With the increasing plasma exposure time, proteins were decomposed to amino acids. Moreau et al. (2007) studied the glidarc discharge effects on various molecular targets of phytopathogens, including membrane proteins and an exoenzyme, pectate lyase (PEL). The effect on membrane proteins was investigated by measuring the total concentration of extractible outer membrane proteins and later were analyzed by SDS-PAGE. A major difference in the electrophoretic pattern of outer membrane proteins, before and after exposure to a glidarc discharge, was observed. However, no variation in the PEL activity was observed after exposure to a glidarc discharge and suggested that oxidation specifically affects membrane-linked proteins due to their close contacts within the membrane. Moisan et al. (2013) reported plasma afterglow particles, reactive species, and UV photons were required for an efficient reduction of the lysozyme enzymatic activity. Additionally, the activity of infectious prion protein was reported to be reduced after exposed to plasma afterglow. Deng et al. (2007b) investigated the application of helium-oxygen APGD to inactivate BSA proteins deposited on stainless steel surfaces. A maximum protein reduction of 4.5 logs was achieved. A possible synergistic effect between atomic oxygen and NO was responsible for destructing and degrading surface protein. Authors further suggested plasmas could potentially be used to inactivate infectious prion proteins. Removal of infective proteins from surgical instruments using RF low-pressure argon/oxygen gas plasma was also previously reported (Baxter et al., 2005, 2006).

4 Kinetics of Microbial Inactivation in Nonthermal Plasma

The physicochemical processes taking place during a plasma sterilization could be identified by analyzing survival curves, which are a plot of the logarithm of the number of surviving microorganisms as a function of plasma exposure time (Moisan et al., 2002). A number of factors dictates the killing process by plasma, such as the type of bacteria, the type of medium, the number of layers in the sample, the type of sample, the working or operating gas type, contribution or no contribution by UV, etc. (Laroussi, 2005).

De Geyter and Morent (2012) reviewed nonthermal plasma sterilization of nonliving and living surfaces, mainly focusing on the inactivation of bacteria. Authors reviewed a number of studies that described different inactivation kinetics of plasma sterilization. Boscariol et al. (2008) investigated *B. atrophaeus* spores inactivation by pure O_2 and O_2-H_2O_2 plasma and obtained linear spore survival curves, where the *D*-values were dependent on discharge power level. Chemical erosion was proposed as a main mechanism for inactivation. Moreau et al. (2000) studied bacterial spore inactivation using low-pressure N_2/O_2 plasma. Spore survival curves with two inactivation phases, with a large number of remnant survivors, were yielded from the flowing afterglow of a microwave-sustained discharge in pure argon. However, addition of O_2 to Ar gas for plasma generation yielded sterilization with three inactivation phases. Moisan et al. (2002) interpreted the three-phase survival curves obtained from plasma sterilization of bacterial spores. It was suggested that the first phase showed the highest killing rate with the smallest *D*-value, due to the action of UV photons on isolated spores. The second slowest kinetics phase was attributed to the UV photons and active species, which induced erosion of materials such as spore coat, debris, and dead spores covering living spores. The third inactivation kinetics phase was started when the last living spores were sufficiently eroded, thus permitting UV photons to act directly, resulting in a *D*-value of the last phase that is very close to first phase *D*-value.

A few authors reported biodecontamination by nonthermal plasma resulted in single-slope curves with *D*-values ranging from 4.5 to 5 min (Choi et al., 2006; Herrmann et al., 1999; Laroussi et al., 2000). Two-slope survival curves in the APP sterilization process were also recorded (Kelly-Wintenberg et al., 1998; Laroussi et al., 2000; Montie et al., 2000). Abramzon et al. (2006) observed biphasic inactivation of bacterial biofilms by RF high-pressure cold plasma jet. The results showed first phase with rapid decline in bacterial survivors, followed by a much slower second phase which was characterized by subsequent decline in microbial level. Authors suggested that an initial rapid decline might be due to the killing of the easily available upper layers of the microorganisms in biofilm form. However, the second phase was slower because the plasma had to penetrate the layers of cell debris and dead cells prior to reaching the inner portion of the biofilm. Authors also discussed earlier reported biphasic and triphasic plasma inactivation kinetics of the free living microorganisms and spores. Kelly-Wintenberg et al. (1998) and Laroussi et al. (2000) reported biphasic plasma inactivation kinetics for *S. aureus* and *E. coli* on polypropylene samples, and *P. aeruginosa* in liquid suspension. It was hypothesized that in a biphasic curve, during the first killing phase, the reaction of active plasma species with the outer membrane of bacterial cells would cause damaging alterations. This results in compromised membrane, where reactive species can easily penetrate through, causing a rapid cell death in the second phase. Similarly, a biphasic trend of inactivation for *P. aeruginosa* biofilms after atmospheric pressure nonthermal plasma exposure was observed by Alkawareek et al. (2012). The rapid decline in the number of surviving cells during the first phase (initial 60 s) of plasma exposure, with the *D*-value of 23.57 s, was observed. After 60 s of plasma exposure, a slower second phase with lower bacterial reduction and a higher *D*-value 128.20 s was recorded. This result was concurred with previous studies that reported biphasic survival curves after APP exposure of *C. violaceum* (Abramzon et al., 2006) and *Neisseria gonorrhoeae* (Xu et al., 2011) biofilms. Authors hypothesized that in deep biofilm layers, polymeric matrix surrounding the cells provides protection; plasma-induced cellular debris may also provide the shielding effect at the surface of biofilms, and this might be the cause of the slower inactivation rate as proposed earlier (Abramzon et al., 2006). Bayliss et al. (2012) investigated microbial responses as a community to a pulsed RF atmospheric plasma plume. Interestingly, plasma treatment beyond 15 s resulted in biphasic inactivation kinetics, although the initial cell population was homogenous. It was shown that plasma treatment rapidly altered the original homogenous bacterial deposition to heterogeneous distribution with cell clumping. The resulting heterogeneity was considered as a key to the biphasic behavior. In this work, for the first time it was shown that, although bacteria were originally present as a single monolayer, the initial plasma treatment introduced significant heterogeneity. The slower inactivation rates after 15 s and appearance of biphasic kinetics was attributed to the shielding effect provided by cell refuge (evolution of cell aggregates referred to as cell refuge) to the viable cells against further plasma treatment. In addition, authors discussed some interesting studies that reported biphasic or triphasic inactivation kinetics as a result of plasma treatment. Biphasic bacterial inactivation kinetics was reported by Heise et al. (2004) and Halfmann et al. (2007), although the initial bacterial deposition was homogenous. Perni et al. (2006) described triphasic inactivation kinetics of *B. subtilis* spores by nonthermal plasma. Inactivation data showed a very short shoulder phase and it was suggested that plasma-generated species overcome the initial threshold of bacterial resistance. Moisan et al. (2001) reported biphasic plasma inactivation kinetics, characterized by an initial rapid reduction phase, followed by a second slower inactivation phase. Hury et al. (1998) reported biphasic inactivation kinetics of *Bacillus* spores in low-pressure, oxygen-based plasmas. It was suggested that the limited diffusion of plasma-generated reactive species within the spore sublayers, and the stacking up of spores, provides a shielding effect against plasma. Cooper et al. (2010) reported a linear inactivation of *B. stratosphericus* when present in fluid or cell suspension form, following plasma treatment. Becker et al. (2005) studied *Bacillus* spore inactivation, using He, He-N_2, He-air as discharge gases, and observed single-phase survival curves that they attributed to the specific operating conditions at atmospheric pressure. Yang et al. (2011) observed triphasic inactivation behavior of *S. mutans* and *L. acidophilus* on PTFE films after plasma treatment. It was suggested that the initial rapid deactivation was the result of the deactivation of cells, residing in the top layers, by plasma. It was followed by a slower rate, because plasma species were blocked by the outer layer of cells and remaining debris of lethal cells, before reaching a layer of cells underneath. Once this shielding effect was eliminated by decomposing the remaining debris of lethal cells, the underlying layer of cells was directly exposed to plasma. Boudam et al. (2006) reported biphasic inactivation for *B. subtilis* spores by flowing afterglow N_2-O_2 discharge. Deng et al. (2005) observed biphasic survival curves with initial D-value for the first phase (D_1) shorter than second phase D-value (D_2). Roth et al. (2010) also reported biphasic inactivation kinetics for *B. subtilis* spores following low-pressure, low-temperature gas plasma sterilization with a rapid inactivation during the first phase, followed by a slower second phase. Recently, Jahid et al. (2015) demonstrated cold oxygen plasma effects on *S. Typhimurium* and cultivable indigenous microorganisms isolated from lettuce. The plasma treatments resulted in nonlinear microbial inactivation curves and the inactivation kinetics were determined by applying GInaFiT 1.6 software (Geeraerd et al., 2005); survivor data was fitted to a modified Weibull model. Jeon et al. (2014) described two-phase inactivation kinetics of bacteria following surface micro discharge (SMD) plasma treatment of 5, 15, 30, 60, and 120 s. The survival curves for the gas compositions containing oxygen showed two phases with a transition between 15 and 30 s. Similar survival curves were observed for O_2+N_2 and O_2+Ar gas plasmas. In the first phase, rapid bacterial inactivation with *D*-values around 5 s was observed, followed by a slower second phase with *D*-values of around 90 s. Reactive species were suggested to be responsible for the bactericidal effect, apart from ozone for SMD plasma-induced two-phase bacterial inactivation. Kostov et al. (2010) observed a bacterial survival curve characterized by a double-slope exponential decay with a rapid first phase and a slower second phase after DBD air plasma treatment. The D_1 value was much shorter than the D_2 value. Kamgang et al. (2007) evaluated efficiency of gliding discharge in humid air against planktonic, adherent, and biofilm cells of *Staphylococcus epidermidis*. For planktonic cells, log-linear kinetics was observed, whereas for plasma treated adherent and biofilm cells, biphasic inactivation kinetics was observed.

Kelly-Wintenberg et al. (1998) demonstrated biphasic inactivation curves for OAUGDP treated *E. coli* and *S. aureus* cells. For *E. coli* and *S. aureus* cells D_1 was 7 and 30 s, respectively. However, D_2 was only 2 s for both cell types. Biphasic inactivation kinetics were described by an initial slow exponential killing phase, followed by a second discrete rapid killing phase. The plausible explanation given was that during the first killing phase, toxic active species were concentrated, which led to alterations at the cell membrane level; this was responsible for the differences in D_1 values of two tested bacteria, considering the difference in their surface and membrane structure. However, once the species concentration was sufficient for lethality, it resulted in a rapid second killing phase with irreversible cell damage and lysis.

The reader is directed to Laroussi (2005, 2009), who explored the inactivation kinetics of APP. Laroussi et al. (2000) reported two-slope bacterial survivor curves with D_2 smaller than D_1. Similarly, biphasic survivor curves were reported by Montie et al. (2000) where the D_1 value was dependent on challenge species under test, and D_2 value was dependent on the type of surface or medium supporting the microorganisms.

Li et al. (2013) observed two-slope survivor curves for *E. coli* and its mutants after direct current atmospheric pressure oxygen plasma treatments, with a minor role of UV. Marchal et al. (2012) observed that the biofilm thickness plays a significant role on the plasma inactivation kinetics. In that study, GInaFiT software was used to determine low-temperature APP inactivation kinetics of Gram-positive bacterial biofilms. Among the proposed microbial survival models by GInaFiT tool, log-linear with tail model was reported as the better fit of the experimental data. Moisan et al. (2013) obtained biphasic survival curve resulted from the exposure of *B. atrophaeus* endospores to the plasma discharge afterglow from a N_2-O_2 gas mixture. The first phase was rapid with shorter D_1 value whereas second phase was much longer with longer D_2 value. Authors suggested that during the first phase the lethal effects were exerted by UV photons on the top spore layers and in the second phase, as spores were within a stack or aggregated, or located within crevices, UV photons could not access them efficiently. The reported remarks on the second phase were in agreement with a previous report (Rossi et al., 2006) which stated stacked spores were a reason for the second phase, although Rossi et al. (2006) accounted both UV irradiation and etching for faster inactivation rates, compared to the UV-only approach suggested by Moisan et al. (2013).

Moisan et al. (2001) interpreted that plasma-induced survival curves involve three basic mechanisms. That includes: direct destruction of microbial genetic material by the action of UV irradiation; erosion of microorganisms and an etching process that results from the adsorption of the plasma-reactive species on the microorganisms; and that is followed by volatile compound formation. Additionally, a number of interesting studies that assessed spore inactivation by plasma, which results into different survivor slopes, were discussed (Hury et al., 1998; Kelly-Wintenberg et al., 1998; Lerouge et al., 2000; Moreau et al., 2000). For a three-phase survival curve reported by Moreau et al. (2000) it was proposed that the highest killing rate during the first step was due to the direct UV irradiation that led to DNA destruction of isolated spores, or the very first layer of spores on a stack. The second slower kinetic step reflected the time required for sufficient erosion of inactivated spores and the debris on top of living spores. A third step with a shorter inactivation time occurs when all the debris is cleared off from the last living spores for the UV photons to hit their genetic material.

Noriega et al. (2011) demonstrated a biphasic inactivation curve for cold atmospheric gas plasma-treated *Listeria innocua* deposited on membrane filters. Park et al. (2003) investigated the decontamination efficiency of microwave-induced atmospheric pressure argon plasma against bacteria and fungi, and observed different survival curves according to the type of microorganisms. Pavlovich et al. (2013b) demonstrated single-slope survival curve of *E. coli* on a stainless steel surface after ambient gas plasma exposure. The first order inactivation kinetics between reactive neutral species and bacteria was suggested; however, particular species reaction with bacteria was not mentioned. Vleugels et al. (2005) observed two or three phases of different kinetics during the APGD plasma inactivation of *Pantoea agglomerans*. For 12 h old biofilm there were three inactivation phases and for 24 h old biofilm, two inactivation phases were observed. It was suggested that in the first phase, naked bacteria on the biofilm surface (or first layer) were directly attacked by plasma species; conversely, for the 24 h old biofilm, the first inactivation phase was slower due to thicker biofilm that contained less naked bacteria than the 12 h old biofilm. The second phase was slower where plasma species can access the bacteria that was not well protected and that lie immediately underneath the surface layer of the biofilm. The third inactivation phase, was a near-constant plateau, when the inactivation rate was slowed down further to nearly zero.

To characterize nonthermal plasma inactivation Perni et al. (2006) proposed three inactivation models, namely: the Baranyi model, the Weibull model, and a third-order polynomial empirical model. They also recommended the Baranyi model as the appropriate basis to unravel physical and biological mechanisms of microbial inactivation by plasma. Experimental inactivation data for *B. subtilis* spores was reliably shown to reproduce using these models. Schnabel et al. (2014) reported three-step inactivation kinetics for *B. atrophaeus* during microwave plasma-processed air treatment. Authors proposed that the first phase with rapid inactivation was due to plasma influence on isolated spores, or the first layer of stacked endospores. It was followed by a slower second inactivation phase which was attributed to the conversion of reactive species $NO^{\cdot}$ to $NO^{\cdot}_2$. The last phase occurred after endospore debris was cleared off and $NO^{\cdot}_2$ interaction was possible (Moisan et al., 2001; Philip et al., 2002). Traba and Liang (2011) reported that inactivation of bacterial biofilm by plasma discharge as a complex process that involves two steps. The second phase with rapid decline in cell viability was mainly attributed to the etching effect of the discharge gases. Tseng et al. (2012) observed a steady and linear inactivation of bacterial cells and spores using gas discharge plasma.

5 Concluding Remarks

Nonthermal plasma is a potential decontamination approach with great advantages in various areas of food, medical, and healthcare sectors. Plasma process involves an intricate sequence of biological interactions in microorganisms due to the action of generated reactive plasma agents. Reactive species play a significant role in dictating antimicrobial effects.

In this chapter, we attempted to compile an overview of the role of different reactive plasma species in microbial inactivation. Further, we discussed plasma effects on different cellular targets. And finally, kinetics of microbial inactivation by plasma was briefly illustrated. Optimization of the decontamination process is highly desirable to achieve successful elimination of the target microbial population, and analyzing plasma effects on cell targets facilitates better understanding of its antimicrobial action. In-depth research on the interaction between active plasma species and various cell macromolecules could help to significantly improve the process efficiency.

References

Abramzon, N., Joaquin, J.C., Bray, J., Brelles-Marino, G., 2006. Biofilm destruction by RF high-pressure cold plasma jet. IEEE Trans. Plasma Sci. 34 (4), 1304–1309.

Akitsu, T., Ohkawa, H., Tsuji, M., Kimura, H., Kogoma, M., 2005. Plasma sterilization using glow discharge at atmospheric pressure. Surf. Coat. Technol. 193 (1–3), 29–34.

Ali, A., Kim, Y.H., Lee, J.Y., Lee, S., Uhm, H.S., Cho, G., Park, B.J., Choi, E.H., 2014. Inactivation of *Propionibacterium acnes* and its biofilm by nonthermal plasma. Curr. Appl. Phys. 14, S142–S148.

Alkawareek, M.Y., Algwari, Q.T., Laverty, G., Gorman, S.P., Graham, W.G., O'Connell, D., Gilmore, B.F., 2012. Eradication of *Pseudomonas aeruginosa* biofilms by atmospheric pressure nonthermal plasma. PLoS ONE 7 (8), e44289.

Alkawareek, M.Y., Gorman, S.P., Graham, W.G., Gilmore, B.F., 2014. Potential cellular targets and antibacterial efficacy of atmospheric pressure nonthermal plasma. Int. J. Antimicrob. Agents 43 (2), 154–160.

Bai, N., Sun, P., Zhou, H.X., Wu, H., Wang, R., Liu, F., Zhu, W., Lopez, J.L., Zhang, J., Fang, J., 2011. Inactivation of *Staphylococcus aureus* in water by a cold, He/O_{-2} atmospheric pressure plasma microjet. Plasma Process. Polym. 8 (5), 424–431.

Baxter, H.C., Campbell, G.A., Whittaker, A.G., Jones, A.C., Aitken, A., Simpson, A.H., Casey, M., Bountiff, L., Gibbard, L., Baxter, R.L., 2005. Elimination of transmissible spongiform encephalopathy infectivity and decontamination of surgical instruments by using radio-frequency gas-plasma treatment. J. Gen. Virol. 86, 2393–2399.

Baxter, H.C., Campbell, G.A., Richardson, P.R., Jones, A.C., Whittle, I., Casey, M., 2006. Surgical instrument decontamination: efficacy of introducing an argon:oxygen RF gas-plasma cleaning step as part of the cleaning cycle for stainless steel instruments. IEEE Trans. Plasma Sci. 34 (4), 1337–1344.

Bayliss, D.L., Walsh, J.L., Iza, F., Shama, G., Holah, J., Kong, M.G., 2012. Complex responses of microorganisms as a community to a flowing atmospheric plasma. Plasma Process. Polym. 9 (6), 597–611.

Becker, K., Koutsospyros, A., Yin, S.M., Christodoulatos, C., Abramzon, N., Joaquin, J.C., Brelles-Marino, G., 2005. Environmental and biological applications of microplasmas. Plasma Phys. Controlled Fusion 47, B513–B523.

Bermudez-Aguirre, D., Wemlinger, E., Pedrow, P., Barbosa-Canovas, G., Garcia-Perez, M., 2013. Effect of atmospheric pressure cold plasma (APCP) on the inactivation of *Escherichia coli* in fresh produce. Food Control 34 (1), 149–157.

Boscariol, M.R., Moreira, A.J., Mansano, R.D., Kikuchi, I.S., Pinto, T.J.A., 2008. Sterilization by pure oxygen plasma and by oxygen-hydrogen peroxide plasma: an efficacy study. Int. J. Pharm. 353 (1–2), 170–175.

Boudam, M.K., Moisan, M., Saoudi, B., Popovici, C., Gherardi, N., Massines, F., 2006. Bacterial spore inactivation by atmospheric-pressure plasmas in the presence or absence of UV photons as obtained with the same gas mixture. J. Phys. D.:Appl. Phys 39 (16), 3494–3507.

Boxhammer, V., Morfill, G.E., Jokipii, J.R., Shimizu, T., Klampfl, T., Li, Y.F., Koritzer, J., Schlegel, J., Zimmermann, J.L., 2012. Bactericidal action of cold atmospheric plasma in solution. New J. Phys. 14, 113042.

Chang, Y.T., Chen, G., 2014. Oral bacterial inactivation using a novel low-temperature atmospheric-pressure plasma device. J. Dent. Sci. http://dx.doi.org/10.1016/j.jds.2014.03.007.

Chiang, M.H., Wu, J.Y., Li, Y.H., Wu, J.S., Chen, S.H., Chang, C.L., 2010. Inactivation of *E. coli* and *B. subtilis* by a parallel-plate dielectric barrier discharge jet. Surf. Coat. Technol. 204 (21–22), 3729–3737.

Choi, J.H., Han, I., Baik, H.K., Lee, M.H., Han, D.W., Park, J.C., Lee, I.S., Song, K.M., Lim, Y.S., 2006. Analysis of sterilization effect by pulsed dielectric barrier discharge. J. Electrost. 64 (1), 17–22.

Colagar, A.H., Sohbatzadeh, F., Mirzanejhad, S., Omran, A.V., 2010. Sterilization of *Streptococcus pyogenes* by afterglow dielectric barrier discharge using O_2 and CO_2 working gases. Biochem. Eng. J. 51 (3), 189–193.

Colagar, A.H., Memariani, H., Sohbatzadeh, F., Omran, A.V., 2013. Nonthermal atmospheric argon plasma jet effects on *Escherichia coli* biomacromolecules. Appl. Biochem. Biotechnol. 171 (7), 1617–1629.

Cooper, M., Fridman, G., Fridman, A., Joshi, S.G., 2010. Biological responses of *Bacillus stratosphericus* to floating electrode-dielectric barrier discharge plasma treatment. J. Appl. Microbiol. 109 (6), 2039–2048.

Cullen, P.J., Misra, N.N., Bourke, P., Keener, K., O'Donnell, C., Moiseev, T., Mosnier, J.P., Milosavljević, V., 2014. Inducing a dielectric barrier discharge plasma within a package. IEEE Trans. Plasma Sci. 42 (10), 2368–2369.

De Geyter, N., Morent, R., 2012. Nonthermal plasma sterilization of living and nonliving surfaces. Annu. Rev. Biomed. Eng. 14, 255–274.

Deng, X.T., Shi, J.J., Shama, G., Kong, M.G., 2005. Effects of microbial loading and sporulation temperature on atmospheric plasma inactivation of *Bacillus subtilis* spores. Appl. Phys. Lett. 87 (15), 153901.

Deng, X.T., Shi, J.J., Kong, M.G., 2006. Physical mechanisms of inactivation of *Bacillus subtilis* spores using cold atmospheric plasmas. IEEE Trans. Plasma Sci. 34 (4), 1310–1316.

Deng, X.T., Shi, J.J., Kong, M.G., 2007a. Protein destruction by a helium atmospheric pressure glow discharge: capability and mechanisms. J. Appl. Phys. 101 (7), 074701.

Deng, X.T., Shi, J.J., Chen, H.L., Kong, M.G., 2007b. Protein destruction by atmospheric pressure glow discharges. Appl. Phys. Lett. 90 (1), 1–3.

Deng, S.X., Cheng, C., Ni, G.H., Meng, Y.D., Chen, H., 2010a. *Bacillus subtilis* devitalization mechanism of atmosphere pressure plasma jet. Curr. Appl. Phys. 10 (4), 1164–1168.

Deng, S.X., Cheng, C., Ni, G.H., Meng, Y.D., Chen, H., 2010b. The interaction of an atmospheric pressure plasma jet using argon or argon plus hydrogen peroxide vapour addition with *Bacillus subtilis*. Chin. Phys. B 19 (10), 105203.

Dolezalova, E., Lukes, P., 2015. Membrane damage and active but nonculturable state in liquid cultures of *Escherichia coli* treated with an atmospheric pressure plasma jet. Bioelectrochemistry 103, 7–14.

Edelblute, C.M., Malik, M.A., Heller, L.C., 2015. Surface-dependent inactivation of model microorganisms with shielded sliding plasma discharges and applied air flow. Bioelectrochemistry 103, 22–27.

Efremov, N.M., Adamiak, B.Y., Blochin, V.I., Dadashev, S.J., Dmitriev, K.I., Gryaznova, O.P., Jusbashev, V.F., 2000. Action of a self-sustained glow discharge in atmospheric pressure air on biological objects. IEEE Trans. Plasma Sci. 28 (1), 238–241.

Elmoualij, B., Thellin, O., Gofflot, S., Heinen, E., Levif, P., Seguin, J., Moisan, M., Leduc, A., Barbeau, J., Zorzi, W., 2012. Decontamination of prions by the flowing afterglow of a reduced-pressure N_2-O_2 cold-plasma. Plasma Process. Polym. 9 (6), 612–618.

Eto, H., Ono, Y., Ogino, A., Nagatsu, M., 2008. Low-temperature sterilization of wrapped materials using flexible sheet-type dielectric barrier discharge. Appl. Phys. Lett. 93, 221502.

Ferrell, J.R., Shen, F., Grey, S.F., Woolverton, C.J., 2013. Pulse-based nonthermal plasma (NTP) disrupts the structural characteristics of bacterial biofilms. Biofouling 29 (5), 585–599.

Fridman, A., 2008. Plasma biology and plasma medicine. In: Plasma Chemistry. Cambridge University Press, New York, pp. 848–961.

Gaunt, L., Beggs, C., Georghiou, G., 2006. Bactericidal action of the reactive species produced by gas-discharge nonthermal plasma at atmospheric pressure: a review. IEEE Trans. Plasma Sci. 34, 1257–1269.

Geeraerd, A.H., Valdramidis, V., Van Impe, J.F., 2005. GInaFiT: a freeware tool to assess non-log-linear microbial survivor curves. Int. J. Food Microbiol. 102 (1), 95–105.

Guo, J., Huang, K., Wang, J., 2015. Bactericidal effect of various nonthermal plasma agents and the influence of experimental conditions in microbial inactivation: a review. Food Control 50, 482–490.

Hähnel, M., Von Woedtke, T., Weltmann, K.D., 2010. Influence of the air humidity on the reduction of *Bacillus* spores in a defined environment at atmospheric pressure using a dielectric barrier surface discharge. Plasma Process. Polym. 7 (3–4), 244–249.

Halfmann, H., Bibinov, N., Wunderlich, J., Awakowicz, P., 2007. A double inductively coupled plasma for sterilization of medical devices. J. Phys. D.: Appl. Phys. 40 (14), 4145–4154.

Han, L., Patil, S., Cullen, P., Keener, K., Bourke, P., 2014. Bacterial inactivation by atmospheric cold plasma: influence of process parameters and effects on cell leakage and DNA. J. Appl. Microbiol. 116 (4), 784–794.

Han, L., Patil, S., Boehm, D., Milosavljevic, V., Cullen, P.J., Bourke, P., 2016. Mechanism of inactivation by high voltage atmospheric cold plasma differs between *Escherichia coli* and *Staphylococcus aureus*. Appl. Environ. Microbiol. 82, 450–458. http://dx.doi.org/10.1128/AEM.02660-15.

Heise, M., Neff, W., Franken, O., Muranyi, P., Wunderlich, J., 2004. Sterilization of polymer foils with dielectric barrier discharges at atmospheric pressure. Plasmas Polym. 9 (1), 23–33.

Helmke, A., Hoffmeister, D., Berge, F., Emmert, S., Laspe, P., Mertens, N., Vioel, W., Weltmann, K.D., 2011. Physical and microbiological characterisation of *Staphylococcus epidermidis* inactivation by dielectric barrier discharge plasma. Plasma Process. Polym. 8 (4), 278–286.

Hernández-Arias, A.N., Rodriguez-Mendez, B.G., Lopez-Callejas, R., Valencia-Alvarado, R., Mercado-Cabrera, A., Pena-Eguiluz, R., Barocio, S.R., Munoz-Castro, A.E., Beneitez, A.D., 2011. Pulsed dielectric barrier discharge for *Bacillus subtilis* inactivation in water. In: Paper Presented at: 14th Latin American Workshop on Plasma Physics (LAWPP), Mar del Plata, Argentina, November 20–25, 2011.

Herrmann, H.W., Henins, I., Park, J., Selwyn, G.S., 1999. Decontamination of chemical and biological warfare (CBW) agents using an atmospheric pressure plasma jet (APPJ). Phys. Plasmas 6 (5), 2284–2289.

Hong, Y.F., Kang, J.G., Lee, H.Y., Uhm, H.S., Moon, E., Park, Y.H., 2009. Sterilization effect of atmospheric plasma on *Escherichia coli* and *Bacillus subtilis* endospores. Lett. Appl. Microbiol. 48 (1), 33–37.

Hou, Y.M., Dong, X.Y., Yu, H., Li, S., Ren, C.S., Zhang, D.J., Xiu, Z.L., 2008. Disintegration of biomacromolecules by dielectric barrier discharge plasma in helium at atmospheric pressure. IEEE Trans. Plasma Sci. 36 (4), 1633–1637.

Hu, M., Guo, Y., 2012. The effect of air plasma on sterilization of *Escherichia coli* in dielectric barrier discharge. Plasma Sci. Technol. 14 (8), 735–740.

Huang, C., Yu, Q.S., Hsieh, F.H., Duan, Y.X., 2007. Bacterial deactivation using a low temperature argon atmospheric plasma brush with oxygen addition. Plasma Process. Polym. 4 (1), 77–87.

Hury, S., Vidal, D.R., Desor, F., Pelletier, J., Lagarde, T.A., 1998. A parametric study of the destruction efficiency of *Bacillus* spores in low pressure oxygen-based plasmas. Lett. Appl. Microbiol. 26 (6), 417–421.

Ikawa, S., Kitano, K., Hamaguchi, S., 2010. Effects of pH on bacterial inactivation in aqueous solutions due to low-temperature atmospheric pressure plasma application. Plasma Process. Polym. 7 (1), 33–42.

Jahid, I.K., Han, N., Zhang, C.Y., Ha, S.D., 2015. Mixed culture biofilms of *Salmonella typhimurium* and cultivable indigenous microorganisms on lettuce show enhanced resistance of their sessile cells to cold oxygen plasma. Food Microbiol. 46, 383–394.

Jayasena, D.D., Kim, H.J., Yong, H.I., Park, S., Kim, K., Choe, W., Jo, C., 2015. Flexible thin-layer dielectric barrier discharge plasma treatment of pork butt and beef loin: effects on pathogen inactivation and meat-quality attributes. Food Microbiol. 46, 51–57.

Jeon, J., Rosentreter, T.M., Li, Y.F., Isbary, G., Thomas, H.M., Zimmermann, J.L., Morfill, G.E., Shimizu, T., 2014. Bactericidal agents produced by surface micro-discharge (SMD) plasma by controlling gas compositions. Plasma Process. Polym. 11 (5), 426–436.

Joaquin, J.C., Kwan, C., Abramzon, N., Vandervoort, K., Brelles-Marino, G., 2009. Is gas-discharge plasma a new solution to the old problem of biofilm inactivation? Microbiology-Sgm 155, 724–732.

Joshi, S.G., Cooper, M., Yost, A., Paff, M., Ercan, U.K., Fridman, G., Friedman, G., Fridman, A., Brooks, A.D., 2011. Nonthermal dielectric-barrier discharge plasma-induced inactivation involves oxidative dna damage and membrane lipid peroxidation in *Escherichia coli*. Antimicrob. Agents Chemother. 55 (3), 1053–1062.

Jung, H., Kim, D.B., Gweon, B., Moon, S.Y., Choe, W., 2010. Enhanced inactivation of bacterial spores by atmospheric pressure plasma with catalyst TiO_2. Appl. Catal. B-Environ. 93 (3–4), 212–216.

Kamgang, J.O., Briandet, R., Herry, J.M., Brisset, J.L., Naitali, M., 2007. Destruction of planktonic, adherent and biofilm cells of *Staphylococcus epidermidis* using a gliding discharge in humid air. J. Appl. Microbiol. 103 (3), 621–628.

Kang, S.K., Choi, M.Y., Koo, I.G., Kim, P.Y., Kim, Y., Kim, G.J., Mohamed, A.A.H., Collins, G.J., Lee, J.K., 2011. Reactive hydroxyl radical-driven oral bacterial inactivation by radio frequency atmospheric plasma. Appl. Phys. Lett. 98, 143702.

Kelly-Wintenberg, K., Montie, T.C., Brickman, C., Roth, J.R., Carr, A.K., Sorge, K., Wadsworth, L.C., Tsai, P.P.Y., 1998. Room temperature sterilization of surfaces and fabrics with a one atmosphere uniform glow discharge plasma. J. Ind. Microbiol. Biotechnol. 20 (1), 69–74.

Kim, S.M., Kim, J.I., 2006. Decomposition of biological macromolecules by plasma generated with helium and oxygen. J. Microbiol. 44 (4), 466–471.

Kim, S.J., Chung, T.H., Bae, S.H., Leem, S.H., 2009a. Bacterial inactivation using atmospheric pressure single pin electrode microplasma jet with a ground ring. Appl. Phys. Lett. 94, 141502.

Kim, S.J., Chung, T.H., Bae, S.H., Leem, S.H., 2009b. Characterization of atmospheric pressure microplasma jet source and its application to bacterial inactivation. Plasma Process. Polym. 6 (10), 676–685.

Korachi, M., Aslan, N., 2011. The effect of atmospheric pressure plasma corona discharge on ph, lipid content and DNA of bacterial cells. Plasma Sci. Technol. 13 (1), 99–105.

Korachi, M., Gurol, C., Aslan, N., 2010. Atmospheric plasma discharge sterilization effects on whole cell fatty acid profiles of *Escherichia coli* and *Staphylococcus aureus*. J. Electrost. 68 (6), 508–512.

Kostov, K.G., Rocha, V., Koga-Ito, C.Y., Matos, B.M., Algatti, M.A., Honda, R.Y., Kayama, M.E., Mota, R.P., 2010. Bacterial sterilization by a dielectric barrier discharge (DBD) in air. Surf. Coat. Technol. 204 (18–19), 2954–2959.

Kuo, S.P., Tarasenko, O., Nourkbash, S., Bakhtina, A., Levon, K., 2006. Plasma effects on bacterial spores in a wet environment. New J. Phys. 8, 1–11.

Kurita, H., Nakajima, T., Yasuda, H., Takashima, K., Mizuno, A., Wilson, J.I.B., Cunningham, S., 2011. Single-molecule measurement of strand breaks on large DNA induced by atmospheric pressure plasma jet. Appl. Phys. Lett. 99 (19), 191504.

Kuzmichev, A.I., Soloshenko, I.A., Tsiolko, V.V., Kryzhanovsky, V.I., Bazhenov, V.Y., Mikhno, I.L., Khomich, V.A., 2000. Feature of sterilization by different type of atmospheric pressure discharges. Proc. Int. Sym. High Press. Low Temp. Plasma Chem. 2, 402–406.

Kvam, E., Davis, B., Mondello, F., Garner, A.L., 2012. Nonthermal atmospheric plasma rapidly disinfects multidrug-resistant microbes by inducing cell surface damage. Antimicrob. Agents Chemother. 56 (4), 2028–2036.

Lackmann, J.W., Bandow, J.E., 2014. Inactivation of microbes and macromolecules by atmospheric-pressure plasma jets. Appl. Microbiol. Biotechnol. 98 (14), 6205–6213.

Lackmann, J., Schneider, S., Narberhaus, F., Benedikt, J., Bandow, J.E., 2012. Characterization of damage to bacteria and bio-macromolecules caused by (V)UV radiation and particles generated by a micro-scale atmospheric pressure plasma jet. In: Machala, Z., Hensel, K., Akishev, Y. (Eds.), Plasma for Bio-Decontamination, Medicine and Food Security. Springer, Dordrecht, pp. 17–29.

Lackmann, J., Schneider, S., Edengeiser, E., Jarzina, F., Brinckmann, S., Steinborn, E., Havenith, M., Benedikt, J., Bandow, J.E., 2013. Photons and particles emitted from cold atmospheric-pressure plasma inactivate bacteria and biomolecules independently and synergistically. J. R. Soc. Interface 10, 20130591.

Laroussi, M., 1996. Sterilization of contaminated matter with an atmospheric pressure plasma. IEEE Trans. Plasma Sci. 24 (3), 1188–1191.

Laroussi, M., 2002. Nonthermal decontamination of biological media by atmospheric-pressure plasmas: review, analysis, and prospects. IEEE Trans. Plasma Sci. 30 (4), 1409–1415.

Laroussi, M., 2005. Low temperature plasma-based sterilization: overview and state-of-the-art. Plasma Process. Polym. 2 (5), 391–400.

Laroussi, M., 2009. Low-temperature plasmas for medicine? IEEE Trans. Plasma Sci. 37 (6), 714–725.

Laroussi, M., Leipold, F., 2004. Evaluation of the roles of reactive species, heat, and UV radiation in the inactivation of bacterial cells by air plasmas at atmospheric pressure. Int. J. Mass Spectrom. 233 (1–3), 81–86.

Laroussi, M., Sayler, G.S., Glascock, B.B., McCurdy, B., Pearce, M.E., Bright, N.G., Malott, C.M., 1999. Images of biological samples undergoing sterilization by a glow discharge at atmospheric pressure. IEEE Trans. Plasma Sci. 27 (1), 34–35.

Laroussi, M., Alexeff, I., Kang, W.L., 2000. Biological decontamination by nonthermal plasmas. IEEE Trans. Plasma Sci. 28 (1), 184–188.

Lee, K.Y., Joo Park, B., Hee Lee, D., Lee, I.S.O., Hyun, S., Chung, K.H., Park, J.C., 2005. Sterilization of *Escherichia coli* and MRSA using microwave-induced argon plasma at atmospheric pressure. Surf. Coat. Technol. 193 (1–3), 35–38.

Lerouge, S., Wertheimer, M.R., Marchand, R., Tabrizian, M., Yahia, L., 2000. Effect of gas composition on spore mortality and etching during low-pressure plasma sterilization. J. Biomed. Mater. Res. 51 (1), 128–135.

Lerouge, S., Wertheimer, M.R., Yahia, L.H., 2001. Plasma sterilization: a review of parameters, mechanisms, and limitations. Plasmas Polym. 6, 175–188.

Li, J., Sakai, N., Watanabe, M., Hotta, E., Wachi, M., 2013. Study on plasma agent effect of a direct-current atmospheric pressure oxygen-plasma jet on inactivation of *E. coli* using bacterial mutants. IEEE Trans. Plasma Sci. 41 (4), 935–941.

Liang, J.L., Zheng, S.H., Ye, S.Y., 2012a. Inactivation of *Penicillium* aerosols by atmospheric positive corona discharge processing. J. Aerosol Sci. 54, 103–112.

Liang, Y.D., Wu, Y., Sun, K., Chen, Q., Shen, F.X., Zhang, J., Yao, M.S., Zhu, T., Fang, J., 2012b. Rapid inactivation of biological species in the air using atmospheric pressure nonthermal plasma. Environ. Sci. Technol. 46 (6), 3360–3368.

Lim, J.P., Uhm, H.S., Li, S.Z., 2007. Influence of oxygen in atmospheric-pressure argon plasma jet on sterilization of *Bacillus atrophaeous* spores. Phys. Plasmas 14, 093504.

Liu, H., Chen, J., Yang, L., Zhou, Y., 2008. Long-distance oxygen plasma sterilization: effects and mechanisms. Appl. Surf. Sci. 254 (6), 1815–1821.

Lu, X., Ye, T., Cao, Y.G., Sun, Z.Y., Xiong, Q., Tang, Z.Y., Xiong, Z.L., Hu, J., Jiang, Z.H., Pan, Y., 2008. The roles of the various plasma agents in the inactivation of bacteria. J. Appl. Phys. 104 (5), 053309-1–053309-5.

Ma, Y., Zhang, G.J., Shi, X.M., Xu, G.M., Yang, Y., 2008. Chemical mechanisms of bacterial inactivation using dielectric barrier discharge plasma in atmospheric air. IEEE Trans. Plasma Sci. 36 (4), 1615–1620.

Machala, Z., Jedlovsky, I., Chladekova, L., Pongrac, B., Giertl, D., Janda, M., Sikurova, L., Polcic, P., 2009. DC discharges in atmospheric air for bio-decontamination: spectroscopic methods for mechanism identification. Eur. Phys. J. D 54 (2), 195–204.

Mai-Prochnow, A., Murphy, A.B., Mclean, K.M., Kong, M.G., Ostrikov, K., 2014. Atmospheric pressure plasmas: infection control and bacterial responses. Int. J. Antimicrob. Agents 43 (6), 508–517.

Marchal, F., Robert, H., Merbahi, N., Fontagne-Faucher, C., Yousfi, M., Romain, C.E., Eichwald, O., Rondel, C., Gabriel, B., 2012. Inactivation of Gram-positive biofilms by low-temperature plasma jet at atmospheric pressure. J. Phys. D-Appl. Phys. 45, 345202.

Mastanaiah, N., Banerjee, P., Johnson, J.A., Roy, S., 2013. Examining the role of ozone in surface plasma sterilization using dielectric barrier discharge (dbd) plasma. Plasma Process. Polym. 10 (12), 1120–1133.

Misra, N.N., Tiwari, B., Raghavarao, K.S.M.S., Cullen, P.J., 2011. Nonthermal plasma inactivation of food-borne pathogens. Food Eng. Rev. 3 (3–4), 159–170.

Mogul, R., Bol'shakov, A.A., Chan, S.L., Stevens, R.M., Khare, B.N., Meyyappan, M., Trent, J.D., 2003. Impact of low-temperature plasmas on *Deinococcus radiodurans* and biomolecules. Biotechnol. Prog. 19 (3), 776–783.

Moisan, M., Barbeau, J., Moreau, S., Pelletier, J., Tabrizian, M., Yahia, L.H., 2001. Low-temperature sterilization using gas plasmas: a review of the experiments and an analysis of the inactivation mechanisms. Int. J. Pharm. 226 (1–2), 1–21.

Moisan, M., Barbeau, J., Crevier, M.C., Pelletier, J., Philip, N., Saoudi, B., 2002. Plasma sterilization: methods mechanisms. Pure Appl. Chem. 74 (3), 349–358.

Moisan, M., Boudam, K., Carignan, D., Keroack, D., Levif, P., Barbeau, J., Seguin, J., Kutasi, K., Elmoualij, B., Thellin, O., Zorzi, W., 2013. Sterilization/disinfection of medical devices using plasma: the flowing afterglow of the reduced-pressure N_2-O_2 discharge as the inactivating medium. Eur. Phys. J. Appl. Phys. 63, 10001.

Montie, T.C., Kelly-Wintenberg, K., Roth, J.R., 2000. An overview of research using the one atmosphere uniform glow discharge plasma (OAUGDP) for sterilization of surfaces and materials. IEEE Trans. Plasma Sci. 28 (1), 41–50.

Moreau, S., Moisan, M., Tabrizian, M., Barbeau, J., Pelletier, J., Ricard, A., Yahia, L., 2000. Using the flowing afterglow of a plasma to inactivate *Bacillus subtilis* spores: influence of the operating conditions. J. Appl. Phys. 88 (2), 1166–1174.

Moreau, M., Feuilloley, M.G.J., Veron, W., Meylheuc, T., Chevalier, S., Brisset, J.L., Orange, N., 2007. Gliding arc discharge in the potato pathogen *Erwinia carotovora* subsp atroseptica: mechanism of lethal action and effect on membrane-associated molecules. Appl. Environ. Microbiol. 73 (18), 5904–5910.

Navabsafa, N., Ghomi, H., Nikkhah, M., Mohades, S., Dabiri, H., Ghasemi, S., 2013. Effect of BCD plasma on a bacteria cell membrane. Plasma Sci. Technol. 15 (7), 685–689.

Nehra, V., Kumar, A., Dwivedi, H.K., 2008. Atmospheric nonthermal plasma sources. Int. J. Eng. 2, 53–68.

Noriega, E., Shama, G., Laca, A., Diaz, M., Kong, M.G., 2011. Cold atmospheric gas plasma disinfection of chicken meat and chicken skin contaminated with *Listeria innocua*. Food Microbiol. 28 (7), 1293–1300.

O'Connell, D., Cox, L.J., Hyland, W.B., Mcmahon, S.J., Reuter, S., Graham, W.G., Gans, T., Currell, F.J., 2011. Cold atmospheric pressure plasma jet interactions with plasmid DNA. Appl. Phys. Lett. 98 (4), 1–3.

Park, B.J., Lee, D.H., Park, J.C., Lee, I.S., Lee, K.Y., Hyun, S.O., Chun, M.S., Chung, K.H., 2003. Sterilization using a microwave-induced argon plasma system at atmospheric pressure. Phys. Plasmas 10 (11), 4539–4544.

Patil, S., Moiseev, T., Misra, N.N., Cullen, P.J., Mosnier, J.P., Keener, K.M., Bourke, P., 2014. Influence of high voltage atmospheric cold plasma process parameters and role of relative humidity on inactivation of *Bacillus atrophaeus* spores inside a sealed package. J. Hosp. Infect. 88 (3), 162–169.

Pavlovich, M.J., Chen, Z., Sakiyama, Y., Clark, D.S., Graves, D.B., 2013a. Ozone correlates with antibacterial effects from indirect air dielectric barrier discharge treatment of water. J. Phys. D.: Appl. Phys. 46 (14), 145202.

Pavlovich, M.J., Chen, Z., Sakiyama, Y., Clark, D.S., Graves, D.B., 2013b. Effect of discharge parameters and surface characteristics on ambient-gas plasma disinfection. Plasma Process. Polym. 10 (1), 69–76.

Perni, S., Deng, X.T.T., Shama, G., Kong, M.G., 2006. Modeling the inactivation kinetics of *Bacillus subtilis* spores by nonthermal plasmas. IEEE Trans. Plasma Sci. 34 (4), 1297–1303.

Philip, N., Saoudi, B., Crevier, M.C., Moisan, M., Barbeau, J., Pelletier, J., 2002. The respective roles of UV photons and oxygen atoms in plasma sterilization at reduced gas pressure: the case of N_2-O_2 mixtures. IEEE Trans. Plasma Sci. 30 (4), 1429–1436.

Pointu, A.M., Ricard, A., Dodet, N., Odic, E., Larbre, J., Ganciu, M., 2005. Production of active species in N_2-O_2 flowing post-discharges at atmospheric pressure for sterilization. J. Phys. D.: Appl. Phys. 38 (12), 1905–1909.

Pompl, R., Jamitzky, F., Shimizu, T., Steffes, B., Bunk, W., Schmidt, H.U., Georgi, M., Ramrath, K., Stolz, W., Stark, R.W., Urayama, T., Fujii, S., Morfill, G.E., 2009. The effect of low-temperature plasma on bacteria as observed by repeated AFM imaging. New J. Phys. 11, 115023-1–115023-11.

Ptasińska, S., Bahnev, B., Stypczynska, A., Bowden, M., Mason, N.J., Braithwaite, N.S., 2010. DNA strand scission induced by a nonthermal atmospheric pressure plasma jet. Phys. Chem. Chem. Phys. 12 (28), 7779–7781.

Rahul, R., Stan, O., Rahman, A., Littlefield, E., Hoshimiya, K., Yalin, A.P., Sharma, A., Pruden, A., Moore, C.A., Yu, Z., Collins, G.J., 2005. Optical and RF electrical characteristics of atmospheric pressure open-air hollow slot microplasmas and application to bacterial inactivation. J. Phys. D.: Appl. Phys. 38 (11), 1750–1759.

Rodriguez-Mendez, B.G., Hernandez-Arias, A.N., Lopez-Callejas, R., Valencia-Alvarado, R., Mercado-Cabrera, A., Pena-Eguiluz, R., Barocio-Delgado, S.R., Munoz-Castro, A.E., De La Piedad-Beneitez, A., 2013. Gas flow effect on *E. coli* and *B. subtilis* bacteria inactivation in water using a pulsed dielectric barrier discharge. IEEE Trans. Plasma Sci. 41 (1), 147–154.

Rossi, F., Kylian, O., Hasiwa, M., 2006. Decontamination of surfaces by low pressure plasma discharges. Plasma Process. Polym. 3 (6–7), 431–442.

Roth, S., Feichtinger, J., Hertel, C., 2010. Characterization of *Bacillus subtilis* spore inactivation in low-pressure, low-temperature gas plasma sterilization processes. J. Appl. Microbiol. 108 (2), 521–531.

Sato, T., Miyahara, T., Doi, A., Ochiai, S., Urayama, T., Nakatani, T., 2006. Sterilization mechanism for *Escherichia coli* by plasma flow at atmospheric pressure. Appl. Phys. Lett. 89 (7), 073902.

Schnabel, U., Andrasch, M., Weltmann, K.D., Ehlbeck, J., 2014. Inactivation of vegetative microorganisms and *Bacillus atrophaeus* endospores by reactive nitrogen species (RNS). Plasma Process. Polym. 11 (2), 110–116.

Schwabedissen, A., Lacinski, P., Chen, X., Engemann, J., 2007. PlasmaLabel: a new method to disinfect goods inside a closed package using dielectric barrier discharges. Contrib. Plasma Phys. 47 (7), 551–558.

Sharma, A., Pruden, A., Yu, Z.Q., Collins, G.J., 2005. Bacterial inactivation in open air by the afterglow plume emitted from a grounded hollow slot electrode. Environ. Sci. Technol. 39 (1), 339–344.

Shimizu, T., Steffes, B., Pompl, R., Jamitzky, F., Bunk, W., Ramrath, K., Georgi, M., Stolz, W., Schmidt, H.U., Urayama, T., Fujii, S., Morfill, G.E., 2008. Characterization of microwave plasma torch for decontamination. Plasma Process. Polym. 5 (6), 577–582.

Sohbatzadeh, F., Colagar, A.H., Mirzanejhad, S., Mahmodi, S., 2010. *E. coli*, *P. aeruginosa*, and *B. cereus* bacteria sterilization using afterglow of nonthermal plasma at atmospheric pressure. Appl. Biochem. Biotechnol. 160 (7), 1978–1984.

Song, H.P., Kim, B., Choe, J.H., Jung, S., Moon, S.Y., Choe, W., Jo, C., 2009. Evaluation of atmospheric pressure plasma to improve the safety of sliced cheese and ham inoculated by 3-strain cocktail *Listeria monocytogenes*. Food Microbiol. 26 (4), 432–436.

Sousa, J.S., Girard, P., Sage, E., Ravanat, J., Puech, V., 2012. DNA oxidation by reactive oxygen species produced by atmospheric pressure microplasmas. In: Machala, Z., Hensel, K., Akishev, Y. (Eds.), Plasma for bio-decontamination, medicine and food security. Springer, Dordrecht, pp. 107–119.

Stoffels, E., Sakiyama, Y., Graves, D.B., 2008. Cold atmospheric plasma: charged species and their interactions with cells and tissues. IEEE Trans. Plasma Sci. 36 (4), 1441–1457.

Sureshkumar, N.S., 2009. Inactivation characteristics of bacteria in capacitively coupled argon plasma. IEEE Trans. Plasma Sci. 37 (12), 2347–2352.

Sureshkumar, A., Sankar, R., Mandal, M., Neogi, S., 2010. Effective bacterial inactivation using low temperature radio frequency plasma. Int. J. Pharm. 396 (1–2), 17–22.

Takai, E., Kitano, K., Kuwabara, J., Shiraki, K., 2012. Protein inactivation by low-temperature atmospheric pressure plasma in aqueous solution. Plasma Process. Polym. 9 (1), 77–82.

Takai, E., Ikawa, S., Kitano, K., Kuwabara, J., Shiraki, K., 2013. Molecular mechanism of plasma sterilization in solution with the reduced pH method: importance of permeation of HOO radicals into the cell membrane. J. Phys. D.: Appl. Phys. 46, 295402.

Thiyagarajan, M., Sarani, A., Gonzales, X., 2013. Atmospheric pressure resistive barrier air plasma jet induced bacterial inactivation in aqueous environment. J. Appl. Phys. 113 (9), 093302.

Traba, C., Liang, J.F., 2011. Susceptibility of *Staphylococcus aureus* biofilms to reactive discharge gases. Biofouling 27 (7), 763–772.

Trompeter, F.J., Neff, W.J., Franken, O., Heise, M., Neiger, M., Liu, S.H., Pietsch, G.J., Saveljew, A.B., 2002. Reduction of *Bacillus Subtilis* and *Aspergillus Niger* spores using nonthermal atmospheric gas discharges. IEEE Trans. Plasma Sci. 30 (4), 1416–1423.

Tseng, S., Abramzon, N., Jackson, J.O., Lin, W.J., 2012. Gas discharge plasmas are effective in inactivating *Bacillus* and *Clostridium* spores. Appl. Microbiol. Biotechnol. 93 (6), 2563–2570.

Uhm, H.S., Lim, J.P., Li, S.Z., 2007. Sterilization of bacterial endospores by an atmospheric-pressure argon plasma jet. Appl. Phys. Lett. 90 (26), 1501–1503.

Van Bokhorst-Van De Veen, H., Xie, H.Y., Esveld, E., Abee, T., Mastwijk, H., Groot, M.N., 2015. Inactivation of chemical and heat-resistant spores of *Bacillus* and *Geobacillus* by nitrogen cold atmospheric plasma evokes distinct changes in morphology and integrity of spores. Food Microbiol. 45, 26–33.

Van Gils, C.A.J., Hofmann, S., Boekema, B., Brandenburg, R., Bruggeman, P.J., 2013. Mechanisms of bacterial inactivation in the liquid phase induced by a remote RF cold atmospheric pressure plasma jet. J. Phys. D.: Appl. Phys. 46, 175203.

Vaze, N.D., Gallagher, M.J., Park, S., Fridman, G., Vasilets, V.N., Gutsol, A.F., Anandan, S., Friedman, G., Fridman, A.A., 2010. Inactivation of bacteria in flight by direct exposure to nonthermal plasma. IEEE Trans. Plasma Sci. 38 (11), 3234–3240.

Venezia, R.A., Orrico, M., Houston, E., Yin, S.M., Naumova, Y.Y., 2008. Lethal activity of nonthermal plasma sterilization against microorganisms. Infect. Control Hosp. Epidemiol. 29 (5), 430–436.

Virto, R., Manas, P., Alvarez, I., Condon, S., Raso, J., 2005. Membrane damage and microbial inactivation by chlorine in the absence and presence of a chlorine-demanding substrate. Appl. Environ. Microbiol. 71 (9), 5022–5028.

Vleugels, M., Shama, G., Deng, X.T., Greenacre, E., Brocklehurst, T., Kong, M.G., 2005. Atmospheric plasma inactivation of biofilm-forming bacteria for food safety control. IEEE Trans. Plasma Sci. 33 (2), 824–828.

Vrajova, J., Chalupova, L., Novotny, O., Cech, J., Krcma, F., Stahel, P., 2009. Removal of paper microbial contamination by atmospheric pressure DBD discharge. Eur. Phys. J. D 54 (2), 233–237.

Wattieaux, G., Yousfi, M., Merbahi, N., 2013. Optical emission spectroscopy for quantification of ultraviolet radiations and biocide active species in microwave argon plasma jet at atmospheric pressure. Spectrochim. Acta. Part B 89, 66–76.

Xu, G.M., Zhang, G.J., Shi, X.M., Ma, Y., Wang, N., Li, Y., 2009. Bacteria inactivation using dbd plasma jet in atmospheric pressure argon. Plasma Sci. Technol. 11 (1), 83–88.

Xu, L., Tu, Y., Yu, Y., Tan, M., Li, J., Chen, H., 2011. Augmented survival of *Neisseria gonorrhoeae* within biofilms: exposure to atmospheric pressure nonthermal plasmas. Eur. J. Clin. Microbiol. Infect. Dis. 30 (1), 25–31.

Yan, X., Zou, F., Lu, X.P., He, G.Y., Shi, M.J., Xiong, Q., Gao, X., Xiong, Z.L., Li, Y., Ma, F.Y., Yu, M., Wang, C.D., Wang, Y.S., Yang, G.X., 2009. Effect of the atmospheric pressure nonequilibrium plasmas on the conformational changes of plasmid DNA. Appl. Phys. Lett. 95 (8), 083702.

Yang, B., Chen, J.R., Yu, Q.S., Li, H., Lin, M.S., Mustapha, A., Hong, L.A., Wang, Y., 2011. Oral bacterial deactivation using a low-temperature atmospheric argon plasma brush. J. Dent. 39 (1), 48–56.

Yasuda, H., Hashimoto, M., Rahman, M.M., Takashima, K., Mizuno, A., 2008. States of biological components in bacteria and bacteriophages during inactivation by atmospheric dielectric barrier discharges. Plasma Process. Polym. 5 (6), 615–621.

Yu, H., Perni, S., Shi, J.J., Wang, D.Z., Kong, M.G., Shama, G., 2006. Effects of cell surface loading and phase of growth in cold atmospheric gas plasma inactivation of *Escherichia coli* K12. J. Appl. Microbiol. 101 (6), 1323–1330.

Yusupov, M., Bogaerts, A., Huygh, S., Snoeckx, R., Van Duin, A.C.T., Neyts, E.C., 2013. Plasma-induced destruction of bacterial cell wall components: a reactive molecular dynamics simulation. J. Phys. Chem. C 117 (11), 5993–5998.

Zhang, Q., Liang, Y.D., Feng, H.Q., Ma, R.N., Tian, Y., Zhang, J., Fang, J., 2013. A study of oxidative stress induced by nonthermal plasma-activated water for bacterial damage. Appl. Phys. Lett. 102 (20), 203701.

Chapter 7

Cold Plasma Interactions With Food Constituents in Liquid and Solid Food Matrices

B. Surowsky*, S. Bußler†, O.K. Schlüter†

*Technische Universität Berlin, Berlin, Germany

†Leibniz Institute for Agricultural Engineering Potsdam-Bornim, Potsdam, Germany

Whereas cold atmospheric pressure plasma (CAPP), with its huge variety of reactive species and resulting interactions, is already complex itself (see Chapter 3), the complexity of reactions occurring becomes even more complex when the gas discharge comes into contact with food matrices.

Food matrices are commonly composed of different amounts of proteins, lipids, carbohydrates, and water. Together with nutrients, such as minerals and vitamins, they are essential for the human diet.

When observing these components isolated from others, and when the composition of reactive species applied is known, their behavior during plasma exposure is comparably easy to predict.

However, food matrices are multicomponent targets with different percentages of constituents. Their interactions influence the impact of the plasma applied. Even the addition of just one component might lead to a totally different behavior. Some of

Cold Plasma in Food and Agriculture. http://dx.doi.org/10.1016/B978-0-12-801365-6.00007-X

them are able to scavenge others against oxidation, while other components act in the opposite way and enhance oxidation reactions. The formation of cross-links between different molecules, particularly proteins, can also be observed. In addition, the degradation and by-products formed often initiate further reactions, leading to multistep chain reactions.

The fact that all of these reactions additionally depend on factors, such as pH, conductivity, state of matter, and macromolecular structure, underlines the challenge behind the application of plasma on food.

The following chapter gives an overview about the most important food constituents and their reactions with reactive oxygen species (ROS), including proteins, lipids, and carbohydrates. The chapter also deals with the differences between liquid- and solid-food matrices and gives some insight into food matrix interactions during plasma exposure.

1 Plasma Treatment of Liquid and Solid Food Systems

There are two major differences between the application of plasma on solid/dry media and liquid media: the penetration depths or contact surface between plasma and food, and the chemistry/physics initiated by ROS.

The application on solid foods is usually limited to a treatment of their surface. The ability of plasma and its reactive species to penetrate into solid foods is dependent on several factors, which are mainly their composition, water content, and porosity. A handful of studies, dealing with the penetration depth of plasma, showed that reactive species including ROS, reactive nitrogen species (RNS), ozone, and UV can only penetrate some μm deep into biofilms. Xiong et al. (2011) studied the penetration depth of a plasma jet into biofilms formed by *Porphyromonas gingivalis* bacteria using a confocal laser scanning microscope. It was found that the plasma was capable of inactivating the bacterial cells up to a depth of 15 μm. Pei et al. (2012) showed that their handheld air plasma jet is capable of inactivating a 25.5 μm *Enterococcus faecalis* biofilm.

The composition of the plasma and its flow rate represent additional limiting factors regarding the penetration depth. While hydrogen peroxide has a half-life of 1 ms and is comparably stable, other ROS such as singlet oxygen and hydroxyl radicals have half-lives of 1 μs and 1 ns, respectively, leading to very limited penetration depths (Table 1; Møller et al., 2007).

However, a limited penetration depth is an advantage in most cases. If the main reason for applying plasma to the surface is to achieve a gentle microbial decontamination, a low penetration depth helps to retain the majority of nutrients inside the food while its surface can be decontaminated properly. This is particularly true for foods with a big surface-to-volume ratio.

TABLE 1 Properties of Selected Reactive Oxygen Species

Property	Hydroxyl Radical $OH^{\cdot}$	Singlet Oxygen $^1O_2$	Superoxide $O_2^{\cdot-}$	Hydrogen Peroxide H_2O_2
Half-life	1 ns	1 μs	1 μs	1 ms
Present in	Plasma, air, and liquid	Plasma, air, and liquid	Plasma, air, and liquid	Air-liquid interface and liquid
Penetration depth (diffusion coefficient 10^{-9} m^2/s)	1 nm	30 nm	30 nm	1 μm

Liquid foods behave very contrary, because every volume element comes into contact with the plasma applied (or at least with subsequent reaction products), so that penetration depth is not that important. Thus, if plasma is applied in order to decontaminate a liquid, not only the microorganisms will be harmed, but also all other surrounding components. Therefore, optimization with regard to a good antimicrobial efficacy and retention of other food constituents at the same time represents a key challenge.

The subsequent chemistry initiated by ROS has several consequences. Once plasma comes into contact with water molecules, water dissociation reactions with electrons occur. The rates of these reactions greatly depend on the water content, as well as on the electron energy and collision cross sections for water molecules with electrons (Locke and Shih, 2011). The major ROS formed in liquids is $OH^{\cdot}$.

$$O + H_2O \rightarrow 2OH^{\cdot}$$

It can be generated by electron dissociation and electron attachment, as well as by thermal dissociation and ion and metastable pathways. OH radicals are able to subsequently react, eg, to hydrogen peroxide.

$$2OH^{\cdot} \rightarrow H_2O_2$$

Depending on the pH of the liquid, hydrogen peroxide can be very stable and might remain active for a much longer time than the plasma exposure itself. It might also form hydroperoxy radicals or superoxide in the presence of OH radicals, which can also be comparably long-lasting.

$$OH^{\cdot} + H_2O_2 \rightarrow OOH^{\cdot} + H_2O$$
$$O_2^{\cdot-} + H^+ \leftrightarrow OOH^{\cdot}$$

The occurrence of these ROS has been reported by several studies (Zhang et al., 2006; Ikawa et al., 2010). In this context, the term "plasma-activated water" is widely used; meaning, the plasma-treated liquid retains its antimicrobial properties for a long time (Kamgang-Youbi et al., 2009; Traylor et al., 2011; Zhang et al., 2013).

While these reactions do not play a role in dry and solid foods, solid surfaces are a place where other fundamental processes such as etching, deposition, recombination, deexcitation, and secondary emission from solids occur. Regarding decontamination, etching is the most important one. Etching processes can be triggered by atomic oxygen, leading to volatile compounds such as CO_2 and H_2O. It is accompanied with a loss of weight, deriving from the ablation of the first (atomic) layer of the surface and also with surface chemical reactions (Williams, 1997).

2 Plasma Effect on Proteins/Enzymes

Proteins consist of a huge number of amino acids (commonly 100–300) and carry out a great variety of biological functions (eg, cell stabilization or catalysis of chemical reactions). They are connected via peptide bonds, forming a polypeptide chain which is the basis (primary structure) of every protein. The secondary structure describes the spatial arrangement of these amino acids. Depending on the hydrogen bonds located between the peptide bonds, mainly four different types of structure are formed: α-helices, β-sheets, turns, and random coils (Fig. 1). The tertiary structure describes the spatial arrangement of the polypeptide chain and is determined by forces and bonds existing between the amino acid side chains, such as disulphide and hydrogen bonds, as well as hydrophobic, ionic, and van der Waals forces. An aggregation of different proteins is called "quaternary structure" and stabilized by, eg, hydrogen or ionic bonds, but also by covalent bonds.

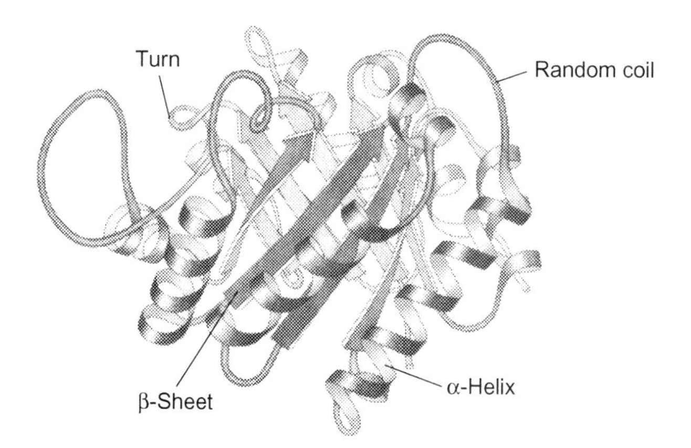

Fig. 1 Typical tertiary structure of a protein with its secondary structure fractions: α-helix, β-sheet, turn, and random coil.

The spatial structure of proteins determines their biological function and, thus, any structural modification of proteins, eg, by oxidation of amino acids, leads to a modification or inhibition of their functionality. Cherry et al. (1999) demonstrated that changing three oxidizable amino acids near the active site of the enzyme peroxidase improved its stability fivefold.

Preferred amino acids for ROS attacks are sulfur-containing and aromatic ones. Two representatives of the former group, cysteine and methionine, are susceptible to reactions with a wide range of ROS, particularly $OH^{\cdot}$ and $^{1}O_2$. Oxidation of the thiol group (–SH) contained in cysteine leads to disulphides such as cystine, but also mixed disulphides can be formed (Fig. 2A). Methionine can be reversibly oxidized to methionine sulfoxide (Fig. 2B) and is supposed to act as an endogenous antioxidant protecting the active site or other sensitive domains in the protein (Levine et al., 1996).

Tryptophan, an aromatic amino acid, can be oxidized under formation of *N*-formylkynurenine. As for methionine, is has been suggested that also tryptophan might act as an antioxidant (Fig. 2C) to a certain extent (Levine et al., 1996).

Another type of oxidative attack results in carbonylation, an irreversible formation of free carbonyl groups (C=O) in amino acids such as arginine, histidine, lysine, proline, threonine, and tryptophan (Fig. 2D). It has been shown that carbonylation can be the reason for the inactivation of enzymes such as catalase (Nguyen and Donaldson, 2005).

All of these reactions are also applicable on enzymes, which are macromolecular biological catalysts and also belong to the group of proteins in most cases.

As naturally occurring compounds in food, they can be either desired or undesired. Most of them belong to the former group, since they catalyze reactions which negatively affect food quality characteristics. In this context, the enzymes polyphenol oxidase (PPO, tyrosinase) and peroxidase (POD) are well known for being involved in enzymatic browning reactions, which also include losses in nutritional value, whereas, eg, lipases are responsible for the formation of off-flavors through the decomposition of lipids.

The inactivation of these and other unwanted enzymes is beneficial and thus part of numerous CAPP-related studies. The majority of them clearly demonstrate that plasma is capable of reducing the activity of enzymes, and some of them also try to explain the mechanisms involved.

Meiqiang et al. (2005) were one of the first groups studying the impact of plasma on the activity of peroxidase. They treated tomato seeds using a magnetized arc-discharge plasma tube and were the also the first (and up to now, only) group who found an increase of peroxidase activity after plasma exposure. However, the study lacks in detail regarding process parameters, such as gas composition and treatment time.

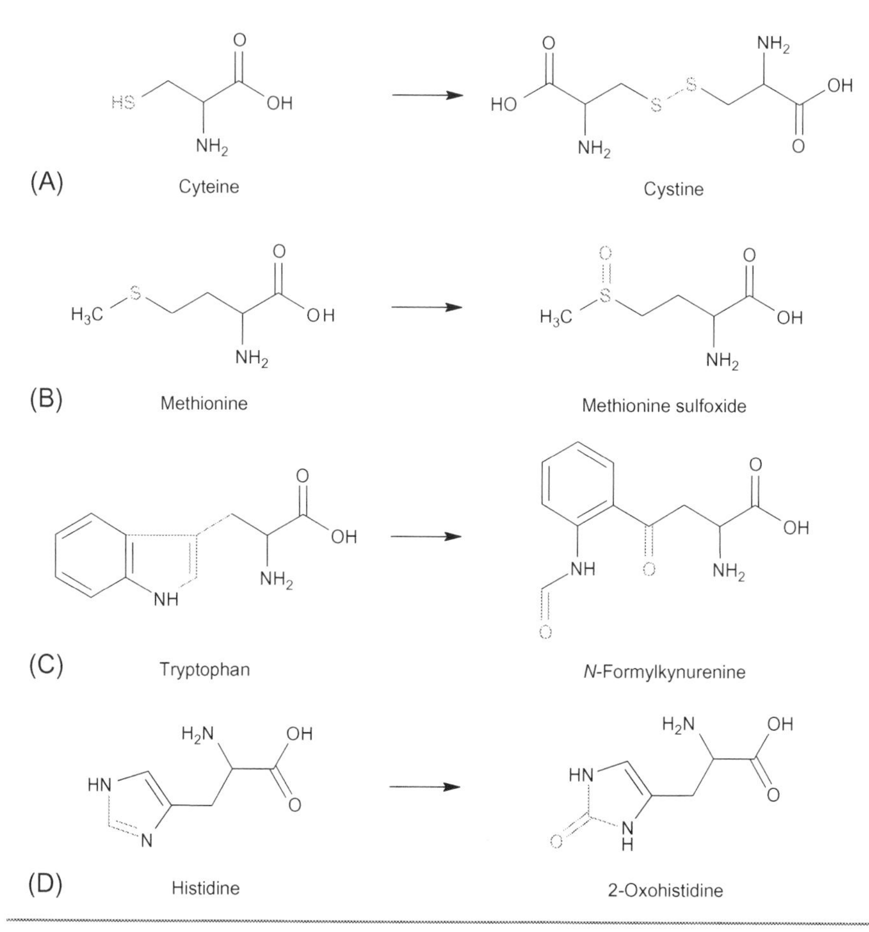

Fig. 2 Common oxidation reactions of amino acids: (A) oxidation of the thiol groups of cysteine and (B) methionine, (C) tryptophan oxidation, and (D) carbonylation of histidine.

The activity of peroxidase was also subject of a study published by Henselová et al. (2012). Significant reductions in activity were achieved for peroxidase derived from maize roots, using a diffuse coplanar surface barrier discharge.

Horseradish peroxidase was subject of a study from Ke and Huang (2013). Using an arc-discharge plasma equipped with argon gas, peroxidase contained in PBS solution was inactivated after a treatment time of 30 min. Fluorescence measurements showed increasing intensities at 450 nm after excitation at 330 nm. Referring to control experiments, it was found that the increasing peak is based on the destruction of heme, the cofactor responsible for the enzyme's activity. Several approaches were used in order to further identify the reactive species involved in peroxidase inactivation. It was concluded that H_2O_2 degraded heme into fluorescent products, while other factors such as OH radicals destroyed the structure of the enzyme. UV was found to be an additional factor, which accelerated the inactivation process in the presence of ROS.

The kinetics of inactivation of tomato peroxidase by atmospheric air dielectric barrier discharge plasma inside a sealed package were part of a further study (Pankaj et al., 2013). The enzyme activity was found to decrease with both treatment time and voltage, following sigmoidal logistic kinetics.

Using fresh-cut apples as a treatment medium, Tappi et al. (2014) investigated the impact of a low-frequency DBD (dielectric barrier discharge) with air as process gas on the activity of native PPO. Its residual activity decreased linearly with increasing treatment times up to about 42% after 30 min. This observation was accompanied with a decreased tendency toward enzymatic browning. The authors conclude that the enzyme's loss in activity might be based on the action of OH and NO radicals on the amino acid structure.

In several studies, single components of CAPP were investigated regarding their impact on enzymes, particularly the effect of UV light is part of numerous studies.

Falguera et al. (2011) studied the effect of UV light on various quality determining characteristics of different apple juices. They found that PPO was completely inactivated after a treatment time of 100 min, whereas POD was already inactivated after 15 min. The same group examined UV light's impact on PPO extracted from *Agaricus bisporus* in sodium phosphate buffer with different melanin concentrations (Falguera et al., 2012). An irradiation time of 35 min led to a complete inactivation of the enzyme in the absence of melanin. However, in the presence of 0.1 mg melanin per ml, even 70 min were not sufficient for a complete inactivation.

Müller et al. (2014) examined the effect of UV-C and UV-B on PPO from mushroom in buffer, apple juice, and grape juice. While UV-B did not affect PPO activity, UV-C decreased its activity depending on the matrix used. Due to the presence of soluble compounds in the juices and a corresponding attenuation of UV-C energy, the highest effect was achieved in buffer.

A further study of the effect of UV-C irradiation on mushroom PPO was conducted by Haddouche et al. (2015). A 5-min exposure with 20 mW/cm^2 resulted in more than 90% reduction of PPO activity. Several typical fruit and vegetable juice components, such as sugars, were added in order to study their impact on PPO inactivation. It was found that fructose accelerated the inactivation of PPO, whereas sucrose and glucose did not show any effect.

Other studies include investigations of the impact of CAPP on dehydrogenase (Meiqiang et al., 2005), malate synthase, isocitrate lyase, catalase and malate dehydrogenase (Nguyen and Donaldson, 2005), pectinolytic enzymes (Falguera et al., 2011), lipase (Li et al., 2011), catalase, dehydrogenase and superoxide dismutase (Henselová et al., 2012), lysozyme (Takai et al., 2012), α-chymotrypsin (Attri and Choi, 2013), and lactate dehydrogenase (Zhang et al., 2015).

Since most of the enzymes existing are proteins, their degradation is mainly based on reactions of ROS with amino acid side chains as described for proteins.

Most of the authors investigating plasma-mediated enzyme degradation try to find explanations for occurring changes in activity and try to identify the reactive species involved. Typical methods used in order to identify changes of the enzymes' structure are circular dichroism spectroscopy and fluorescence spectroscopy. While the former is an excellent tool for determining changes of the different structure fractions, fluorescence spectroscopy is widely used, eg, for the quantification of aromatic amino acids such as tyrosine and tryptophan. The intensity as well as location of their characteristic peaks has been shown to be a helpful indicator for occurring enzyme modifications.

Table 2 shows the mechanisms and ROS proposed to be involved in CAPP mediated enzyme inactivation. As partially described previously, Ke and Huang (2013)

TABLE 2 Overview of Studies Investigating the Mechanisms Behind Enzyme Inactivation by CAPP

Enzyme	Matrix	Enzyme Activity	Reason/ Mechanism	Reactive Species (Major Role)	References
Peroxidase from horseradish	PBS buffer	Decrease	Heme destruction by H_2O_2 and structural modifications	H_2O_2, $OH^{\cdot}$, UV	Ke and Huang (2013)
Lysozyme from egg white	PBS buffer	Decrease	Change of secondary structure and increased molecular weight	ROS (no effect by UV)	Takai et al. (2012)
α-Chymotrypsin from bovine pancreas	PBS buffer	Decrease	Loss of secondary structure, conformational change, and unfolding	not indicated	Attri and Choi (2013)
Lipase from *Candida rugose*	PBS buffer	Increase	Change of secondary structure	$OH^{\cdot}$, $O^{\cdot}$: activity increase; ozone alone: activity decrease; UV: no effect	Li et al. (2011)
Lactate dehydrogenase from rabbit muscle	PBS buffer	Decrease	Modification of secondary molecular structure (reduced α-helix, increased β-sheet) and polymerization of peptide chains	Long-living ROS: H_2O_2, ozone, and NO_3^-	Zhang et al. (2015)

observed an increasing peak at 450 nm (excitation 330 nm) after exposing horseradish peroxidase to CAPP. After conducting several control experiments with enzymes with and without heme contained as cofactor, they conclude that its origin is a fluorescent product formed as a result of heme destruction. They furthermore postulate that monitoring the intensity change for this peak could be a good method for assessing the CAPP-induced effects on horseradish peroxidase. According to the study, H_2O_2 is the key factor involved in heme destruction (Fig. 3), while other compounds such as $OH^{\cdot}$ and UV radiation could destroy the structure of the enzyme and, thus, accelerate the degradation process.

It is noticeable that all of the studies shown are highlighting structural changes as being responsible for the activity changes. However, those changes are not necessarily accompanied with activity losses. While the majority of studies found decreased enzyme activities, Li et al. (2011) have shown that lipase activity can increase during plasma exposure due to changes of the enzymes' secondary structure initiated by ROS. However, which reactive species are dominant and whether an increase or a decrease of activity occurs has been shown to hold importance. The action of ozone, as well as UV alone, resulted in decreasing activities. Zhang et al. (2015) give detailed information on the changes of the different fractions. According to their results, the α-helix content decreased upon plasma exposure, whereas β-sheet regions, as well as random coils, increased.

These studies support our own findings, which were the result of studies using polyphenloxidase from mushroom as well as peroxidase from horseradish in a (solid) model food system (Surowsky et al., 2013). It had been shown that CAPP is capable

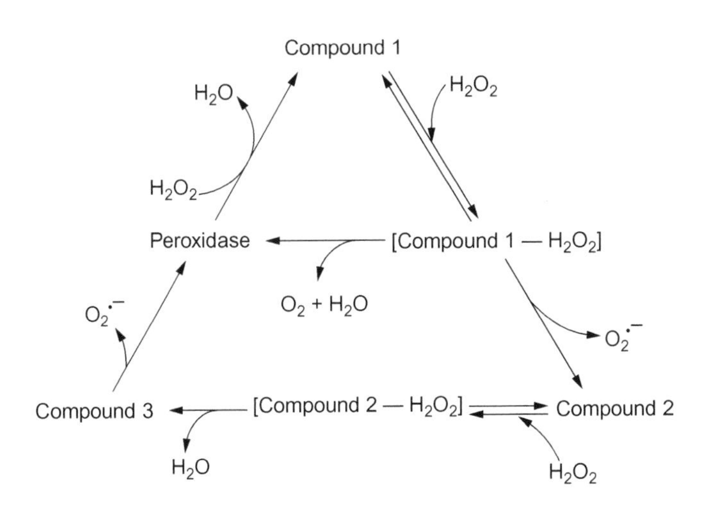

Fig. 3 Proposed reaction pathway of peroxidase with H_2O_2 in the absence of reducing substrate. *(From Ke, Z., Huang, Q., 2013. Inactivation and heme degradation of horseradish peroxidase induced by discharge plasma. Plasma Process. Polym. 10 (8), 731–739.)*

of reducing the activity of both PPO and POD, as a result of changes of their secondary structure fractions. In both cases, decreasing α-helix contents were accompanied by decreasing β-sheet contents. The results of fluorescence measurements supported the occurrence of structural changes. Decreasing fluorescence intensities of tryptophan, as well as a red shift, indicated a change of tryptophan surroundings to a more polar environment.

3 Plasma Effect on Lipids

Lipids consist of fatty acids, which can be saturated, monounsaturated, or polyunsaturated, depending on the number of double bonds existing between their carbon atoms. The primary targets for ROS (particularly $OH^{\cdot}$ and $^{1}O_2$) are C–H-bonds (methyl groups), preferably those lying between double bonds, because the energy input needed in order to abstract a hydrogen atom is much lower there compared to CH-bonds bound elsewhere (272 vs. 422 kJ/mol; Wollrab, 2009; Fig. 4). Thus, the more double bonds a fatty acid contains, the more susceptible it is against homolytic ROS attacks. Typical ROS sensitive fatty acids are, eg, linoleic acid (18:2) and α-linolenic acid (18:3), containing two and three double bounds, respectively.

Lipid oxidation is accompanied with the formation of peroxy radicals, which in turn also break CH-bonds under formation of hydroperoxide and another radical. This chain reaction can be divided into three steps: initiation, propagation, and termination (Vatansever et al., 2013). It can be stopped either by a recombination of radicals or by the effect of antioxidants, which act as radical scavengers. Hydroperoxides can initiate various subsequent reactions, such as decomposition of lipids or chain-linking of fatty acids via ether or peroxide bridges. Since they are comparably labile, hydroperoxides can also dissociate to aldehydes.

The initial oxidation of only a few lipid molecules can result in significant tissue damage of bacterial cells (Mylonas and Kouretas, 1999).

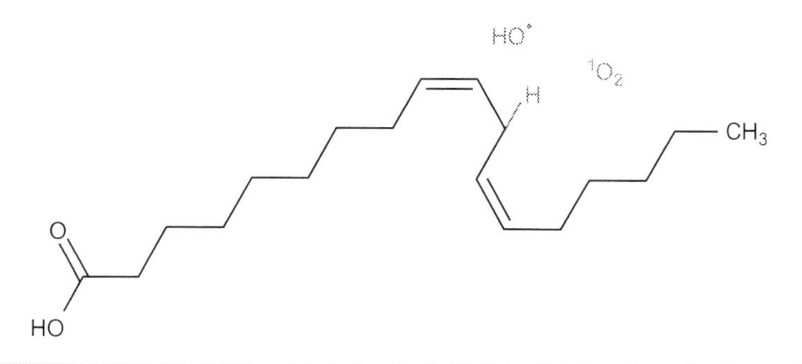

Fig. 4 ROS attack on methyl group located between the double bonds of linoleic acid.

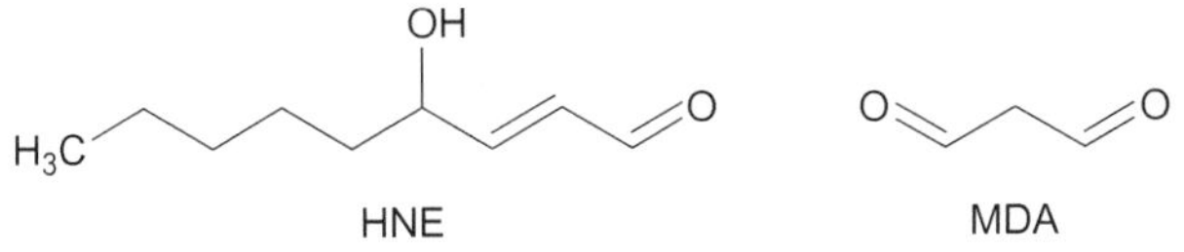

HNE MDA

Fig. 5 Two reactive lipid oxidation products: HNE (4-hydroxy-2-nonenal) and MDA (malondialdehyde).

In case of the oxidation of linoleic acid and linolenic acid, the compounds 4-hydroxy-2-nonenal and malondialdehyde are finally formed (Fig. 5). These two aldehydes are important biomarkers of oxidative stress, highly reactive and able to form conjugates with DNA and proteins (Esterbauer et al., 1991).

Since lipid peroxidation is accompanied with the generation of off-flavors, it has a huge impact on sensory attributes of food. Depending on the fatty acid oxidized, rancid, fishy, fatty, or metallic flavors can occur. Volatile carbonyl compounds such as pentanal, *cis*-4-heptanal and *trans*-6-nonenal are responsible for this.

Korachi et al. (2015) investigated the biochemical changes to milk following a treatment by CAPP with emphasis on changes to free fatty acids. Whereas the total free fatty acid concentration did not change significantly, an effect on polyunsaturated fatty acids was observed (3–2.5% after 20 min). The amount of the predominant fatty acid contained in milk, hexadecanoic acid (C16:0), was found to decrease after 3 min of plasma exposure; however, longer treatment times resulted in increasing amounts. The authors conclude that the occurring changes may be attributed to dehydrogenation caused by oxygen radicals. They also found that the levels of long-chain fatty acids such as stearic acid (C18:0) decreased, while the levels of short-chain fatty acids (C10:0 and C12:0) increased. According to the study, this could be attributed to hydrolytic effects.

The generation of shorter chain fatty acids as a result of the impact of reactive species has also been reported in other studies. According to these studies, reactive species can initiate lipid peroxidation and produce hydroperoxide, which may be subsequently converted into secondary oxidation products such as aldehydes and shorter chain fatty acyl compounds (Benedetti et al., 1984; Kappus, 1985). Other studies showed that exposure of polyunsaturated fatty acids to reactive species (hydroperoxyl radicals, superoxide radicals, and singlet oxygen) can also result in the generation of shorter fatty acids (Farr and Kogoma, 1991; Doroszkiewicz et al., 1994).

Whereas plasma-based lipid oxidation is usually unwanted, some studies are dealing with cold plasma as a tool to accelerate lipid oxidation in order to simulate slow alteration processes (Van Durme et al., 2014; Vandamme et al., 2015). Van Durme et al., concluded that plasma application on vegetable oil leads to the formation of several secondary volatile lipid oxidation products. Aldehydes were mainly formed by the action of atomic oxygen, while singlet oxygen induced the formation of 2-pentyl furan. Vandamme et al., used fish oil instead of vegetable oil. They also found different secondary lipid oxidation products such as 2-propenal, 2-pentenal, and heptanal in the plasma-treated samples. Both studies conclude that, in comparison to thermally based tests, plasma exposure is capable of accelerating lipid oxidation in a realistic manner.

4 Plasma Effect on Carbohydrates

Carbohydrates are particularly susceptible to the action of $OH^{\cdot}$. Sugar alcohols such as mannitol and sorbitol have shown to scavenge OH and thus protect, eg, more vital cellular components from being oxidized (Smirnoff and Cumbes, 1989; Shen et al., 1997). Their hydroxyl radical scavenging activities were shown to increase with increasing sugar alcohol concentration. It was also found that sucrose and different polyols were the most effective scavengers. Using a xanthine oxidase-malate dehydrogenase system, Smirnoff and Cumbes further demonstrated that myo-inositol, sorbitol, and mannitol are capable of protecting malate dehydrogenase from being oxidized. 20 mM mannitol scavenged ~60% of the hydroxyl radicals.

Isbell and Frush (1973) and Isbell et al. (1973) showed that the hydrogen peroxide mediated oxidation of aldoses (glucose, mannose, galactose, allose, altrose, talose, xylose, arabinose, lyxose, and ribose), as well as ketoses (fructose and sorbose) leads to the formation of formic acid (Fig. 6). They found that aqueous alkaline hydrogen peroxide solutions degrade aldohexoses almost quantitatively to 6 mol of formic

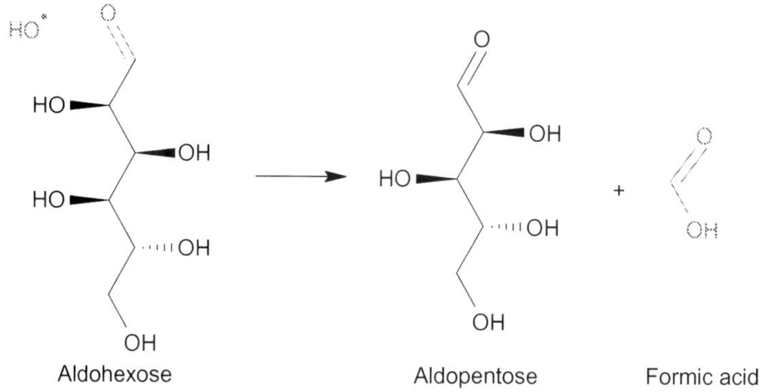

Fig. 6 General mechanism of the oxidation of aldohexose (here: glucose): formation of aldopentose and formic acid. *(From Isbell, H. S., Frush, H. L., Martin, E. T. 1973. Reactions of carbohydrates with hydroperoxides: part I. Oxidation of aldoses with sodium peroxide. Carbohydr. Res. 26 (2), 287–295.)*

acid, and aldopentoses to 5 mol. The mechanism behind formic acid formation is explained as stepwise degradation of aldoses: the addition of a hydroperoxide anion to the aldehydo modification of the sugar and subsequent decomposition of the adduct to formic acid and the next lower aldose. In addition, it was found that iron salts accelerate this reaction. The reactivity of the different hexoses and pentoses can be directly linked to their tendency to exist in the aldehyde form. Thus, glucose, which has the lowest proportion of the aldehyde form, showed the lowest reaction rate of the six aldohexoses studied. In the pentose series, ribose has the highest reaction rate because it has the highest proportion of the aldehyde form.

5 Matrix Interactions During Plasma Exposure: Radical Scavengers and Protective Effects

Once plasma comes into contact with the food matrix, the reactions occurring are dependent on the composition of both the plasma as well as the food matrix.

Molecule-radical interactions include reactions directly occurring between plasma-immanent species and molecules, as well as reactions between subsequently formed radicals and molecules. Particularly important are those reactions which involve antioxidants and water (see Section 7.1), since these compounds determine the progress of the reaction cascade initiated by ROS. Table 3 gives an overview about the scavenging ability of various molecules against hydroxyl radicals, singlet oxygen, superoxide, and hydrogen peroxide.

Minor components such as ascorbic acid (vitamin C), tocopherol (vitamin E), flavonoids, and carotenoids are crucial antioxidative components of the food matrix. All of them are protecting the matrix against ROS, however, their mechanisms of action differ.

Radical scavengers contain sterically hindered phenolic groups which help to interrupt the radical transfer occurring during oxidation reactions by forming other, inert radicals. This finally stops the reactions cascade and protects other molecules from getting oxidized. One important compound belonging to this group is vitamin E.

It has been shown that vitamin E is most efficient in scavenging peroxyl radicals in vivo (Niki, 2014). Other sources also report that it is capable of scavenging hydroxyl radicals ($OH^{\cdot}$), singlet oxygen ($^1O_2$), and superoxide ($O_2^{\cdot-}$) (Wang and Jiao, 2000; Halliwell, 2006). Vitamin E particularly inhibits lipid peroxidation by breaking chain propagation both in vitro and in vivo (Niki, 2014).

Reducing agents can act as an antioxidant as well. These molecules have very low redox potentials, so they can be oxidized much easier than other compounds. Thus, if representatives of this group are present in the food matrix, ROS will preferably oxidize them. Ascorbic acid is one of the most widespread reducing agents in food.

TABLE 3 Scavenging Abilities of Different Molecules

Antioxidant	Hydroxyl Radical $OH^\cdot$	Singlet Oxygen $^1O_2$	Superoxide $O_2^{\cdot-}$	Hydrogen Peroxide H_2O_2
Histidine	■	■	□	□
Mannitol	■	□	□	□
Sorbitol	■	□	□	□
Myo-Inositol	■	□	□	□
Proline	■	□	□	□
Vitamin E	■	■	■	□
Dimethyl sulfoxide	■	□	□	□
Chlorogenic acid	■	□	□	■
Sodium azide	□	■	□	□
Vitamin C	□	■	■	■
β-carotene	□	■	□	□
ADPA	□	■	□	□
Superoxide dismutase	□	□	■	□
Glutathione	□	□	■	■
Catalase	□	□	□	■
References	Smirnoff and Cumbes (1989), Wang and Jiao (2000), Møller et al. (2007), Takamatsu et al. (2015), and Vatansever et al. (2013)	Wang and Jiao (2000), Møller et al. (2007), Matsumara et al. (2013), and Vatansever et al. (2013)	Wang and Jiao (2000), Møller et al. (2007), and Nakamura et al. (2012)	Wang and Jiao (2000), Vakrat-Haglili et al. (2005), and Møller et al. (2007)

■—high antioxidant activity and □—low/no antioxidant activity.

If ascorbic acid is exposed to oxygen, it generally follows the Michaeles concept of a reversible two-step oxidation involving a free radical intermediate, resulting in the formation of dehydroascorbic acid (DHA; Fig. 7; Michaelis, 1932; Bielski et al., 1975).

DHA can be subsequently converted into 2,3-diketo-L-gulonic acid by hydrolysis (Fig. 8; Deutsch and Santhosh-Kumar, 1996). The 2,3-diketo-L-gulonic acid has no

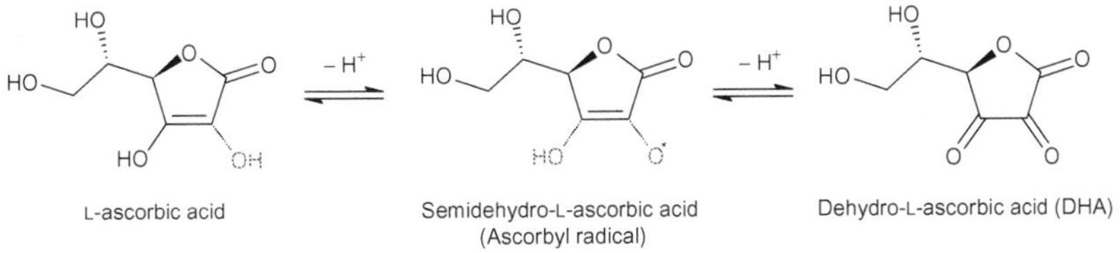

Fig. 7 Aerobic two-step degradation pathway of L-ascorbic acid.

biological function and cannot be reconverted back to DHA. Deutsch (1998) also reports that DHA and 2,3-diketo-L-gulonic acid are more susceptible to oxidation by ROS than ascorbic acid.

Besides these well-known antioxidants, several studies show that various carbohydrates are also capable of protecting other molecules from getting oxidized. Joslyn and Miller (1949) and Kyzlink and Curda (1970) reported that carbohydrates such as fructose, glucose, and sucrose are capable of protecting ascorbic acid—although it has a very low redox potential—from getting oxidized by ambient oxygen. Birch and Pepper (1983) confirmed these results for glucose, sorbitol, fructose, maltose as well as maltitol.

Sugars have also been shown to protect lipids from being oxidized (Faraji and Lindsay, 2004). Antioxidant activity was confirmed for fructose, sucrose, raffinose, sorbitol, and mannitol at concentrations of 16% in the aqueous phase in model fish-oil-in-water emulsions. Combinations of sugars and phenolic antioxidants such as tocopherol decreased the antioxidant activity, compared to that for individual phenolic antioxidants. This phenomenon is explained with the H-bonding activity of sugars/polyols, which hinders the H-donating activity of phenolic antioxidants. Protective effects of ions on enzyme structure and activity against plasma-based degradation have also been observed (Attri and Choi, 2013). It was found that chymotrypsin contained in buffer was protected by triethylammonium sulfate, as well as by triethylammonium dihydrogen phosphate (TEAP). Particularly the addition of TEAP resulted in retention of the enzymes' structure and activity.

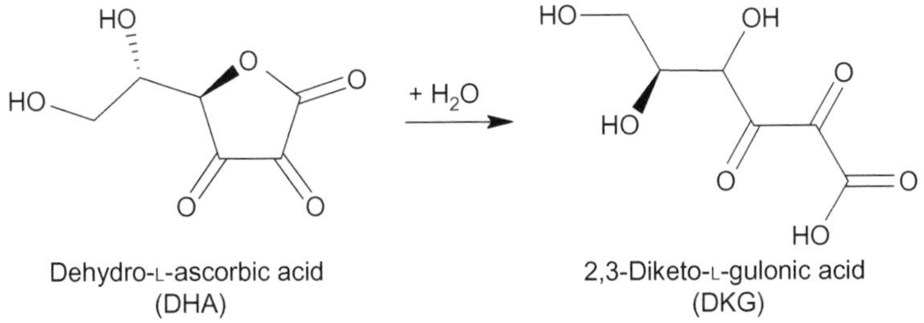

Fig. 8 Degradation of dehydro-L-ascorbic acid in aqueous solution.

These findings show that the result of applying plasma on multicomponent food is hard to predict, especially when it is applied on liquid ones, where basically every component can react with each other. This underlines the need for two particular things: a detailed knowledge of the composition and properties of the food to be treated and, as a consequence, a tailor-made plasma process in terms of setup and ROS composition.

6 Plasma Effects on Functional Properties of Food Systems

It is well-known that CAPP also modifies the surface structure of materials at the micro- to nanometer range (Pankaj et al., 2014a,b) and hereby allows to chemically and physically modify surface characteristics of polymeric materials without affecting their bulk properties. With regard to the modification of surface properties of polymeric materials, CAPP treatment is advantageous for several reasons: its use is free of hazardous solvents; there is uniformity of treatment; and there is no generation of thermal damage when in contact with materials (Desmet et al., 2009; Misra et al., 2015; Morent et al., 2011). Extended plasma treatment results in the ablation of surfaces caused by UV radiation and bombardment of energetic particles (eg, electrons, ions, radicals, and excited atoms/molecules) on the polymer surface, consequently increasing surface roughness (De Geyter et al., 2007). Because of the high etch rate, argon and oxygen plasmas are particularly useful in ablating polymers (Fricke et al., 2011). Besides ablating surfaces, CAPP can be used for the surface functionalization of polymers, polymer degradation, and cross-linking (Pankaj et al., 2014a). Surface functionalization refers to the formation of functional groups on the polymer surface (eg, oxygen- and nitrogen-containing groups) occurring when hydrogen atoms on polymer chains form carbon radicals, subsequently causing oxidation or nitration. Plasma-specific effects on polymer surfaces may offer an innovative approach for the modification of biopolymers in the food sector. Given its diversity of application possibilities, the plasma technology is applicable to moist and dry surfaces from animal and vegetable origin.

The number of reports on plasma treatment of plant surfaces is limited and different plasma sources were used in the studies. In order to elucidate whether ion bombardment and subsequent oxidation reactions cause etching of the upper epidermal layer and cell ablation, Grzegorzewski et al. (2010) analyzed the surface morphology of plasma-treated lamb's lettuce by means of SEM. While untreated lettuce leaves exhibited wide areas with thick platelets and small-sized granular structures, the surface of plasma-treated leaves became rough and granular structures disappeared with increasing exposure time. Consequently, the surface wettability of leaves was increased. Whereas the surface of the pristine leaves was hydrophobic (contact angle of 88), exposure to direct plasma treatment gradually reduced the contact angle until a value of 34 was reached (180 s exposure) and the surfaces became more hydrophilic.

Bormashenko et al. (2012) reported the possibility to modify the surfaces wetting properties of a diversity of seeds, including lentils, beans and wheat, by cold radiofrequency air plasma treatment. Here again, air plasma treatment led to the dramatic decrease in the apparent contact angle, in turn changing the wettability of beans and lentils and giving rise to a change in the water absorption (imbibition) of the seeds.

However, besides the resulting accelerated germination rates of seeds, post-harvest application of plasma on plant materials may be used as a targeted tool for the stress-induced modification and intensification of plants' secondary metabolism and, hence, concluding for the possibility of eliciting phytochemical synthesis without damaging the plant tissue. Further, plasma processing of plant material for food may change the structure of the food matrix, which is of great importance for the bioavailability of phytochemicals ingested in the matrix. Yet, the interactions of plasma-immanent species with dietary bioactive compounds in foods are not clearly understood. The health-promoting effects of flavonoids are attributed to a high antioxidant activity and metal-binding properties, by which cells are protected against the damaging effects of ROS and RNS such as singlet oxygen, superoxide, peroxyl radicals, hydroxyl radicals, and peroxynitrite (Haenen et al., 1997; Hu et al., 1995). Since plasma might generate numerous reactive species, flavonoids are ideal target compounds to elucidate the interactions and effects of plasma-immanent reactive species on bioactive molecules due to their antioxidant activity protecting cells against the damaging effects of ROS.

Grzegorzewski et al. (2010) exposed 1,4-benzopyrone derivates to different CAPP discharges using different feed gases (argon, oxygen) and excitation sources (radio frequency, microwave) at both atmospheric and low pressure. A structure-dependent degradation upon plasma-chemical reactions were observed. The plasma-induced effects on the flavonol glycoside profile of pea seeds, sprouts, and seedlings, reported by Bußler et al. (2015a,b) may be a consequence of (i) the impact on photosynthetic efficiency and on related signaling pathway(s) involved in the synthesis of plant secondary metabolites and (ii) the protection against oxidative stress resulting from excessive strain by ROS and UV radiation from plasma.

During post-harvest processing of plant-based food materials, the plasma technology offers several approaches to modify their techno-functional properties.

Chen (2013) investigated the impact of low-pressure plasma treatment on the properties of long-grain brown rice and determined the microstructure of the brown rice surface, and the cooking, textural, and pasting properties of plasma-treated brown rice. Indica brown rice was treated for 30 min at voltage settings of 1, 2, and

3 kV in air. Exposure to plasma treatment resulted in etching of the brown rice surface, allowing water to be easily absorbed by the rice kernel during soaking. This led to a reduction in cooking time, elongation ratio, width expansion ratio, water absorption, and cooking loss of brown rice. Plasma further modified the pasting properties of the rice, and a significant decrease in peak viscosity and breakdown were detected. It was shown that the starch structure of brown rice was influenced by low-pressure plasma; In addition, decreases in enthalpy and crystallinity were measured by a differential scanning calorimeter and X-ray diffractometer. Similar plasma-induced effects were reported for basmati rice (Thirumdas et al., 2015). Hence, plasma treatment can be used to improve cooking quality of long-grain and basmati rice.

Different researchers evaluated the potential of CAPP treatment for the functionalization of dry, bulky, and powdery food materials. Misra et al. (2015) explored the possible effects of CAPP as a means to change the structural and functional properties on strong and weak wheat flours. Considering that previous studies have demonstrated that ozone modifies the functional properties of wheat flour (Chittrakorn, 2008), they generated high in ozone concentration using a dielectric barrier discharge plasma device and air as the working gas. Plasma treatments were found to result in a voltage and treatment time-dependent increase in the viscoelasticity of the dough produced from the wheat flour. Those effects were attributed to the alteration of the secondary structure of gluten proteins, following measurements via FTIR spectroscopy.

Bußler et al. (2015a,b) found that plasma treatment was capable of modifying protein- and techno-functional properties of different flour fractions from grain pea (*Pisum sativum* "Salamanca"). Experiments using a pea protein isolate indicated that the reason for the increase in water and fat binding capacities in protein rich pea flour to 113% and 116%, respectively, was based on plasma-induced modifications of the proteins. With increasing exposure to plasma, the fluorescence emission intensity of soluble proteins increased at 328 nm and decreased at 355 nm, indicating structural and/or compositional changes of the proteins. The results indicate that the application of CAPP can be exploited as a means to modulate the functionality of dry bulk materials in the food sector.

Because the result of applying plasma on multicomponent food is hard to predict, a wide range of interactions between different components need to be taken into consideration and are presently the subject of intense research. Plasma treatment of single-component or model food systems is promising in order to elucidate the underlying plasma-induced mechanisms.

Lii et al. (2002) exposed granular starches of nine botanical origins to low-pressure glow plasma generated in air. Starches were partly oxidized to carboxylic starches, and partly depolymerized when their affinity to plasma depended on their botanical origin. The authors claimed that treatment with glow plasma may offer an alternative method of wasteless dextrinization of starches.

The interaction between CAPP and whey protein isolate model solutions was investigated by Segat et al. (2015). They found an increase in yellow color and a minor reduction in pH value, which they attributed to the reactions of reactive oxygen and nitrogen plasma species. Mild oxidation in the proteins occurred upon exposure to plasma for 15 min, accompanied by an increase in carbonyl groups and surface hydrophobicity, besides the reduction of free SH groups. Those results point to the effects on amino acid residues. Moreover, the authors correlated the reduction of free SH groups to the aggregation among proteins or a strong oxidative effect on cysteine. Dynamic light scattering revealed a certain degree of unfolding, as confirmed by high performance liquid chromatography profiles. Those plasma-induced protein structure modifications may be responsible for the improvement in foaming and emulsifying capacity. Reported results demonstrate that cold plasma can be successfully applied in order to selectively modify the structure of proteins and starches and therefore, improve their functionality; hence, this offers the opportunity to use plasma-treated food components as an ingredient in different formulated food to express targeted functionality.

Another potential application of the plasma technology in the food sector may further aim at the targeted modification or functionalization of product surfaces. Up to now, little literature is available dealing with these application-oriented utilization of plasma. As many of the plasma sources relevant for plasma application in the food sector can be easily integrated into continuous industrial production or finishing lines, surface treatment of processed food using plasma seems to be promising. One possible application may offer the surface modification of bakery products. Misra et al. (2014) demonstrated the potential of plasma in enhancing the surface hydrophobicity of freshly baked biscuits, evident from the increased spread area of vegetable oil. For a given volume of oil, up to 50% more spreading of oil could be achieved within a few seconds of the process, as compared to currently used methods. For this particular application it could be advantageous that the induced effects that are time-dependent fade over time. Nevertheless, detailed investigations regarding the effects of cold plasma on the chemical constituents of the biscuits and on the very probable oxidation of fat is urgently required.

7 Summary

Compared to traditional food preservation methods such as pasteurization, an efficient application of CAPP requires a much more detailed knowledge of the process and the product in order to achieve the desired goal. This includes knowledge of the plasma source and the proportions of different reactive species contained in the resulting plasma, as well as of the type and composition of the food matrix.

Fig. 9 gives a final overview of the major effects of CAPP on food constituents described in this chapter. On their way to the food surface, the reactive species

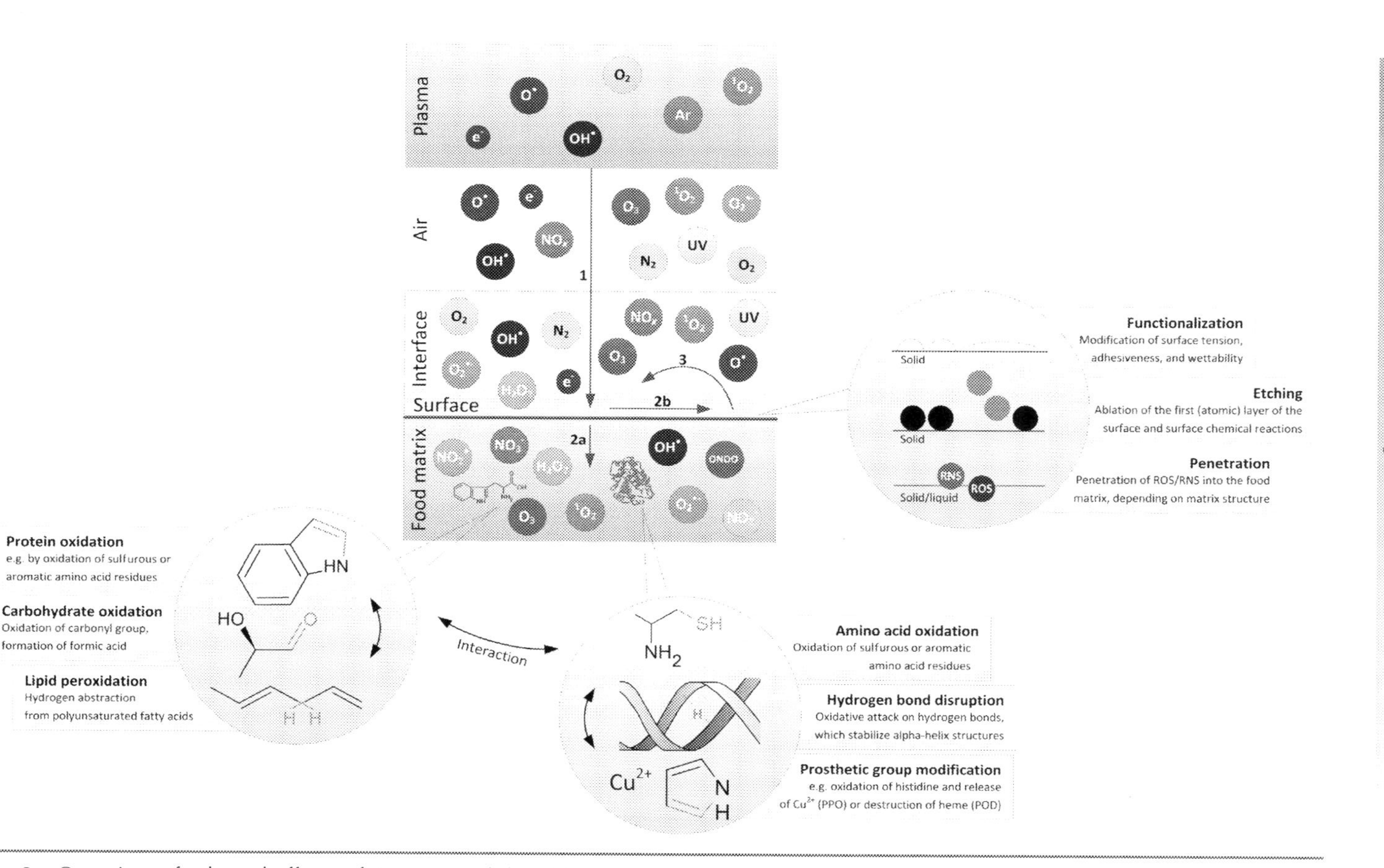

Fig. 9 Overview of selected effects of CAPP on solid and liquid food systems.

produced by a plasma source come into contact with molecules of the surrounding air, resulting in the generation of further ROS and RNS (1). After arriving at the surface, they can act in different ways: they can be absorbed there, they can migrate through the surface (2a) or they can initiate chemical reactions with the surface (2b). Finally, they can also be removed by desorption and return back to the gas phase, where recombination between desorbed species and new reactive species can occur (3) (D'Angelo, 2010).

Whereas the application of CAPP on solid foods mostly affects their surface, where, eg, functionalization and etching can occur and penetration of reactive species is limited, liquid food matrices represent a further reactive environment. Due to their interaction with water and other molecules, ROS and RNS initiate multiple chain reactions, resulting in a great variety of different species. This can finally result in the oxidation of constituents such as proteins, carbohydrates, and lipids. The activity of enzymes can also be influenced as a consequence of amino acid oxidation, hydrogen bond disruption, and prosthetic group modification. In addition, interactions between oxidation products and other compounds such as antioxidants need to be taken into account.

The complexity of these molecule-molecule interactions underlines the importance of a tailor-made process design. In order to achieve the goal, which might be a gentle microbial inactivation in most cases, plasma sources need to be adapted to the specific purpose of application. This will be one of the major tasks in the field of CAPP and food before large-scale processes can be implemented in industry.

References

Attri, P., Choi, E.H., 2013. Influence of reactive oxygen species on the enzyme stability and activity in the presence of ionic liquids. PLoS ONE. 8, e75096.

Benedetti, A., Comporti, M., Fulceri, R., Esterbauer, H., 1984. Cytotoxic aldehydes originating from the peroxidation of liver microsomal lipids: identification of 4, 5-dihydroxydecenal. Biochim. Biophys. Acta Lipids Lipid Metab. 792 (2), 172–181.

Bielski, B.H., Richter, H.W., Chan, P.C., 1975. Some properties of the ascorbate free radical. Ann. N. Y. Acad. Sci. 258 (1), 231–237.

Birch, G.G., Pepper, T., 1983. Protection of vitamin C by sugars and their hydrogenated derivatives. J. Agric. Food Chem. 31 (5), 980–985.

Bormashenko, E., Grynyov, R., Bormashenko, Y., Drori, E., 2012. Cold radiofrequency plasma treatment modifies wettability and germination speed of plant seeds. Sci. Rep. 2, 741. http://dx.doi.org/10.1038/srep00741.

Bußler, S., Herppich, W.B., Neugart, S., Schreiner, M., Ehlbeck, J., Rohn, S., Schlüter, O., 2015a. Impact of cold atmospheric pressure plasma on physiology and flavonol glycoside profile of peas (*Pisum sativum* 'Salamanca'). Food Res. Int. 76 (Part 1), 132–141. http://dx.doi.org/10.1016/j.foodres.2015.03.045.

Bußler, S., Steins, V., Ehlbeck, J., Schlüter, O., 2015b. Impact of thermal treatment versus cold atmospheric plasma processing on the techno-functional protein properties from *Pisum sativum* 'Salamanca'. J. Food Eng. 167 (B), 166–174. http://dx.doi.org/10.1016/j.jfoodeng.2015.05.036.

Chen, H.H., 2013. Investigation of properties of long-grain brown rice treated by low-pressure plasma. Food Bioprocess Technol. 7 (9), 2484–2491. http://dx.doi.org/10.1007/s11947-013-1217-2.

Cherry, J.R., Lamsa, M.H., Schneider, P., Vind, J., Svendsen, A., Jones, A., Pedersen, A.H., 1999. Directed evolution of a fungal peroxidase. Nat. Biotechnol. 17 (4), 379–384. http://dx.doi.org/10.1038/7939.

Chittrakorn, S., 2008. Use of Ozone as an Alternative to Chlorine for Treatment of Softwheat Flours. Kansas State University, Manhatthan.

D'Angelo, D., 2010. Plasma-surface Interaction. In: Rauscher, H., Perucca, M., Buyle, G. (Eds.), Plasma technology for Hyperfunctional Surfaces: Food, Biomedical and Textile Applications. Wiley, Weinheim.

De Geyter, N., Morent, R., Leys, C., Gengembre, L., Payen, E., 2007. Treatment of polymer films with a dielectric barrier discharge in air, helium and argon at medium pressure. Surf. Coat. Technol. 201 (16), 7066–7075.

Desmet, T., Morent, R., Geyter, N.D., Leys, C., Schacht, E., Dubruel, P., 2009. Nonthermal plasma technology as a versatile strategy for polymeric biomaterials surface modification: a review. Biomacromolecules 10 (9), 2351–2378.

Deutsch, J.C., 1998. Ascorbic acid oxidation by hydrogen peroxide. Anal. Biochem. 255 (1), 1–7.

Deutsch, J.C., Santhosh-Kumar, C.R., 1996. Dehydroascorbic acid undergoes hydrolysis on solubilization which can be reversed with mercaptoethanol. J. Chromatogr. A. 724 (1), 271–278.

Doroszkiewicz, W., Sikorska, I., Jankowski, S., 1994. Studies on the influence of ozone on complement-mediated killing of bacteria. FEMS Immunol. Med. Microbiol. 9 (4), 281–285.

Esterbauer, H., Schaur, R.J., Zollner, H., 1991. Chemistry and biochemistry of 4-hydroxynonenal, malonaldehyde and related aldehydes. Free Radic. Biol. Med. 11 (1), 81–128.

Falguera, V., Pagán, J., Ibarz, A., 2011. Effect of UV irradiation on enzymatic activities and physicochemical properties of apple juices from different varieties. LWT-Food Sci. Technol. 44 (1), 115–119.

Falguera, V., Pagán, J., Garza, S., Garvín, A., Ibarz, A., 2012. Inactivation of polyphenol oxidase by ultraviolet irradiation: protective effect of melanins. J. Food Eng. 110 (2), 305–309.

Faraji, H., Lindsay, R.C., 2004. Characterization of the antioxidant activity of sugars and polyhydric alcohols in fish oil emulsions. J. Agric. Food Chem. 52 (23), 7164–7171.

Farr, S.B., Kogoma, T., 1991. Oxidative stress responses in *Escherichia coli* and *Salmonella typhimurium*. Microbiol. Rev. 55 (4), 561–585.

Fricke, K., Steffen, H., Von Woedtke, T., Schröder, K., Weltmann, K.D., 2011. High rate of etching of polymers by means of an atmospheric pressure plasma jet. Plasma Process. Polym. 8, 51–58.

Grzegorzewski, F., Rohn, S., Kroh, L.W., Geyer, M., Schlüter, O., 2010. Surface morphology and chemical composition of lamb's lettuce (*Valerianella locusta*) after exposure to a low-pressure oxygen plasma. Food Chem. 122, 1145–1152.

Haddouche, L., Phalak, A., Tikekar, R.V., 2015. Inactivation of polyphenol oxidase using 254 nm ultraviolet light in a model system. LWT-Food Sci. Technol. 62 (1), 97–103.

Haenen, G.R., Paquay, J.B., Korthouver, R.E., Bast, A., 1997. Peroxynitrite scavenging by flavonoids. Biochem. Biophys. Res. Commun. 236, 591.

Halliwell, B., 2006. Reactive species and antioxidants. Redox biology is a fundamental theme of aerobic life. Plant Physiol. 141 (2), 312–322.

Henselová, M., Slováková, L., Martinka, M., Zahoranová, A., 2012. Growth, anatomy and enzyme activity changes in maize roots induced by treatment of seeds with low-temperature plasma. Biologia 67, 490–497.

Hu, J.P., Calomme, M., Lasure, A., De Bruyne, T., Pieters, A., Vlietinck, A., Vanden Berghe, D.A., 1995. Structure-activity relationship of flavonoids with superoxide scavenging activity. Biol. Trace Elem. Res. 47, 327.

Ikawa, S., Kitano, K., Hamaguchi, S., 2010. Effects of pH on bacterial inactivation in aqueous solutions due to low-temperature atmospheric pressure plasma application. Plasma Process. Polym. 7 (1), 33–42.

Isbell, H.S., Frush, H.L., 1973. Reaction of carbohydrates with hydroperoxides: part II. Oxidation of ketoses with the hydroperoxide anion. Carbohydr. Res. 28 (2), 295–301.

Isbell, H.S., Frush, H.L., Martin, E.T., 1973. Reactions of carbohydrates with hydroperoxides: part I. Oxidation of aldoses with sodium peroxide. Carbohydr. Res. 26 (2), 287–295.

Joslyn, M.A., Miller, J., 1949. Effect of sugars on oxidation of ascorbic acid. J. Food Sci. 14 (4), 325–339.

Kamgang-Youbi, G., Herry, J.M., Meylheuc, T., Brisset, J.L., Bellon-Fontaine, M.N., Doubla, A., Naitali, M., 2009. Microbial inactivation using plasma-activated water obtained by gliding electric discharges. Lett. Appl. Microbiol. 48 (1), 13–18.

Kappus, H., 1985. Lipid peroxidation: mechanisms, analysis, enzymology and biological relevance. In: Sies, H. (Ed.), Oxidative Stress. Acadamic Press, London, pp. 273–310.

Ke, Z., Huang, Q., 2013. Inactivation and heme degradation of horseradish peroxidase induced by discharge plasma. Plasma Process. Polym. 10 (8), 731–739.

Korachi, M., Ozen, F., Aslan, N., Vannini, L., Guerzoni, M.E., Gottardi, D., Ekinci, F.Y., 2015. Biochemical changes to milk following treatment by a novel, cold atmospheric plasma system. Int. Dairy J. 42, 64–69.

Kyzlink, V., Curda, D., 1970. Einfluss der Saccharose und der Zuganglichkeit des Sauerstoffs auf den Oxydationsverlauf der L-Ascorbinsaure im flussigen medium. Z. Lebensm. Unters. Forsch. 143 (4), 263–273.

Levine, R.L., Mosoni, L., Berlett, B.S., Stadtman, E.R., 1996. Methionine residues as endogenous antioxidants in proteins. Proc. Natl. Acad. Sci. 96 (26), 15036–15040.

Li, H.-P., Wang, L.-Y., Li, G., Jin, L.-H., Le, P.-S., Zhao, H.-X., Xing, X.-H., Bao, C.-Y., 2011. Manipulation of lipase activity by the helium radio-frequency, atmospheric-pressure glow discharge plasma jet. Plasma Process. Polym. 8 (3), 224–229. http://dx.doi.org/10.1002/ppap.201000035.

Lii, C.Y., Liao, C.D., Stobinski, L., Tomasik, P., 2002. Behaviour of granular starches in low-pressure glow plasma. Carbohydr. Polym. 49 (4), 499–507. http://dx.doi.org/10.1016/S0144-8617(01)00365-4.

Locke, B.R., Shih, K.-Y., 2011. Review of the methods to form hydrogen peroxide in electrical discharge plasma with liquid water. Plasma Sources Sci. Technol. 20 (3), 034006.

Matsumura, Y., Iwasawa, A., Kobayashi, T., Kamachi, T., Ozawa, T., Kohno, M., 2013. Detection of high-frequency ultrasound-induced singlet oxygen by the ESR spin-trapping method. Chem. Lett. 42 (10), 1291–1293.

Meiqiang, Y., Mingjing, H., Buzhou, M., Tengcai, M., 2005. Stimulating effects of seed treatment by magnetized plasma on tomato growth and yield. Plasma Sci. Technol. 7 (6), 3143.

Michaelis, L., 1932. Theory of the reversible two-step oxidation. J. Biol. Chem. 96 (3), 703–715.

Misra, N.N., Sullivan, C., Pankaj, S.K., Alvarez-Jubete, L., Cama, R., Jacoby, F., Cullen, P.J., 2014. Enhancement of oil spreadability of biscuit surface by nonthermal barrier discharge plasma. Innovative Food Sci. Emerg. Technol. 26, 456–461. http://dx.doi.org/10.1016/j.ifset.2014.10.001.

Misra, N.N., Kaur, S., Tiwari, B.K., Kaur, A., Singh, N., Cullen, P.J., 2015. Atmospheric pressure cold plasma (ACP) treatment of wheat flour. Food Hydrocoll. 44, 115–121. http://dx.doi.org/10.1016/j.foodhyd.2014.08.019.

Møller, I.M., Jensen, P.E., Hansson, A., 2007. Oxidative modifications to cellular components in plants. Annu. Rev. Plant Biol. 58, 459–481.

Morent, R., De Geyter, N., Desmet, T., Dubruel, P., Leys, C., 2011. Plasma surface modification of biodegradable polymers: a review. Plasma Process. Polym. 8 (3), 171–190.

Müller, A., Noack, L., Greiner, R., Stahl, M.R., Posten, C., 2014. Effect of UV-C and UV-B treatment on polyphenol oxidase activity and shelf life of apple and grape juices. Innovative Food Sci. Emerg. Technol. 26, 498–504.

Mylonas, C., Kouretas, D., 1999. Lipid peroxidation and tissue damage. In Vivo 13, 295–309.

Nakamura, K., Kanno, T., Mokudai, T., Iwasawa, A., Niwano, Y., Kohno, M., 2012. Microbial resistance in relation to catalase activity to oxidative stress induced by photolysis of hydrogen peroxide. Microbiol. Immunol. 56 (1), 48–55.

Niki, E., 2014. Role of vitamin E as a lipid-soluble peroxyl radical scavenger: in vitro and in vivo evidence. Free Radic. Biol. Med. 66, 3–12.

Nguyen, A.T., Donaldson, R.P., 2005. Metal-catalyzed oxidation induces carbonylation of peroxisomal proteins and loss of enzymatic activities. Arch. Biochem. Biophys. 439 (1), 25–31.

Pankaj, S.K., Misra, N.N., Cullen, P.J., 2013. Kinetics of tomato peroxidase inactivation by atmospheric pressure cold plasma based on dielectric barrier discharge. Innovative Food Sci. Emerg. Technol. 19, 153–157.

Pankaj, S.K., Bueno-Ferrer, C., Misra, N.N., Milosavljević, V., O'Donnell, C.P., Bourke, P., 2014a. Applications of cold plasma technology in food packaging. Trends Food Sci. Technol. 35 (1), 5–17.

Pankaj, S.K., Bueno-Ferrer, C., Misra, N.N., O'Neill, L., Jiménez, A., Bourke, P., Cullen, P.J., 2014b. Characterization of polylactic acid films for food packaging as affected by dielectric barrier discharge atmospheric plasma. Innovative Food Sci. Emerg. Technol. 21, 107–113. http://dx.doi.org/10.1016/j.ifset.2013.10.007.

Pei, X., Lu, X., Liu, J., Liu, D., Yang, Y., Ostrikov, K., Pan, Y., 2012. Inactivation of a 25.5 μm *Enterococcus faecalis* biofilm by a room-temperature, battery-operated, handheld air plasma jet. J. Phys. D. Appl. Phys. 45 (16), 165205.

Segat, A., Misra, N.N., Cullen, P.J., Innocente, N., 2015. Atmospheric pressure cold plasma (ACP) treatment of whey protein isolate model solution. Innovative Food Sci. Emerg. Technol. 29, 247–254. http://dx.doi.org/10.1016/j.ifset.2015.03.014.

Shen, B.O., Jensen, R.G., Bohnert, H.J., 1997. Increased resistance to oxidative stress in transgenic plants by targeting mannitol biosynthesis to chloroplasts. Plant Physiol. 113 (4), 1177–1183.

Smirnoff, N., Cumbes, Q.J., 1989. Hydroxyl radical scavenging activity of compatible solutes. Phytochemistry 28 (4), 1057–1060.

Surowsky, B., Fischer, A., Schlueter, O., Knorr, D., 2013. Cold plasma effects on enzyme activity in a model food system. Innovative Food Sci. Emerg. Technol. 146–152.

Takai, E., Kitano, K., Kuwabara, J., Shiraki, K., 2012. Protein inactivation by low-temperature atmospheric pressure plasma in aqueous solution. Plasma Process. Polym. 9 (1), 77–82.

Takamatsu, T., Uehara, K., Sasaki, Y., Hidekazu, M., Matsumura, Y., Iwasawa, A., et al., 2015. Microbial inactivation in the liquid phase induced by multigas plasma jet. PLoS One 10 (7), e0132381.

Tappi, S., Berardinelli, A., Ragni, L., Dalla Rosa, M., Guarnieri, A., Rocculi, P., 2014. Atmospheric gas plasma treatment of fresh-cut apples. Innovative Food Sci. Emerg. Technol. 21, 114–122.

Thirumdas, R., Deshmukh, R.R., Annapure, U.S., 2015. Effect of low temperature plasma processing on physicochemical properties and cooking quality of basmati rice. Innovative Food Sci. Emerg. Technol. 31, 83–90. http://dx.doi.org/10.1016/j.ifset.2015.08.003.

Traylor, M.J., Pavlovich, M.J., Karim, S., Hait, P., Sakiyama, Y., Clark, D.S., Graves, D.B., 2011. Long-term antibacterial efficacy of air plasma-activated water. J. Phys. D. Appl. Phys. 44 (47), 472001.

Vakrat-Haglili, Y., Weiner, L., Brumfeld, V., Brandis, A., Salomon, Y., McIlroy, B., et al., 2005. The microenvironment effect on the generation of reactive oxygen species by Pd-bacteriopheophorbide. J. Am. Chem. Soc. 127 (17), 6487–6497.

Van Durme, J., Nikiforov, A., Vandamme, J., Leys, C., De Winne, A., 2014. Accelerated lipid oxidation using non-thermal plasma technology: evaluation of volatile compounds. Food Res. Int. 62, 868–876.

Vandamme, J., Nikiforov, A., Dujardin, K., Leys, C., De Cooman, L., Van Durme, J., 2015. Critical evaluation of non-thermal plasma as an innovative accelerated lipid oxidation technique in fish oil. Food Res. Int. 72, 115–125.

Vatansever, F., De Melo, W.C., Avci, P., Vecchio, D., Sadasivam, M., Gupta, A., Tegos, G.P., 2013. Antimicrobial strategies centered around reactive oxygen species-bactericidal antibiotics, photodynamic therapy, and beyond. FEMS Microbiol. Rev. 37 (6), 955–989.

Wang, S.Y., Jiao, H., 2000. Scavenging capacity of berry crops on superoxide radicals, hydrogen peroxide, hydroxyl radicals, and singlet oxygen. J. Agric. Food Chem. 48 (11), 5677–5684.

Williams, P.F., 1997. Plasma Processing of Semiconductors, vol. 336. Springer, New York.

Wollrab, A., 2009. Organische Chemie, vol. 3. Springer, Heidelberg, Germany. ISBN 978-3-642-00780-4.

Xiong, Z., Du, T., Lu, X., Cao, Y., Pan, Y., 2011. How deep can plasma penetrate into a biofilm? Appl. Phys. Lett. 98 (22), 221503.

Zhang, R., Wang, L., Wu, Y., Guan, Z., Jia, Z., 2006. Bacterial decontamination of water by bipolar pulsed discharge in a gas-liquid-solid three-phase discharge reactor. IEEE Trans. Plasma Sci. 34 (4), 1370–1374.

Zhang, Q., Liang, Y., Feng, H., Ma, R., Tian, Y., Zhang, J., Fang, J., 2013. A study of oxidative stress induced by non-thermal plasma-activated water for bacterial damage. Appl. Phys. Lett. 102 (20), 203701.

Zhang, H., Xu, Z., Shen, J., Li, X., Ding, L., Ma, J., Lan, Y., Xia, W., Cheng, C., Sun, Q., Zhang, Z., Chu, P.K., 2015. Effects and mechanism of atmospheric-pressure dielectric barrier discharge cold plasma on lactate dehydrogenase (LDH) enzyme. Sci. Rep. 5. http://dx.doi.org/10.1038/srep10031.

Chapter 8

Plasma in Agriculture

T. Ohta
Meijo University, Nagoya, Japan

1 Introduction

The world's population is projected to rise from about 7 billion in 2012 to 9.6 billion people in 2050, as predicted by the World Resources Institute. With an accelerating growth of the world's population, the demand for foods is poised to grow significantly. In order to meet this demand, the global agriculture and food industry will likely face more intense pressure than ever before. Subsequently, the burden on the environment and natural resources will also rise significantly. A concern of growing importance within the agriculture sector is the menace of pesticide, fumigant, and agrochemical usage which, despite increasing the agricultural production, are harmful to the health of most living beings and the environment. In addition, many insect pests have developed resistance to the fumigants and pesticides, resulting in their usage at increased concentrations by the farmers; this occurs in developing as well as developed countries. Therefore, it is essential that scientists focus on technologies for controlling the preharvest pest population in fields, as well as during postharvest storage of grains. At the same time, the new technologies developed should not leave any residual chemistry on the food grains, and have the agility to modify the mode of

Cold Plasma in Food and Agriculture. http://dx.doi.org/10.1016/B978-0-12-801365-6.00008-1

action, when needed, without incurring marked economic pressure. In yet another context, a number of reports have highlighted the problem posed by fungi of *Fusarium* species and the associated mycotoxins, which are major contaminants in grains and fruits. In short, to reduce/arrest fungal growth, mycotoxin contamination, and pesticide residues in crops or harvested grains—while mitigating postharvest losses, germination abilities, and safety issues—effective technologies are increasingly needed. In addition, keeping the limited resources in mind, it is important to develop sustainable food production and processing technologies, which consume less water and energy and are environment friendly.

Cold plasma processing methods possess many advantages in agriculture, owing to their operation at low-temperatures and short processing times, without inducing damage to crops, foods, seeds, humans, and the environment. Plasma discharges produce reactive neutral species, charged species (electrons, ions), electric fields, and ultraviolet radiation. These factors cause the change in density of reactive oxygen species (ROS), reactive nitrogen species (RNS), pH, oxidation-reduction potential, electrical conductivity, and so on, and affect seed germination, plant growth, and the quality of agricultural product. Fig. 1 provides a summary of the plasma applications in agricultural processing operations, which is broadly classified into preharvest and postharvest. At this point of time, the contribution of plasma technologies in agricultural operations are limited to the decontamination of seeds or crops intended for sowing or storage, disinfection of processing surfaces or tools, the enhancement of seed germination or growth, production of nitrogen-based fertilizers, soil remediation, reduction of pathogen invasion, and the removal of ethylene from air to reduce rate of aging (Filimonova and Amirov, 2001; Ito et al., 2014).

Within this chapter, a review of the research in the area of cold plasma technology-based decontamination of seed material and plants, germination enhancement, induction of plant growth, and other aspects of importance to the evolving field of "plasma agriculture" have been discussed. While the chapter

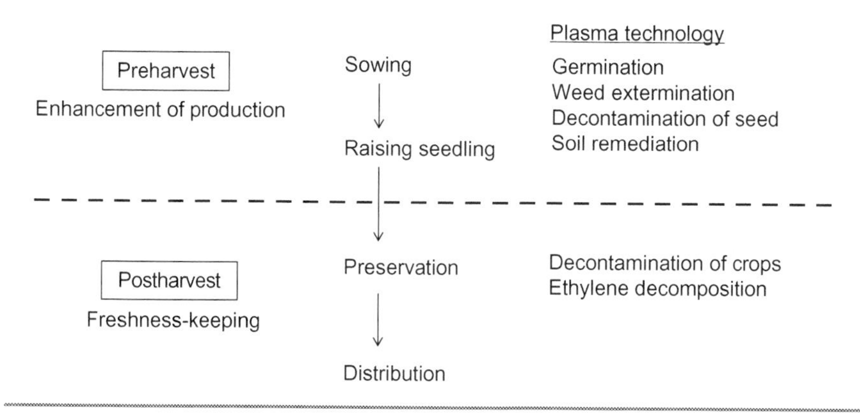

Fig. 1 Plasma applications to agricultural production operations.

provides a brief presentation of the studies conducted so far, the salient results of each study have been captured throughout. There are other important, yet underresearched applications of plasma in agriculture, which have not been discussed in this chapter for reasons of brevity; some examples include air cleaning, ethylene removal from storehouse, and removal of volatile organic compounds from greenhouse facilities or poultry farms (Andersen et al., 2013; Ye et al., 2013).

2 Decontamination of Seeds Using Cold Plasma

At the first instance, it is worthwhile noting the subtleties of the terminology; while "grains" are intended for human consumption, "seeds" are meant for sowing and agriculture use. Nevertheless, the terms are often used interchangeably. A summary of the research studies to date, pertinent to the decontamination of seeds using cold plasma technology is tabulated in Table 1. A description of the results of these studies is provided within this section. However, some details regarding decontamination of grains using plasma will also be provided later within this book (Chapter 9).

In one of the earliest studies, Selcuk et al. (2008) investigated the inactivation efficacy of low-pressure cold plasma using air or SF_6 for two pathogenic fungi (*Aspergillus* and *Penicillum*) on the seed surface. The study involved testing a range of seeds, including seeds of tomato, wheat, bean, chick pea, soy bean, barley, oat, rye,

TABLE 1 Examples from Literature on Decontamination of Seeds Using Cold Plasma

Microorganisms/ Pest	Seeds	Sources	References
Aspergillus paraciticus 798 and *Penicillum* MS1982	Tomato (*Lycopersicon esculentum*), wheat (*Triticum durum*), bean (*Phaseolus vulgaris*), chick pea (*Cicer arietinum*), soybean (*Glycine max* cv.), barley (*Hordeum vulgare*), oat (*Avena sativa*), rye (*Secale cereale*), lentil (*Lens culinaris*), and corn (*Zea mays*)	Low pressure cold plasma	Selcuk et al. (2008)
Bacillus atrophaeus	Rapeseed (*Brassica napus*)	DBD plasma	Schnabel et al. (2012)
Native microflora	Chickpea (*Cicer arietinum*)	Surface microdischarge	Mitra et al. (2013)
Insect (aphids)		Ozone-mist spray	Ebihara et al. (2013)
Penicillium digitatum	*Citrus unshiu*	Atmospheric-pressure oxygen radical source	Hashizume et al. (2013, 2014, 2015), Iseki et al. (2010, 2011), and Ishikawa et al. (2012)

lentil, and corn. The plasma decontamination process was performed by batch process in vacuum chamber, using gas injection followed by plasma discharge for the duration of 5–20 min. The plasma treatment was found to reduce the fungal counts on seeds to levels $<1\%$ of the initial counts, depending on the original contamination, without affecting the seed germination quality. The SF_6 plasma treatment for 15 min was found to decrease the count for both species by 3-log. Therefore, this study confirmed plasma treatment to be a rapid, functional decontamination method which enables elimination of aflatoxin-producing fungus from surfaces of seeds and nuts.

In another study, the seeds of *Brassica napus*, inoculated with *Bacillus atrophaeus*, were subjected to direct dielectric barrier discharge (DBD) plasma treatments or indirect microwave plasma-processed air. Reduction rates of up to 0.5 or 5.2 log were achieved after 15 min of treatment time (Schnabel et al., 2012). In addition, the overall seed viability was not affected by plasma treatment.

Chickpea seeds have been exposed to atmospheric pressure cold plasma in air using a surface microdischarge (Mitra et al., 2013). A 2-log reduction of the natural microbiota has been observed after treatment for 5 min. An increase in seed vigor is also noted after 1 min of treatment, with higher overall seed germination (89.2%) and germination rate. The seed membrane permeability analysis has revealed that water conductivity increases with an increase in treatment time; also, the roughness profile of the seed cotyledon is altered significantly with longer treatment times.

While fungal decontamination is a major safety issue in stored seeds and grains, insect infestation in seeds is yet another issue of high importance. Reports have indicated the insecticidal activity of atmospheric pressure plasma against larval and pupal stages of *Plodia interpunctella* (Indian meal moth) (Abd El-Aziz et al., 2014). The activities of antioxidant enzymes, catalase and glutathione *S*-transferase, were found to increase in the body tissue of larval instar, indicating the oxidizing effects of plasma. The effects of ozone-mist spray using the surface DBD on insects and agricultural plants was reported (Ebihara et al., 2013). Aphids (viz. red, chrysanthermum, and cotton aphids) were exterminated in 30 s without noticeable damages to the leaves of tomato and eggplant. The ROS such as hydroxyl radical ($^{\cdot}OH$), hydroperoxide radical ($^{\cdot}HO_2$), the superoxide anion radical ($^{\cdot}O_2^{-}$), and ozonide radical ion ($^{\cdot}O_3^{-}$) have been attributed to the rapid sterilization of aphids. Mechanistically, it was suggested that ozone and its derivative radicals were transported into the aphid's body through spiracles, thereafter via the trachea, and finally reached the cells to act on the proteins and organelles (eg, nucleus), ultimately resulting in death.

In order to inactivate the microorganisms on the seed, it is essential to investigate the reaction between plasma chemical species, the seed surface, as well as the intracellular structures of the seeds and microorganisms. Otto et al. (2011) have reviewed the chemical and physical methods for cleaning and disinfection. Iseki et al. (2010) reported the inactivation of *Penicillium digitatum* spores, which causes green mold on citrus, using high density atmospheric pressure plasma. Unlike most other studies, they reported that the contribution of ozone and ultraviolet radiation to the inactivation was relatively minor. *P. digitatum* spores were inactivated by a nonequilibrium atmospheric-pressure oxygen radical source for evaluating the contribution of ROS, such as $O(^3P)$, $O_2(^1D)$, and O_3 in the gas phase, to the inactivation (Hashizume et al., 2013; Iseki et al., 2011). The densities of ROS were measured by using vacuum ultraviolet or ultraviolet absorption spectroscopy, and the inactivation rate constant was estimated on the basis of the Chick-Watson law using the values of density and inactivation curve. They observed that $O(^3P)$ is the crucial species responsible for inactivation of *P. digitatum* spores, with only minor roles played by $O_2(^1D)$ and O_3. The free radicals in spores of *P. digitatum* generated from plasma electric discharge have been measured using real time in situ electron spin resonance (ESR) spectroscopy (Ishikawa et al., 2012). The ESR signal from the spores may be assigned to a stable free radical as intracellular semiquinone and has been found to decrease during O atom irradiation. To investigate the inactivation process mediated by neutral oxygen species, the intracellular organelles in the spores treated with a nonequilibrium atmospheric pressure plasma and an atmospheric-pressure oxygen radical source have been observed with a fluorescent confocal-laser microscope (Hashizume et al., 2014, 2015; Hiroshi et al., 2013). The results suggest that neutral oxygen species, in particular atomic oxygen, induce a minor structural change or functional inhibition of cell membranes, which results in the oxidation of the intracellular organelles via penetration of ROS into the cell. Furthermore, through transmission electron microscopy, the intracellular nanostructures have also been visualized to degrade by excess oxygen radicals: that is, $O(^3P)$ dose of $>1.0 \times 10^{20}$ cm^{-2}.

3 Enhancement of Seed Germination

A considerable amount of effort has been taken up by the research community to develop seed priming techniques which could enhance seed germination. Seed priming aims at an alleviation of the stresses associated with the soil, environment, and those inherent to the seeds, thereby promoting the growth process and crop performance (Harris et al., 2001). Several approaches have been evaluated for enhancing the germination of seeds, including the use of magnetic fields (Vashisth and Nagarajan, 2008, 2010), ultrasound treatment (Yaldagard et al., 2007), and exposure to electric fields (Sidaway, 1966; Tkalec et al., 2009). In recent times, cold plasma technology has also attracted the interest of scientists owing to its germination enhancement capabilities. Particularly, the germination rate and yield of germination seeds has been demonstrated to be enhanced by plasma treatments. A summary of the research studies supporting the seed germination enhancement by cold plasma is tabulated in Table 2.

TABLE 2 Examples from Literature on Enhancement of Seed Germination

Seeds	Sources	References
Lentils (*Lens culinaris*), beans (*Phaseolus vulgaris*), wheat (*Triticum*, species C9),	Cold radio-frequency air plasma	Bormashenko et al. (2012)
Oat (*Avena sativa*) and wheat (*Triticum eastivum*)	Low pressure cold plasma	Sera et al. (2010)
Lamb's quarters (*Chenopodium album* agg.)	Low pressure microwave discharge	Šerá et al. (2008, 2009)
Blue lupine (*Lupinus angustifolius*), catgut (*Galega virginiana*), honey clover and soy (*Melilotus albus*)	5.28 MHz capacitively coupled plasma	Filatova et al. (2010, 2011)
Tomato (*Lycopersicon esculentum* L. Mill. cv. zhongshu No. 6)	Arc discharge plasma	Yin et al. (2005)
Tomato (*Lycopersicon esculentum* cv. PKM1)	Ozone generator	Sudhakar et al. (2011)
Radish sprout (*Raphanus sativus* var. *longipinnatus*)	Atmospheric discharge plasma	Hayashi et al. (2015)
Wheat (*Triticum aestivum*)	Atmospheric pressure surface discharge	Dobrin et al. (2015)
Oat (*Avena savita*)	Low pressure cold plasma	Dubinov et al. (2000)
Barley (*Hordeum vulgare*), radish (*Raphanus sativus*), pea cultivar (*Pisum sativum* cv. Little Marvel, *P. sativum* cv. Alaska), soybean (*Glycine max* (L.) Merr.), corn (*Zea mays* L.), and bean (*Phaseolus vulgaris* L.)	RF rotating plasma, low pressure	Volin et al. (2000)
Herbaceous plant (*Andrographis paniculata*)	DBD plasma	Tong et al. (2014)
Corn	Ozone generator	Violleau et al. (2008)
Soybean (*Glycine max* L. Merr cv. Zhongdou 40)	Low pressure RF plasma	Ling et al. (2014)
Pea (*Pisum sativum* L. var. Prophet)	Diffuse coplanar surface DBD plasma	Stolárik et al. (2015)
Lamb's quarters (*Chenopodium album* agg.)	Surfatron discharge	Straňák et al. (2007)

Mechanistic insights into the germination enhancement have been provided by several studies. The wetting properties of the surfaces of lentils, beans (*Phaseolus vulgaris*) and wheat were found to alter under the influence of cold plasma in air (Bormashenko et al., 2012, 2015). In particular, a dramatic decrease in the apparent contact angle has been noted, which is partially attributed to the oxidation of seed surface by active species in plasma. A growth in the nitrogen containing groups of air plasma treated seeds has also been registered using selective ion mass spectrometry (SIMS).

Wheat and oat caryopses have been observed to be stimulated by cold plasma discharge (Sera et al., 2010). The wheat seed coat showed an eroded surface after plasma treatment. Plasma treatment restricts any germination enhancement of wheat on the first day, but elongation of the footstalk does occur in plants grown from treated seeds. In the context of oats, plasma treatment does not affect germination; however, it does accelerate the rootlet generation in plants grown from treated seeds. In addition, exposure to plasma also brings about changes in the metabolism in oats, as well as wheat. For example, differences in the content of phenolic compounds between sprouts of control and treated seeds has been observed. These phenomena indicate penetration of active species from plasma through the porous seed coat into the seed, where they react with cellular components.

The seeds of lamb's quarters (*Chenopodium album* agg.) have also been reported to be stimulated by low-pressure microwave discharge using Ar/O_2 or Ar/N_2 mixture gases (Šerá et al., 2008). The treated seeds exhibit structural changes on the surface of the seat coat. The germination time and the growth of sprout length also improve. Germination rate for the untreated seeds was 15%, while it increased approximately three times (max 55%) for seeds treated by plasma from 12 to 48 min. In addition, germination of two seed lots of lamb's quarters were stimulated with differences in the starting percentage germinations (17%, 8%) (Šerá et al., 2009). Significant differences among dispersion of population parameters have confirmed the erratic dormancy and gradual germination of lamb's quarters.

The effect of 5.28 MHz capacitively coupled RF discharge plasma and radiowave (magnetic and electric components of the electromagnetic field strength were 590 A/m and 12,700 V/m) treatments on seed sowing quality has been investigated (Filatova et al., 2010, 2011). It has been shown that presowing plasma and electromagnetic treatments for 10–15 min results in enhanced germination of seeds, viability, and crop capacity.

In yet another study, tomato seeds were treated with magnetized arc plasma before they were sown. The dehydrogenase activity and activity of peroxidase isoenzyme increased with the increase in the current, and decreased when the current increased beyond 1.5 A. The researchers did not observe any difference in germination percentage between treatments and control under laboratory conditions. However, large differences were observed in germination percentage in the pot experiments (Yin et al., 2005). Particularly, in the pot experiment, the sprouting rate for the treatment with a 1.5 A current was 32.75%, whereas the untreated was only

4.75% on the 11th day. The time to germination initiation in treated seeds was much less compared to the control (by more than a day). Interestingly, the 1.5 A treatment also increased the tomato yield by up to 20.7%.

The role of ozone (O_3) gas treatment produced by a corona discharge on seeds of *Lycopersicon esculentum* cv. PKM1 (tomato) to release dormancy has been investigated (Sudhakar et al., 2011). A faster start of germination on exposure to O_3 concentration of 0.01 g O_3 g^{-1} seeds (98–99%) was observed compared to control seeds. O_3 treatment loses its effect steadily when the stored time of treated seeds prolongs, and the germination efficiency of all the treatments are more or less same, 6 months after treatment. A reduced level of abscisic acid in O_3-treated seeds has been associated with acceleration of seed germination. Abscisic acid is a plant hormone that suppresses germination.

The impact of atmospheric plasma treatment of seeds on growth regulation characteristics of radish sprout (*Raphanus sativus* var. *longipinnatus*) has been investigated by Hayashi et al. (2015). This research group observed an enhanced germination, with up to 1.6 times longer stem and root development in plants after a cultivation period of 72 h (see Fig. 2). The "reduction type" thiol in radish sprouts,

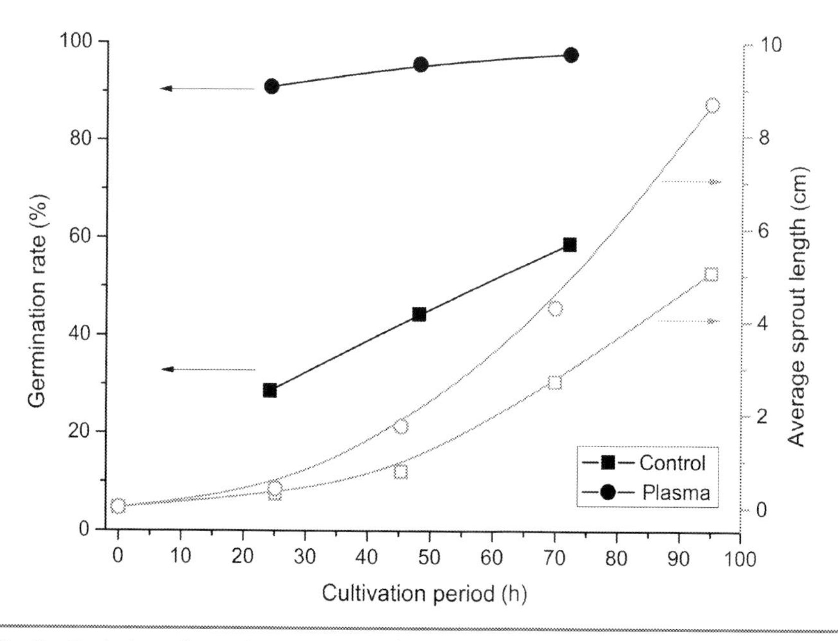

Fig. 2 Evolution of germination rate and sprout length of control and oxygen plasma-treated radish sprouts, as a function of cultivation time. *(From Hayashi, N., Ono, R., Shiratani, M., Yonesu, A., 2015. Antioxidative activity and growth regulation of* Brassicaceae *induced by oxygen radical irradiation. Jpn. J. Appl. Phys. 54, 06GD01.)*

responsible for maintaining the redox state of plant cells and the activation of growth factors, is modified by cold plasma. Not only the levels of reduction-type thiols, but also the average length of sprouts increase as a function of the plasma treatment period. The growth regulation originates from the change in the antioxidative activity of plant cells induced by active oxygen species generated in the oxygen plasma, which leads to the production of growth factor in plants.

The effect of atmospheric pressure cold plasma from a surface discharge on wheat seeds has been investigated (Dobrin et al., 2015). The plasma treatment had little effect on the germination rate, but influenced growth parameters. Higher root length was achieved by the plasma treatment as compared with the untreated samples. On the other hand, the sprout's length was about two times narrower for the treated samples as compared with the control seeds. The sprouts and roots of the plasma-treated seeds were heavier than those of the control samples.

Oat and barley seeds have been exposed to both continuous and pulsed glow discharge plasmas, with a pulse repetition rate of 0.5 Hz and a pulse duration of 150–200 ms in air under 0.1–0.2 Pa (Dubinov et al., 2000). Stimulating effects of plasma on the germination yield and sprout length were noticed. The observed effects exhibited a strong dependence on whether the discharge was continuous or pulsed.

A study was aimed at evaluating the effects of 13.56 MHz RF plasma rotating reactor on seed germination (Volin et al., 2000). The seeds of radish and two pea cultivars treated with CF_4 expressed a significant delay in germination compared to control. The plasma treatment with octadecafluorodecalin significantly delayed germination in soybean, corn, and bean seeds. Notably, it has been observed that an increase in coating thickness results in a stronger degree of delay in germination. Other chemical treatments could have different effects. For example, (1) seeds treated with cyclohexane accelerate soybean germination but not corn seeds; (2) hydrazine treatment accelerates corn seed germination slightly; (3) aniline treatment enhances soybean and corn seed germination percentage. Therefore, it can be concluded that plasma coatings play an important role in the rate of imbibition.

On treating *Andrographis paniculata* seedlings with an atmospheric pressure DBD air plasma, an improvement in the permeability has been observed (Tong et al., 2014). This study documented an acceleration of seed germination and seedling emergence after only 10 s of treatment at ~6 kV. The catalase activity and catalase isoenzyme expression have also been reported to improve, while the malondialdehyde content in the seedlings decreases. Although, post-treatment, the seed germination enhanced, there were no obvious changes observed in seedling emergence. These facts lead to the conclusion that air plasma could change the physiological and biochemical phenomena in seeds through the action of active plasma species on seed coat.

An increase in the germination rate and faster start of germination in corn seeds treated with high purity oxygen O_2 ($[O_3]=0$ g/m^3) and oxygen mixed with ozone O_2/O_3 ($[O_3]=20$ g/m^3) has also been recorded (Violleau et al., 2008). This early germination start results in higher germination yield with longer roots on the fourth and fifth days. Nevertheless, too long of an ozone treatment seems to be unfavorable for seed growth.

Plasma treatments with a radio frequency source operating at 13.56 MHz frequency has been shown to enhance seed germination and seedling growth of soybeans, with treatments at 80 W power exhibiting the highest stimulatory effect (Ling et al., 2014). Germination and vigor indices significantly increased by 14.66% and 63.33%, respectively. Seed's water uptake improved by 14.03%, and apparent contact angle decreased by 26.19%. Characteristics of seedling growth, including shoot length, shoot dry weight, root length, and root dry weight, significantly increased by 13.77%, 21.95%, 21.42%, and 27.51%, respectively, compared with control. In addition, soluble sugar and protein contents were 16.51% and 25.08% higher than those of the control. From the results, the plasma treatment had a greater stimulatory effect on plant roots.

In a contemporary study, pea (*Pisum sativum* L. var. Prophet) seeds have been treated using a diffuse coplanar surface barrier discharge for various duration (ranging from 60 to 600 s) (Stolárik et al., 2015). The plasma treatment has been observed to induce significant changes in the seeds' surfaces, which has been associated with water permeability into the seeds. In addition to the effects of plasma, the germination percentage of pea seeds, the growth parameters (root and shoot length, dry weight), and the vigor of seedlings have been observed to depend also on the duration of exposure. The plasma treatment causes changes in endogenous hormones (auxins and cytokinins and their catabolites and conjugates), which correlates with increased growth of the pea seedlings. The results suggest an interaction among the modification of seed structure demonstrated by cold plasma in the induction of rapid germination and hormonal activities associated with plant signaling and development, during the early growth of pea seedlings.

A summary of the mechanism of action of plasma treatment resulting in germination enhancement in seeds is provided in Fig. 3. It should be noted that the seed germination enhancement action of plasma is not only beneficial for agriculture applications, but also in food processing sector. For example, the malting, brewing, weaning foods, and specialty flours industry, also rely on germination as a major operation; technologies resulting in rapid germination could significantly reduce energy usage, thereby contributing toward sustainability in food processing.

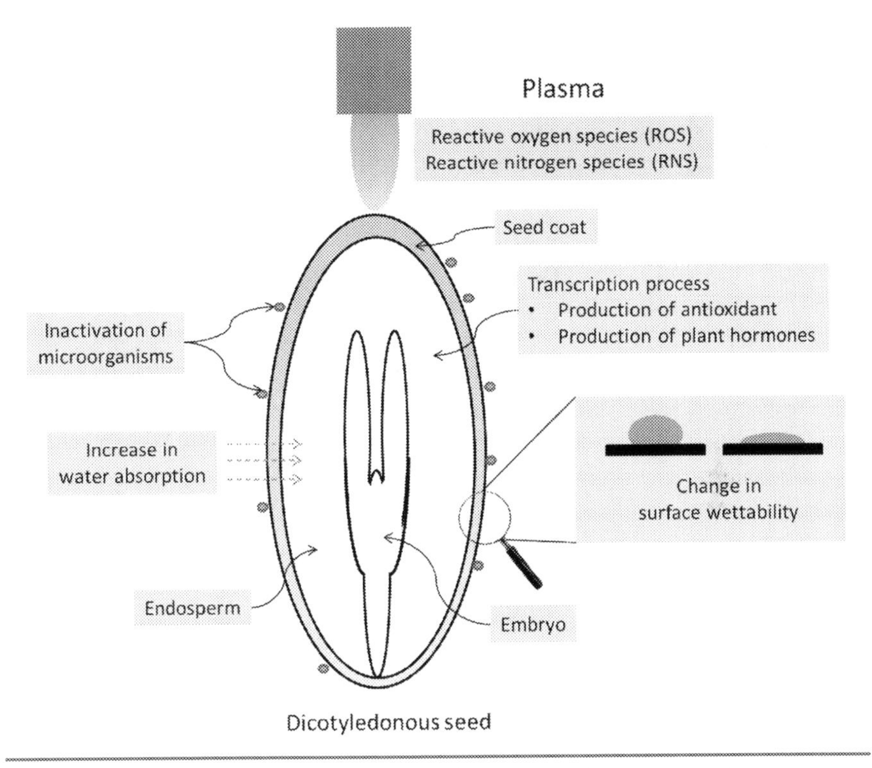

Fig. 3 A pictorial summary of the mode of action of plasma on seeds, resulting in germination enhancement.

4 Growth of Plants

The effect of water treated with different atmospheric plasmas, viz. thermal spark discharge, gliding arc discharge, and transferred arc discharge, on germination, growth rates, and overall nutrition of several plants has been reported by Park et al. (2013). The authors found that nonthermal gliding arc discharge results in acidic pH, and the production of a significant amount of oxidizing species (eg, H_2O_2), while gliding arc discharge causes significant acidification of water, due to RNS (NO, NO_2, and NO_3). Spark discharge treatment results in neutral or higher (basic) pH, depending on initial water composition, and the production of RNS. Park et al. (2013) noted that depending on the plasma source, the use of plasma-activated water resulted in an increase in plant growth (watermelon, zinnia [*Zinnia peruviana*], alfalfa, pole beans, and shade champ grass), and enhanced the action of chemical fertilizers. The possibility of using plasma as a means to capture atmospheric nitrogen in water has also been proposed by this group of researchers. In another study, the drainage water from

Brassica rapa (Chinese cabbage) pots has been exposed to plasma and then recycled to irrigate the plants over 28 days of cultivation (Takaki et al., 2013). The growth rate was found to have increased significantly with plasma irradiation of the drainage water. The nitrate (NO_2^-) and nitrous (NO_3^-) radicals produced by plasma irradiation was observed to have increased with the treatment time. Furthermore, the nitrogen concentration of the leaves also increased due to plasma treatment, while the bacterial count in the drainage water significantly decreased.

An investigation has been carried out to evaluate the long-term effects of DBD air plasma irradiation of radish seeds on the subsequent sprout growth (*Raphanus sativus* L.) (Sarinont et al., 2014). Plasma treatment for a duration of 3 min led to the growth enhancement of radish sprouts and the maximum average length was 3.7 times that of control. However, the response of the plants to the plasma treatment becomes gradually weak with time.

A comprehensive study by Kitazaki et al. (2014) employing a combinatorial plasma irradiation approach to study the growth enhancement of radish sprouts, has revealed up to 250% longer seedling post-plasma exposure for 180 s. Treatment for 180 s also induced the growth enhancement of radish sprouts for 7 days. They also identified that the temperature rise and etching of the seeds during the plasma irradiation have little or no effect on growth enhancement.

At present, the studies related to plasma-led growth enhancement in plants are very limited. Clearly, for plant growth enhancement applications, controlling the dose and flux of radicals is important for clarifying the growth enhancement mechanisms and optimizing the effects of the plasma. Therefore, standardization of plasma processes could help in accelerating research in this area.

5 Soil Remediation Using Cold Plasma

Poor agricultural practices from the overuse of pesticides and growth of heavy industries have laid waste to much of the arable land and water resources. Decontaminating polluted land and water is a complex and costly challenge, even for the most developed countries. Several technologies exist for the remediation of contaminated soils, including soil washing, phytoextraction, electrokinetic remediation, stabilization/solidification, phytostabilization, in situ chemical stabilization, and bioremediation (Chiang et al., 2016). However, the search for low-cost technologies for soil remediation and decontamination of water bodies has motivated scientists to explore plasma technologies for such applications. The decontamination of waste water using plasma technologies has been discussed later within this book (Chapter 13). What follows hereafter is a summary of the research studies investigating the applications of cold plasma technologies for soil remediation.

The remediation of Acid scarlet GR, a model azo-dye contaminated soil, using silent discharge plasma in atmospheric pressure DBD has recently been reported (Lu et al., 2014). It was observed that the decolorization rate of Acid scarlet GR in the soil increases with the applied voltage and discharge frequency. Treatment for up to 25 min was found to result in 93% degradation of Acid scarlet GR and 74% removal of chemical oxygen demand.

Lou et al. (2012) investigated the remediation of chloramphenicol-contaminated soil using an atmospheric pressure DBD plasma source (Lou et al., 2012). The degradation efficiency of the process was found to depend on the moisture content of the soil, the inducer gas, and the applied voltage, with up to 81% degradation after 20 min of treatment at an optimal soil moisture content of 10%. Using different gases, viz. oxygen, air, nitrogen, and argon, Lou et al. (2012) concluded that O_3 has a dominant role in the degradation process. In addition, it was observed that the presence of elemental iron favored the degradation and improved the efficiency by 8%, at a usage level of only 2%.

A combination of pulsed discharge plasma and TiO_2 catalysis was proposed by Wang et al. (2011) to remediate *p*-nitrophenol (PNP) contaminated soil. It was observed that 88.8% of PNP was removed within 10 min of treatment. The active species (eg, O_3 and H_2O_2) enhanced PNP degradation and mineralization. The main intermediates were identified as hydroquinone, benzoquinone, catechol, phenol, benzo[d][1, 2, 3]-trioxole, acetic acid, formic acid, NO_2, NO_3, and oxalic acid. In a subsequent study, the PNP degradation by DBD plasma in four different soils (quartz sand, sand, sandy soil, clay soil) was studied by Wang et al. (2014). The research group observed that the rate constant of PNP degradation, approximated by a first-order kinetics, varied with the depth of the soil layers. Furthermore, the degradation was higher in moist soil compared to dry soil, which was attributed to the formation of hydrogen peroxide and OH radicals in the presence of moisture. The depth dependence was attributed to the different diffusion behavior of active species and the differences in porosity of the packed soils. This observation of the spatial variance in the plasma chemical effects has far outreaching consequences and implications for validation of spatial homogeneity of plasma processes.

Chlorinated organic compounds are frequently encountered soil contaminants, and persist for extended periods of time (Wang et al., 2010a,b). The remediation of pentachlorophenol contaminated soil using pulsed corona discharge plasma was studied by Wang et al. (2010a). It was observed that with greater energy input to the plasma (controlled by peak pulsed voltage or pulse frequency), the degradation increased. On investigating the action of different species, it was noted that ozone was the key species, followed by others, such as N, N_2^+, N^+, and $^{\bullet}OH$. Through chromatographic separation and mass-based identification, tetrachlorocatechol, tetrachlorohydroquinone, acetic acid, formic acid, and oxalic acid were identified as the main degradation pathway intermediates.

The application of a DBD plasma source to dissipate kerosene components in the soil matrix has also been investigated (Redolfi et al., 2010). Experiments have revealed that up to 90% of the total kerosene components can be eliminated at an energy density of 960 J/g of soil. The main mechanism of dissipation has been attributed to the partial oxidation of kerosene in the soil matrix, whereas negligible transfer of organic contaminants from solid phase to gas phase has been observed.

In a distinct study, DBD plasma has been explored for the production of ozone in gas phase to treat the soil (Stryczewska et al., 2005). It was observed that the soil treatment with ozone resulted in lower bacterial counts in the soil. Therefore, in principle, plant diseases due to soil bacteria can be controlled through plasma treatment. It is important however, to preserve the action of nitrogen-fixing bacteria in soil and this aspect remains unaddressed. Besides a sterilization effect, the nitrogen content (NH_4-N, NO_3-N) in soil has also been reported to increase after DBD treatment, while the mineral content does not change significantly.

Overall, based on the above studies, one can note that the treatment time required for appreciable recovery of soil quality remains higher in most cases. The process of cold plasma-assisted soil remediation is in itself very complex, as it involves both soil properties (composition, nutritive value, contaminants, and microbiology), plasma species (electrons, excited atoms of oxygen and nitrogen, ozone, nitrogen oxides, radicals, and UV radiation), and parameters (pressure, temperature, moisture content, and time of exposure), all of which play a role in the process (Stryczewska et al., 2005). In addition, the sheer scale of operations required for soil remediation could be a huge challenge. Tackling these techno-economic challenges requires significant advances with respect to the development of efficient plasma sources operating in air.

6 Concluding Remarks

With the growing world population, the global agriculture is under pressure to produce more food from the limited land and natural resources. Plasma technologies have recently been explored for their possible contributions to the agricultural production operations. Several applications of plasma technologies have been demonstrated in literature, including the decontamination of seeds, the germination enhancement in seeds, an enhancement of plant growth, insect and fungal control in stored seeds, and reclamation of contaminated soil. The decontamination of seeds and enhancement of plant growth in greenhouse agriculture using plasma technology offers the opportunity to practically realize organic farming without the use of pesticides. The remediation of soil using plasma treatment has been found successful at the laboratory scale, in most studies reported so far. However, an upscaling of the plasma technologies for practical agricultural operations requires significant efforts from the plasma research community.

References

Abd El-Aziz, M.F., Mahmoud, E.A., Elaragi, G.M., 2014. Non thermal plasma for control of the Indian meal moth, *Plodia interpunctella* (*Lepidoptera*: *Pyralidae*). J. Stored Prod. Res. 59, 215–221.

Andersen, K.B., Beukes, J.A., Feilberg, A., 2013. Non-thermal plasma for odour reduction from pig houses—a pilot scale investigation. Chem. Eng. J. 223, 638–646.

Bormashenko, E., Grynyov, R., Bormashenko, Y., Drori, E., 2012. Cold radiofrequency plasma treatment modifies wettability and germination speed of plant seeds. Sci. Rep. 2, 741.

Bormashenko, E., Shapira, Y., Grynyov, R., Whyman, G., Bormashenko, Y., Drori, E., 2015. Interaction of cold radiofrequency plasma with seeds of beans (*Phaseolus vulgaris*). J. Exp. Bot. 66, 4013–4021.

Chiang, P.-N., Tong, O.-Y., Chiou, C.-S., Lin, Y.-A., Wang, M.-K., Liu, C.-C., 2016. Reclamation of zinc-contaminated soil using a dissolved organic carbon solution prepared using liquid fertilizer from food-waste composting. J. Hazard. Mater. 301, 100–105.

Dobrin, D., Magureanu, M., Mandache, N.B., Ionita, M.-D., 2015. The effect of non-thermal plasma treatment on wheat germination and early growth. Innovative Food Sci. Emerg. Technol. 29, 255–260.

Dubinov, A.E., Lazarenko, E.M., Selemir, V.D., 2000. Effect of glow discharge air plasma on grain crops seed. IEEE Trans. Plasma Sci. 28, 180–183.

Ebihara, K., Mitsugi, F., Ikegami, T., Nakamura, N., Hashimoto, Y., Yamashita, Y., Baba, S., Stryczewska, H.D., Pawlat, J., Teii, S., Sung, T.-L., 2013. Ozone-mist spray sterilization for pest control in agricultural management. Eur. Phys. J. Appl. Phys. 61, 201324322.

Filatova, I.I., Azharonok, V.V., Kadyrov, M., Beljavsky, V., Sera, B., Hruskova, I., Spatenka, P., Sery, M., 2010. RF and microwave plasma applications for pre-sowing caryopsis treatments. Publ. Astron. Obs. Belgrade 89, 289–292.

Filatova, I., Azharonok, V., Kadyrov, M., Beljavsky, V., Gvozdov, A., Shik, A., Antonuk, A., 2011. The effect of plasma treatment of seeds of some grain and legumes on their sowing quality and productivity. Rom. J. Phys. 56, 139–143.

Filimonova, E.A., Amirov, R.K., 2001. Simulation of ethylene conversion initiated by a streamer corona in an air flow. Plasma Phys. Rep. 27, 708–714.

Harris, D., Pathan, A.K., Gothkar, P., Joshi, A., Chivasa, W., Nyamudeza, P., 2001. On-farm seed priming: using participatory methods to revive and refine a key technology. Agric. Syst. 69, 151–164.

Hashizume, H., Ohta, T., Fengdong, J., Takeda, K., Ishikawa, K., Hori, M., Ito, M., 2013. Inactivation effects of neutral reactive-oxygen species on *Penicillium digitatum* spores using non-equilibrium atmospheric-pressure oxygen radical source. Appl. Phys. Lett. 103, 153708.

Hashizume, H., Ohta, T., Takeda, K., Ishikawa, K., Hori, M., Ito, M., 2014. Oxidation mechanism of *Penicillium digitatum* spores through neutral oxygen radicals. Jpn. J. Appl. Phys. 53, 010209.

Hashizume, H., Ohta, T., Takeda, K., Ishikawa, K., Hori, M., Ito, M., 2015. Quantitative clarification of inactivation mechanism of *Penicillium digitatum* spores treated with neutral oxygen radicals. Jpn. J. Appl. Phys. 54, 01AG05.

Hayashi, N., Ono, R., Shiratani, M., Yonesu, A., 2015. Antioxidative activity and growth regulation of *Brassicaceae* induced by oxygen radical irradiation. Jpn. J. Appl. Phys. 54, 06GD01.

Hiroshi, H., Takayuki, O., Takumi, M., Sachiko, I., Masaru, H., Masafumi, I., 2013. Inactivation process of *Penicillium digitatum* spores treated with non-equilibrium atmospheric pressure plasma. Jpn. J. Appl. Phys. 52, 056202.

Iseki, S., Ohta, T., Aomatsu, A., Ito, M., Kano, H., Higashijima, Y., Hori, M., 2010. Rapid inactivation of *Penicillium digitatum* spores using high-density nonequilibrium atmospheric pressure plasma. Appl. Phys. Lett. 96, 153704.

Iseki, S., Hashizume, H., Jia, F., Takeda, K., Ishikawa, K., Ohta, T., Ito, M., Hori, M., 2011. Inactivation of *Penicillium digitatum* spores by a high-density ground-state atomic oxygen-radical source employing an atmospheric-pressure plasma. Appl. Phys. Express 4, 116201.

Ishikawa, K., Mizuno, H., Tanaka, H., Tamiya, K., Hashizume, H., Ohta, T., Ito, M., Iseki, S., Takeda, K., Kondo, H., Sekine, M., Hori, M., 2012. Real-time in situ electron spin resonance measurements on fungal spores of *Penicillium digitatum* during exposure of oxygen plasmas. Appl. Phys. Lett. 101, 013704.

Ito, T., Kawamura, T., Nakagawa, A., Yamazaki, S., Syuto, B., Takaki, K., 2014. Preservation of fresh food using AC electric field. J. Adv. Oxid. Technol. 17, 249–253.

Kitazaki, S., Sarinont, T., Koga, K., Hayashi, N., Shiratani, M., 2014. Plasma induced long-term growth enhancement of *Raphanus sativus* L. using combinatorial atmospheric air dielectric barrier discharge plasmas. Curr. Appl. Phys. 14, S149–S153.

Ling, L., Jiafeng, J., Jiangang, L., Minchong, S., Xin, H., Hanliang, S., Yuanhua, D., 2014. Effects of cold plasma treatment on seed germination and seedling growth of soybean. Sci. Rep. 4, 5859.

Lou, J., Lu, N., Li, J., Wang, T., Wu, Y., 2012. Remediation of chloramphenicol-contaminated soil by atmospheric pressure dielectric barrier discharge. Chem. Eng. J. 180, 99–105.

Lu, N., Lou, J., Wang, C., Li, J., Wu, Y., 2014. Evaluating the effects of silent discharge plasma on remediation of acid scarlet GR-contaminated soil. Water Air Soil Pollut. 225, 1–7.

Mitra, A., Li, Y.-F., Klämpfl, T.G., Shimizu, T., Jeon, J., Morfill, G.E., Zimmermann, J.L., 2013. Inactivation of surface-borne microorganisms and increased germination of seed specimen by cold atmospheric plasma. Food Bioprocess Technol. 7 (2014), 645–653.

Otto, C., Zahn, S., Rost, F., Zahn, P., Jaros, D., Rohm, H., 2011. Physical methods for cleaning and disinfection of surfaces. Food Eng. Rev. 3, 171–188.

Park, D.P., Davis, K., Gilani, S., Alonzo, C.A., Dobrynin, D., Friedman, G., Fridman, A., Rabinovich, A., Fridman, G., 2013. Reactive nitrogen species produced in water by non-equilibrium plasma increase plant growth rate and nutritional yield. Curr. Appl. Phys. 13, S19–S29.

Redolfi, M., Makhloufi, C., Ognier, S., Cavadias, S., 2010. Oxidation of kerosene components in a soil matrix by a dielectric barrier discharge reactor. Process Saf. Environ. Prot. 88, 207–212.

Sarinont, T., Amano, T., Kitazaki, S., Koga, K., Uchida, G., Shiratani, M., Hayashi, N., 2014. Growth enhancement effects of radish sprouts: atmospheric pressure plasma irradiation vs. heat shock. J. Phys. Conf. Ser. 518, 012017.

Schnabel, U., Niquet, R., Krohmann, U., Winter, J., Schlüter, O., Weltmann, K.-D., Ehlbeck, J., 2012. Decontamination of microbiologically contaminated specimen by direct and indirect plasma treatment. Plasma Process. Polym. 9, 569–575.

Selcuk, M., Oksuz, L., Basaran, P., 2008. Decontamination of grains and legumes infected with *Aspergillus* spp. and *Penicillum* spp. by cold plasma treatment. Bioresour. Technol. 99, 5104–5109.

Será, B., Stranák, V., Serý, M., Tichý, M., Spatenka, P., 2008. Germination of *Chenopodium album* in response to microwave plasma treatment. Plasma Sci. Technol. 10, 506–511.

Šerá, B., Šerý, M., Štraňák, V., Špatenka, P., Tichý, M., 2009. Does cold plasma affect breaking dormancy and seed germination? A study on seeds of Lamb's Quarters (*Chenopodium album* agg.). Plasma Sci. Technol. 11 (6), 750–754. http://dx.doi.org/10.1088/1009-0630/11/6/22.

Sera, B., Spatenka, P., Sery, M., Vrchotova, N., Hruskova, I., 2010. Influence of plasma treatment on wheat and oat germination and early growth. IEEE Trans. Plasma Sci. 38, 2963–2968.

Sidaway, G.H., 1966. Influence of electrostatic fields on seed germination. Nature 211, 303.

Stolárik, T., Henselová, M., Martinka, M., Novák, O., Zahoranová, A., Černák, M., 2015. Effect of low-temperature plasma on the structure of seeds, growth and metabolism of endogenous phytohormones in pea (*Pisum sativum* L.). Plasma Chem. Plasma Process. 35, 659–676.

Straňák, V., Tichý, M., Kříha, V., Scholtz, V., Šerá, B., Špatenka, P., 2007. Technological applications of surfatron produced discharge. J. Optoelectron. Adv. Mater. 9, 852–857.

Stryczewska, H.D., Ebihara, K., Takayama, M., Gyoutoku, Y., Tachibana, M., 2005. Non-thermal plasma-based technology for soil treatment. Plasma Process. Polym. 2, 238–245.

Sudhakar, N., Nagendra-Prasad, D., Mohan, N., Hill, B., Gunasekaran, M., Murugesan, K., 2011. Assessing influence of ozone in tomato seed dormancy alleviation. Am. J. Plant Sci. 2, 443–448.

Takaki, K., Takahata, J., Watanabe, S., Satta, N., Yamada, O., Fujio, T., Sasaki, Y., 2013. Improvements in plant growth rate using underwater discharge. J. Phys. Conf. Ser. 418, 012140.

Tkalec, M., Malarić, K., Pavlica, M., Pevalek-Kozlina, B., Vidaković-Cifrek, Ž., 2009. Effects of radiofrequency electromagnetic fields on seed germination and root meristematic cells of *Allium cepa* L. Mutat. Res., Genet. Toxicol. Environ. Mutagen. 672, 76–81.

Tong, J., He, R., Zhang, X., Zhan, R., Chen, W., Yang, S., 2014. Effects of atmospheric pressure air plasma pretreatment on the seed germination and early growth of *Andrographis paniculata*. Plasma Sci. Technol. 16, 260–266.

Vashisth, A., Nagarajan, S., 2008. Exposure of seeds to static magnetic field enhances germination and early growth characteristics in chickpea (*Cicer arietinum* L.). Bioelectromagnetics 29, 571–578.

Vashisth, A., Nagarajan, S., 2010. Effect on germination and early growth characteristics in sunflower (*Helianthus annuus*) seeds exposed to static magnetic field. J. Plant Physiol. 167, 149–156.

Violleau, F., Hadjeba, K., Albet, J., Cazalis, R., Surel, O., 2008. Effect of oxidative treatment on corn seed germination kinetics. Ozone Sci. Eng. 30, 418–422.

Volin, J.C., Denes, F.S., Young, R.A., Park, S.M.T., 2000. Modification of seed germination performance through cold plasma chemistry technology. Crop Sci. 40, 1706–1718.

Wang, T.C., Lu, N., Li, J., Wu, Y., 2010a. Degradation of pentachlorophenol in soil by pulsed corona discharge plasma. J. Hazard. Mater. 180, 436–441.

Wang, T.C., Lu, N., Li, J., Wu, Y., 2010b. Evaluation of the potential of pentachlorophenol degradation in soil by pulsed corona discharge plasma from soil characteristics. Environ. Sci. Technol. 44, 3105–3110.

Wang, T.C., Lu, N., Li, J., Wu, Y., 2011. Plasma-TiO_2 catalytic method for high-efficiency remediation of *p*-nitrophenol contaminated soil in pulsed discharge. Environ. Sci. Technol. 45, 9301–9307.

Wang, T.C., Qu, G., Li, J., Liang, D., Hu, S., 2014. Depth dependence of *p*-nitrophenol removal in soil by pulsed discharge plasma. Chem. Eng. J. 239, 178–184.

Yaldagard, M., Mortazavi, S., Tabatabaie, F., 2007. The effectiveness of ultrasound treatment on the germination stimulation of barley seed and its alpha-amylase activity. World Acad. Sci. Eng. Technol. 34, 154–157.

Ye, S.-Y., Fang, Y.-C., Song, X.-L., Luo, S.-C., Ye, L.-M., 2013. Decomposition of ethylene in cold storage by plasma-assisted photocatalyst process with TiO_2/ACF-based photocatalyst prepared by gamma irradiation. Chem. Eng. J. 225, 499–508.

Yin, M., Huang, M., Ma, B., Ma, T., 2005. Stimulating effects of seed treatment by magnetized plasma on tomato growth and yield. Plasma Sci. Technol. 7, 3143.

Chapter 9

Cold Plasma for Food Safety

D. Ziuzina*, N.N. Misra†
*Dublin Institute of Technology, Dublin, Ireland, †General Mills India Pvt Ltd., Mumbai, India

1 Introduction

The risk associated with the transmission of foodborne pathogenic infections has become a global issue of concern to the entire food industry. In the European Union alone, in 2011, a total of 5648 foodborne outbreaks were reported, resulting in 69,553 human cases, 7125 hospitalizations, and 93 deaths (EFSA, 2013). Health-associated outbreaks, which have been linked to the consumption of food produce, demands research for more efficient decontamination techniques since current sanitation procedures show limited efficacy against bacterial pathogens (Olaimat and Holley, 2012; Ölmez and Temur, 2010; Warning and Datta, 2013).

Despite the beneficial health effects, minimally processed food products can be a vehicle for the transmission of bacterial, parasitic, and viral pathogens capable of causing human illness (Abadias et al., 2008). Foodborne human illnesses resulting from consumption of contaminated fresh produce have been widely reported throughout the world. In the EU in 2009 and 2010, respectively, 4.4% and 10% of the foodborne-verified outbreaks were linked with the consumption of vegetables, fruits, and berries (Van Boxstael et al., 2013). Most reporting countries identified

Cold Plasma in Food and Agriculture. http://dx.doi.org/10.1016/B978-0-12-801365-6.00009-3

Escherichia coli O157:H7, *Listeria monocytogenes*, and *Salmonella* spp. as the target pathogens of concern (Olaimat and Holley, 2012; Raybaudi-Massilia et al., 2009). Fresh produce implicated in illness outbreaks in the first decades of the 21st century, caused by contamination with *E. coli*, *Salmonella*, and *L. monocytogenes*, included spinach, lettuce, radish, alfalfa sprouts, tomatoes, peppers, cantaloupe, strawberries, and fruit and vegetable salads (Olaimat and Holley, 2012). In the United States from 1973 to 1997, *Salmonella* spp. was responsible for nearly half of the outbreaks due to bacteria, and it remains a suspect in a number of recently documented outbreaks of illnesses (Berger et al., 2010; Olaimat and Holley, 2012).

In the EU *Salmonella* Enteritidis and Typhimurium serovars are recognized as the pathogens most frequently associated with outbreaks of foodborne illness (Fernandez et al., 2013). *E. coli* O157 foodborne outbreaks have become increasingly common: from 1982 to 2002, a total of 350 outbreaks were reported in the United States (Rangel et al., 2005). In 2006, in the United States there was a multistate outbreak of *E. coli* 0157:H7 transmitted via spinach, with 276 cases of illness and three deaths (Goodburn and Wallace, 2013). In 2011, a large *E. coli* associated outbreak occurred in Germany, which involved 3911 cases of illness with 47 deaths (Olaimat and Holley, 2012). *Listeria* is an important cause of human illness in the United States, which in 2011 caused a lethal outbreak with a total of 147 infected persons, with a reported 30 deaths (CDC, 2011).

Similar to fresh plant produce, the rich pool of nutrients in meat makes it an ideal environment for the growth and proliferation of spoilage microorganisms, particularly foodborne human pathogens. Of particular significance in this context are *E. coli* and *L. monocytogenes*, which although inactivated by traditional processing, often transmits through ready-to-eat (RTE) foods such as soft cheese, poultry products, RTE meat products, smoked fish, and seafood sources (Martín et al., 2014). *Campylobacter* spp. are another class of bacteria often associated with poultry products. With the increased demand for high quality, convenience, safety, fresh appearance and an extended shelf life of fresh meat products, alternative nonthermal preservation technologies such as high pressure, super-chilling, biopreservatives, and active packaging have been extensively investigated in recent years (Zhou et al., 2010). In general, each of these technologies have their own merits and demerits; however, most of them are not effective against spores.

Deterioration of fresh produce caused by bacteria, namely *Erwinia* spp., *Enterobacter*, *Propionibacterium chlohexanicum*, *Pseudomonas* spp., and lactic acid bacteria may also constitute a hazard for consumers by the possible presence of microbial toxins. In line with the produce spoilage microorganisms, fungi—including yeast and molds—are largely responsible for the spoilage and quality loss of fresh produce. Molds, such as *Penicillium* spp., *Aspergillus* spp., *Eurotium* spp., *Alternaria* spp., *Cladosporium* spp., *Paecilomyces* spp., and *Botrytis* spp., are commonly involved in the spoilage of fresh fruits and some processed fruit derivatives, including the thermally processed. Yeasts, such as *Saccharomyces* spp. *Cryptococcus* spp., and *Rhodotorula* spp. are found in fresh fruits, and *Zygosaccharomyces* spp., *Hanseniaspora* spp., *Candida* spp., *Debaryomyces* spp., and *Pichia* spp. are found in dried fruits (Raybaudi-Massilia et al., 2009). In beef, pork, mutton, turkey, and chicken meat products *Salmonella*, *E. coli*, *Staphylococcus aureus*, *L. monocytogenes*, *Shigella flexneri*, *Bacillus subtilis*, yeasts, and molds are commonly encountered foodborne pathogens (Hygreeva et al., 2014).

Over the last decades, it has become clear that many human pathogens grow predominantly as biofilms on the surfaces of most of their habitats, rather than in planktonic mode (Giaouris et al., 2014; Sharma et al., 2014). Formation of bacterial biofilms on food-contact surfaces, on food-processing equipment and in potable water distribution systems contributes to food spoilage, cross-contamination of food products and the spread of foodborne pathogens (Kim and Wei, 2012). Moreover, biofilms are more resistant to various environmental stresses; therefore, these represent a major challenge to the food industry (Borges et al., 2014).

Although current decontamination approaches—such as chemical preservation (ascorbic acid and calcium salts), the addition of biopreservatives, mild heat treatments, microwave processing, reduction of water activity, ionizing irradiation, the use of disinfectants (electrolyzed water treatment, chlorination, hydrogen peroxide), high hydrolytic-pressure technology, a high-intensity pulsed electric field, pulsed light, ozone technology, vacuum/hypobaric packaging, and hurdle technologies—provide good results in a number of cases, firm adhesion of pathogens to surfaces, internalization of pathogens, and biofilm formation limits the usefulness and effectiveness of most processing and chemical-sanitizing methods. Other drawbacks include the high initial costs of equipment, an adverse alteration of food properties (such as color, taste and smell, or damage to their structure) and formation of toxic byproducts (Scholtz et al., 2015). Failure to eradicate foodborne pathogens, bacterial biofilms, and increase in the number of food-related human illnesses has initiated a search for new or terminal-control decontamination approaches, which could prevent contamination during processing, while maintaining produce quality characteristics and extending shelf life.

To date, among the various physical and chemical food decontamination techniques evaluated, cold plasma demonstrates high efficiency for the reduction of bacterial contaminants in fresh produce (Baier et al., 2014; Fernandez et al., 2013; Lacombe et al., 2015; Misra et al., 2014; Ziuzina et al., 2014); milk and milk products (Kim et al., 2015; Song et al., 2009; Yong et al., 2015a,b); and meat and meat products (Jayasena et al., 2015; Kim et al., 2011, 2014a). The versatility of cold plasma technology for decontamination of foods by acting upon a range of microorganisms can be appreciated from the discussion, which will follow in this chapter. It will also become clear that the chemical species of plasma are effective against bacteria, bacterial spores, and fungi. The sterilization of food packaging materials using cold plasma technology is another mature area of research, with commercial applications under development. However, this topic is covered in Chapter 12, where it fits well.

Microbiological issues associated with food production and achievements in the current plasma research are described in this chapter. The chapter provides an overview of the effects of plasma process parameters on decontamination efficacy. In addition to the variables of the plasma process, factors such as the type of bacteria, cell concentration, and substrate or food also influence the decontamination efficacy; therefore, the focus is also set on these factors during the discussions.

2 Microbiological Safety of Plant Origin Foods

2.1 Fruits and Vegetables

In recent years, cold plasma treatment of fresh fruits and vegetables has been the subject of much research, which has demonstrated that plasma technology offers a good alternative to conventional methods within food production settings. Table 1 presents a summary of the results from research activities focused on inactivation of pathogens associated with fresh produce using cold plasma technology. What follows next is a discussion of selected works from this table in order to highlight the intricacies and criticalities arising from process conditions, the produce characteristics, and the nature of the pathogen.

2.1.1 The Effects of Plasma Source and Process Parameters

In one of the early works, Critzer et al. (2007) investigated the efficacy of one atmosphere uniform glow discharge plasma generated at 9 kV, and reported >2 log reduction of *E. coli* O157:H7 inoculated on apples after 2 min of treatment, while 1 min of treatment resulted in similar inactivation levels for *Salmonella* inoculated on cantaloupe. An extended treatment for 3 and 5 min reduced *L. monocytogenes* populations on lettuce by >3 and >5 log reductions, respectively. Significant reductions by up to 3.7 log of *E. coli* and *Salmonella* populations inoculated on apples were achieved following 3 min of treatment with cold atmospheric plasma generated in a gliding arc discharge by Niemira and Sites (2008).

An interesting approach of plasma application was demonstrated by Klockow and Keener (2009), where prepackaged whole spinach leaves inoculated with *E. coli* O157:H7 were exposed for 5 min to cold plasma and subsequently stored for 0.5–24 h. The largest reductions of 3–5 log CFU/leaf were achieved after 24 h of storage. In this study an extended posttreatment storage time emerged as a critical treatment parameter for the maximal efficiency of bacterial inactivation with this type of system. In general,

TABLE 1 A Summary of the Research on Inactivation of Microorganisms on Fresh Produce by Cold Plasma Treatment

Plasma Source	Gas	Bacteria	Produce	Maximal log Reduction	Reference
One atmosphere uniform glow discharge plasma	Air	*E. coli*, *Salmonella*, *Listeria*	Apples, cantaloupe, lettuce	*E. coli* > 2, *Salmonella* > 3, *Listeria* > 5	Critzer et al. (2007)
Gliding arc discharge	Air	*E. coli*, *Salmonella Stanley*	Apples	*E. coli* ~ 3.6, *Salmonella* ~ 3.7	Niemira and Sites (2008)
DBD	Air, oxygen	*E. coli*	Spinach	*E. coli* ~ 5	Klockow and Keener (2009)
Plasma micro jet	Air	*Salmonella* spp.	Carrots, cucumbers, pears	*Salmonella* ~90–100% (4–5 log)	Wang et al. (2012b)
Plasma jet	Argon	*E. coli*	Corn salad leaves	*E. coli* ~ 3.6	Baier et al. (2013)
Needle array plasma	Argon	*E. coli*	Lettuce, carrots, tomatoes	*E. coli* ~ 1.7	Bermúdez-Aguirre et al. (2013)
Plasma jet	Nitrogen	*Salmonella*	Lettuce, strawberry, potato	*Salmonella* ~ 2.7	Fernandez et al. (2013)
Radio-frequency plasma/chamber	Oxygen	*Salmonella*	Spinach, lettuce, tomato, potato	*Salmonella* <3	Zhang et al. (2013)
Plasma jet, DBD	Argon/oxygen mix	*E. coli*	Corn salad, cucumber, apples, tomato	*E. coli* ~ 4.7	Baier et al. (2014)
Long wavelength UV light Photoplasma, Model: Induct, ID 60/chamber	Air	*Aeromonas hydrophila*	Lettuce	*Aeromonas hydrophila* ~ 5	Jahid et al. (2014)
Microwave-powered atmospheric cold plasma/chamber	Nitrogen, Nitrogen/oxygen mix, Helium, Helium-oxygen mix	*Aspergillus flavus*, *B. cereus* spores, Background microflora	Red pepper powder	*Aspergillus flavus* ~ 2.5 *Bacillus cereus* ~ 3.4, Background microflora ~ 1	Kim et al. (2014b)

Continued

TABLE 1 A Summary of the Research on Inactivation of Microorganisms on Fresh Produce by Cold Plasma Treatment—cont'd

Plasma Source	Gas	Bacteria	Produce	Maximal log Reduction	Reference
Long wavelength UV light Photoplasma, Model: Induct, ID 60/chamber	Air	*Listeria*	Lettuces, cabbages	*Listeria*~4	Srey et al. (2014)
Dielectric barrier discharge, high-voltage atmospheric cold plasma	Air	*E. coli*, *Listeria*, *Salmonella*, Background microflora	Cherry tomatoes, Strawberries	Cherry tomatoes: *E. coli*~3.1 *Salmonella*~6.3 *Listeria*~6.7 Strawberries: *E. coli*~3.5 *Salmonella*~3.8 *Listeria*~4.2	Ziuzina et al. (2014)
Plasma jet	Air	Background microflora	Blueberries	Background microflora~2	Lacombe et al. (2015)
Long wavelength UV light Photoplasma, Model: Induct, ID 60/chamber	Air	*Salmonella* Background microflora	Lettuce	*Salmonella*~3.7 Background microflora~4.1	Jahid et al. (2015)
Radio-frequency plasma jet	Argon	*Salmonella*, *B. subtilis* spores, *B. atrophaeus* spores	Whole black pepper	*Salmonella*~4.1, *B. subtilis* spores~2.4, *B. atrophaeus* spores~2.8	Hertwig et al. (2015)
Plasma jet	Argon	*E. coli* O157:H7 *E. coli* O104:H4	Corn salad leaves	*E. coli O157:H7*~3.3 *E. coli O104:H4*~3.3	Baier et al. (2015b)
Microwave plasma torch	Air	*E. coli*, Background microflora	Apples, cucumbers, tomatoes, carrots	*E. coli*~6 Background microflora~5.2	Baier et al. (2015a)
DBD plasma	Air	*E. coli* *Listeria*	Radicchio leaves	*E. coli*~1.3 *Listeria*~2.2	Pasquali et al. (2016)

researchers studying decontamination effects of cold plasma have demonstrated that antimicrobial efficacy of plasma is influenced by the amount and type of reactive species generated during plasma processing, which are governed by experimental parameters, such as working gas, voltage levels, frequency, distance between samples and plasma emitter, electrodes geometry, and the treatment time.

High variability in the maximal bacterial inactivation achieved by cold plasma largely originates from the treatment conditions and the equipment used. For example, gliding arc discharge plasma operating in air was more effective at the higher flow rates against bacteria inoculated on apples (Niemira and Sites, 2008). Higher antimicrobial efficiencies were achieved with plasma treatment against *E. coli* inoculated on corn salad leaves at higher input power and in conjunction with lower initial bacterial loads. However, treatment at higher power levels was shown to have detrimental impact on the tissue quality characteristics, thereby limiting the use of this setup as a "high-intensity-short-time" treatment (Baier et al., 2013). Similarly, substantial reduction of *E. coli* on spinach was associated with notable color degradation of produce (Klockow and Keener, 2009). The operating gas could have a profound impact on the efficacy of plasma process. Baier et al. (2015a) demonstrated a high potential of argon nonthermal plasma for decontamination of perishable fresh produce. In this work, an effective yet gentle plasma application at a higher distance between the tip of the glowing plasma and the sample surface reduced shiga toxin-producing *E. coli* strains inoculated on corn salad leaves by up to 3.3 log in 1–2 min.

2.1.2 The Effects of Plasma-Activated Water

Some recent studies have demonstrated that plasma-activated water (PAW) has outstanding antibacterial ability. Plasma treatment causes the change in water acidity; however, in line with the acidic environment in the presence of hydrogen peroxide and ozone, these may be considered as key species responsible for the antimicrobial action of PAW (Scholtz et al., 2015). Ma et al. (2015) demonstrated the efficacy of PAW in terms of reduction of monoculture *S. aureus* inoculated on strawberry fruits. Using a single electrode nonthermal atmospheric pressure plasma jet, Ar/O_2-based plasma was generated at 18 kV over sterile deionized water for either 10 or 20 min. Plasma-generated chemically reactive species tend to penetrate the water phase and change the chemical composition of water, thereby inducing bactericidal activity. Inoculated fruits were immersed in 10 or 20 min-PAW for 5, 10, or 15 min and stored for either 0 or 4 days at 20°C. PAW achieved substantial reduction of *S. aureus* inoculated on strawberries ranging from 1.6 to 2.3 log. Interestingly, bacterial numbers tended to decrease further by a log after 4 days of storage. The inactivation efficiency depended on the plasma-activation time for PAW generation and PAW-treatment time of strawberries inoculated with *S. aureus*. Importantly, no significant change was found in quality of the PAW-treated strawberries (details in Chapter 10).

A complete picture of the chemical species and the set of reactions in plasma-liquid interactions is yet to be obtained; once this is established, cold plasma could lend itself as an alternative sanitizer for the fresh produce industry.

2.1.3 The Effect of Produce Surface and Topology

In order to reflect wide range of substrates and infection causative agents implicated in foodborne related human illnesses, most studies utilize more than one type of produce and bacterial type, which in conjunction with the plasma source has been demonstrated to have an impact on overall antimicrobial efficacy of treatment. Fernandez et al. (2013) demonstrated that efficacy of a nitrogen plasma jet against *Salmonella* inoculated on different substrates was influenced by the substrate surface features, with higher inactivation rates achieved on abiotic surface compared with the surface of fresh produce. Thus, 2 min of cold plasma treatment reduced the population of cells attached on membrane filters by 2.7 log; however, an extended 15 min of treatment was required to reduce bacterial populations on lettuce, strawberry, and potato by 2.7, 1.7, and 0.9 log, respectively. The roughness of food surfaces as compared to the membrane filters and the possibility of bacterial penetration inside internal areas of produce were demonstrated to be key factors associated with the decrease in antimicrobial efficacy of treatment.

A wider range of produce was investigated by Baier et al. (2015a); however, no correlation between produce surface characteristics and the initial microbial count and plasma decontamination efficacy was found in this work. For instance, after 5 min of treatment, the highest inactivation levels in total microbial counts (5.2 log) was recorded on the rough surface of carrots with higher initial microbial levels, whereas this treatment reduced bacterial numbers by 3.4 and 3.3 log on the smooth surface of apples and tomatoes, respectively, associated with lower levels of initial mesophilic counts.

2.1.4 The Effects of Microbial Growth Phase, Concentration, and Cell Type

Fernandez et al. (2013) reported that intrinsic parameters, such as bacterial growth phase and growth temperature, play a minor role in the antimicrobial efficacy of cold plasma. In contrast, Yu et al. (2006) found that inactivation efficacy of plasma treatment was a function of *E. coli* cell concentration and the cell growth phase. Following 2.5 min of treatment, a decrease in the number of cells deposited on the surface of membrane filters by 7 log was achieved at the lowest cell concentration of 10^7 CFU/cm^2, compared to a value of 1 log reduction obtained at the highest cell density of 10^{11} CFU/cm^2. However, unexpected results were recorded with respect to the effect of bacterial growth phase on antimicrobial efficacy of treatment; the exponential phase cells were not more susceptible than those from the stationary phase. Similar trends in decreasing inactivation efficiency of plasma at higher initial cell concentrations were reported by Baier et al. (2015b). Direct treatment for 2 min and lower (5 mm) distance between plasma discharge and gel discs inoculated with *E. coli* led to the most efficient and complete inactivation at lower initial cell concentrations of 10^5 CFU/cm^2 as compared with concentrations of 10^7 CFU/cm^2 in conjunction with an extended 4 min of treatment. An improved plasma processing time was recorded by Wang et al. (2012a) with 90–100% inactivation of *Salmonella* inoculated on carrots, cucumber, and pear slices achieved after application of 4 s of treatment.

In addition to the challenges associated with bacterial concentration and the produce surface, bacterial cell wall characteristics may also play a significant role in the general bacterial resistance to various disinfection agents, and may therefore interfere with the optimum application of cold plasma. Thus, population of *E. coli* O157:H7 inoculated on radicchio leaves was significantly reduced after 15 min of 15 kV air-generated plasma treatment (1.35 log MPN/cm^2). However, 30 min of treatment was necessary to achieve a significant reduction of *L. monocytogenes* counts (2.2 log CFU/cm^2) (Pasquali et al., 2016). Nonetheless, despite the range of various target intrinsic parameters, high-voltage in-package dielectric barrier discharge (DBD) cold plasma treatment utilized for decontamination of cherry tomatoes and strawberries demonstrated high efficiency against key pathogenic microorganisms. In this study, short treatments for 10, 60, and 120 s in conjunction with 24 h of post-treatment storage reduced the population of *Salmonella*, *E. coli* and *L. monocytogenes* on cherry tomatoes to undetectable levels (6.3, 6.7, and 3.1 $\log_{10}$ CFU/sample, respectively), whereas background microflora of cherry tomatoes was not detected after 120–300 s. The effect of substrate complexity on cold plasma treatment was evident in the case of the treatment of inoculated strawberries, where an extended treatment for 5 min was required to achieve considerable reductions of *E. coli*, *Salmonella* and *L. monocytogenes* by 3.5, 3.8, and 4.2 $\log_{10}$ CFU/sample, respectively (Ziuzina et al., 2014). The study conducted by Misra et al. (2014) revealed that 5 min of high-voltage in-package DBD air-generated plasma was very effective against the natural microflora of strawberries, resulting in 2 log reductions within 24 h of post-plasma treatment, which was achieved without adversely affecting produce quality.

2.1.5 The Inactivation of Internalized Bacteria and Biofilms

Bacterial pathogens can rapidly and irreversibly attach to fruits and vegetables while persisting for long periods of time. For example, within 30 s of exposure, 30% of *Salmonella* inoculum could firmly attach to green pepper slices, which cannot easily be removed by washing (Solomon and Sharma, 2009). Warning and Datta (2013) described that plant cuts, lenticels, trichomes, locations around the veins and stomata are the preferential places for bacterial cell attachment. A notable factor that could have a potential contribution to elevated resistance to antimicrobial agents is the fact that bacterial pathogens can become internalized inside plant tissue. Jahid et al. (2014) visualized bacteria associated with the stomatal wall, as well as inside the stomata. Gu et al. (2013) reported that following leaf colonization, bacteria could enter tomato leaves through hydathodes (plant water pores), resulting in the internal translocation of the bacteria inside plants. It should be noted that technologies such as ultraviolet and pulsed white light fail to eradicate internalized pathogens due to shadowing effects (Gómez-López et al., 2007).

The occurrence of highly tolerant bacterial biofilms on food, food-contact surfaces, on food-processing equipment and in potable water distribution systems contributes to food spoilage, cross-contamination of food products, and the spread of foodborne pathogens (Kim and Wei, 2012) and represents a great challenge in food production settings. In light of the issues associated with biofilm resistance, cold plasma technology has received increased attention as an alternative approach for eradication of bacterial biofilms in the food-processing sector.

In one of the foremost studies, Perni et al. (2008a) evaluated the efficacy of treatments using a cold plasma pen for decontamination of various microorganisms on honeydew melon and mango skin. They reported a 3-log reduction of *E. coli*. On exposing the cut surfaces of mangoes and cantaloupes to the same plasma source, a 2.5-log reduction in the population of *E. coli* and *L. monocytogenes* on mangoes and 1.5- and 2-log reductions, respectively, on melons were observed (Perni et al., 2008b). The lower inactivation with cut fruits compared to whole fruits was attributed to the migration of bacteria into the interior of the tissues (ie, internalization).

Jahid et al. (2015) reported increased resistance of *Salmonella* Typhimurium bacterial biofilms on lettuce leaves to plasma treatment due to internalization and extensive colonization in produce stomatal wells. The antimicrobial potential of in-package atmospheric pressure cold plasma treatment with subsequent 24 h of storage was effective for inactivation of pathogens in the form of monoculture biofilms commonly implicated in foodborne-associated human infections, *Salmonella*, *L. monocytogenes* and *E. coli* biofilms, developed on lettuce (Ziuzina et al., 2015). In this study, plasma treatment was challenged with associated bacteria internalized in the lettuce tissue. Indeed, cold plasma treatment in conjunction with 24 h of post-treatment storage had detrimental effects on unprotected bacteria, eliminating most of the cells from the surface of lettuce. However, inside the stomata, where high concentrations of cells were noted and strong biofilm formation is anticipated, bacterial cells remained intact, suggesting that chemical species of plasma could not penetrate colonized stomata through the complex biofilm matrices. The presence of some uncolonized stomata in control often pose difficulties in coming to conclusions about the decontamination efficacy of plasma against internalized bacteria (Fig. 1).

Although significant biofilm reductions by a maximum of 5 $\log_{10}$ CFU/sample have been achieved, inactivation levels are strongly dependent on the experimental storage conditions; thus, plasma treatment requires further research to address the

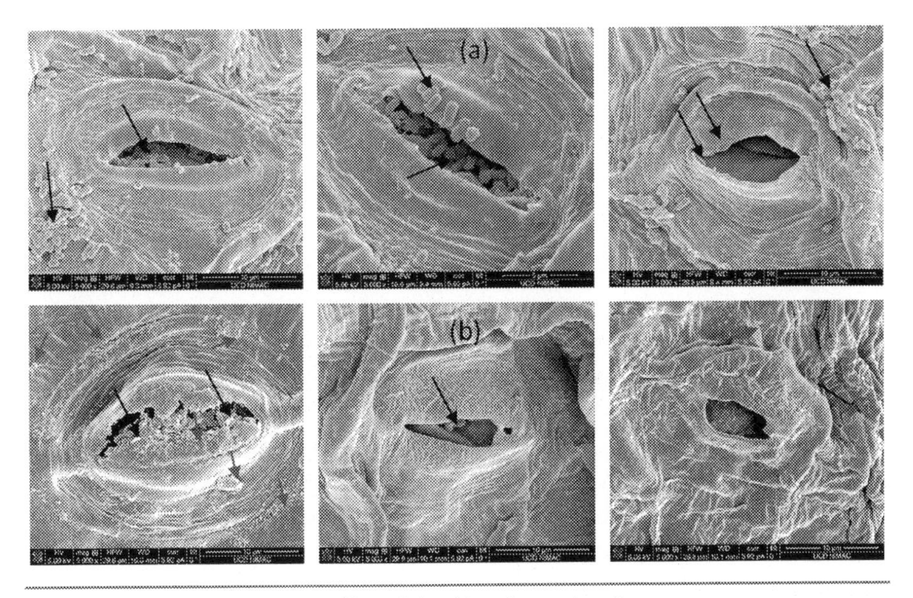

Fig. 1 SEM images of *Salmonella* 48 h biofilms formed on lettuce at room temperature (A) untreated control and (B) atmospheric cold plasma-treated sample. Black arrows indicate intact bacterial cells and white arrows indicate cell debris. *(From Ziuzina, D., Han, L., Cullen, P.J., Bourke, P., 2015. Cold plasma inactivation of internalised bacteria and biofilms for* Salmonella enterica *serovar Typhimurium,* Listeria monocytogenes *and* Escherichia coli. *Int. J. Food Microbiol., 210, 53–61, with permission from Elsevier.)*

influence of various produce storage conditions, the bacterial stress response, produce commodity, and the anticipated levels of initial bacterial colonization. The incidence of bacterial internalization and challenges associated with biofilm formation, which may significantly impact inactivation efficacy of cold plasma and indeed other antimicrobial treatments, highlight the importance of informed, effective microbiological control required for the consistent inactivation of pathogens in produce where enhanced produce quality characteristics are also attained. How these food safety challenges may be effectively addressed through utilization of cold plasma technology is an important question, the answers to which need further work.

2.2 Fruit Juices

While regular consumption of fresh fruits and fruit juices helps to prevent many degenerative diseases, such as cardiovascular problems, diabetes, and cancers, they could also constitute a hazard for consumers by the possible presence of pathogenic microorganisms or microbial toxins (Raybaudi-Massilia et al., 2009). The presence of *E. coli*, *Salmonella*, and *S. aureus* in fruit juices is of primary concern because these pathogens have been associated with a number of outbreaks associated with fruit juices (Aneja et al., 2014). In a recent study, 25 microbial species, including 9 bacterial isolates, 5 yeast isolates, and 11 molds, were isolated from a total of 30 juice samples with *Aspergillus flavus* and *Rhodotorula mucilaginosa* observed in many. Among the bacteria, *Bacillus cereus* and *Serratia* were dominant, but *E. coli* and *S. aureus* were also detected (Aneja et al., 2014). Sanitation methods in use may adversely affect the sensory characteristics of fruit juices, since the concentration of antimicrobials that are necessary to ensure microbiological safety of the produce do not match consumers expectations from a sensory point of view (Raybaudi-Massilia et al., 2009). Moreover, compounds recognized as "possibly carcinogenic" to humans, such as furan, can be formed in fruit juices as a result of pasteurization (Palmers et al., 2015).

It is well-recognized that the ionization occurring during plasma discharge allows substantial levels of ozone to be formed and to dissolve in liquids (Espie et al., 2001). Ozone in turn is characterized by a high oxidation potential that conveys bactericidal and virucidal properties (Patil et al., 2009). Gaseous ozone treatment has been evaluated in a number of studies for its antimicrobial properties and has been found to be an effective alternative to traditional thermal pasteurization for the control of *E. coli* populations in orange juice (Patil et al., 2009), and apple juice (Patil et al., 2010). Reductions in bacterial numbers by 5 log cycles (within the order recommended by FDA) with relatively short treatments for 1–18 min, depending on the type of juice (pulp content) and pH of the juice have been recorded. However, information on plasma-mediated inactivation of bacterial pathogens associated with fruit juices is relatively sparse. The ability of cold plasma to ensure microbiological safety and shelf life of freshly squeezed orange juice has been reported. The effective killing action of a DBD against *S. aureus*, *E. coli*, and *Candida albicans* inoculated in orange juice by >5 log after 12, 8, and 25 s of treatment, respectively, at peak voltage of 30 kV has been documented by Shi et al. (2011). In an earlier study, direct current corona discharges obtained using a pulsating 0–15 kV DC power supply allowed the generation of NTP in apple juice (Montenegro et al., 2002). A reduction in the number of *E. coli* O157:H7 in apple juice by more than 5 log CFU/mL in 40 s was noted in this study.

Surowsky et al. (2014) investigated inactivation of *Citrobacter freundii* in apple juice using an atmospheric plasma jet operating at a gas flow rate of 5 L/min, with 10 mm distance between nozzle outlet and juice sample. Reductions by about 5 log cycles of *C. freundii* in apple juice were achieved using longer plasma exposure times (8 min) and higher oxygen concentrations (0.1% oxygen) in the argon gas mixture. Interestingly, higher reduction levels were associated with longer posttreatment storage of the samples (24 h). The authors indicated that increasing concentrations of oxygen in the process gas led to the formation of hydrogen peroxide in the juice, which was responsible for an increased antimicrobial behavior of the plasma. The impact of plasma on the juice quality was not addressed in the study. In general, most studies have concluded that cold plasma does not adversely affect the quality (see Chapter 10).

Because cold plasma is a relatively novel technology for fruit juice decontamination, extensive research is still required to establish: (i) the inactivation efficacy of treatment toward reduction of a wider range of microorganisms associated with different fruit-based products; (ii) any possible long-term effects on physicochemical characteristics; (iii) the safety of plasma-treated juices for human consumption.

2.3 Food Grains and Nuts

Dried seeds are popular as a ready-to-eat food, particularly due to the positive perception of their health benefits among consumers. However, a number of outbreaks associated with low-moisture products, including nuts and seeds, were documented in the first decades of the 21st century. For example, in 2001, *Salmonella* Typhimurium DT 104 caused an international outbreak with 52 cases in Sweden, Norway, Germany, United Kingdom, and Australia, which was due to the consumption of a sesame seed product (Harris et al., 2015). The pathogens had been associated with a wide range of products, including almonds, cashew, coconut, hazelnut, pine nut, pistachio, walnut, peanut, sesame seed, etc. (Harris et al., 2015). Fungal contamination in nuts due to *Penicillium* spp., *Aspergillus*, *Fusarium* spp., *Trichoderma* spp., and *Cladosporium* spp. is also very common. As a result of fungal contamination, mycotoxins are produced during production, harvest, storage, and processing, which causes intoxication and serious consequences for human health following ingestion. Cereal grains constitute an important part of a staple and healthy diet. However, cereal grains just after being harvested may also contain microbial contamination due to the families, such as *Pseudomonadaceae*, *Micrococcaceae*, *Lactobacillaceae*, and *Bacillaceae*, and molds, namely *Alternaria*, *Fusarium*, *Aspergillus*, *Helminthosporium*, and *Cladosporium*, where an increased resistance of mycotoxins to decontamination is one of the most important subjects of health safety concerns. Aflatoxin B1 (AFB1) is a particular example of a stable carcinogenic mycotoxin produced by the species of *Aspergillus* genus, which cannot be destroyed with the traditional food-processing operations (Pleadin et al., 2015). In general, foodborne pathogens in low-water activity foods often exhibit an increased tolerance to heat and other treatments that are lethal to cells in high-water activity environments (Beuchat et al., 2013). The safety risks associated with nuts and grains have initiated the search for alternative decontamination procedures, where cold plasma has been demonstrated to possess reasonable effectiveness for microbiological control of low-water activity products.

In an early study, Deng et al. (2007) demonstrated an effective and rapid inactivation of *E. coli* inoculated on almonds, using DBD source for generation of plasma with almost 5 log reduction achieved after 30 s of treatment. In this study, higher inactivation rates were associated with higher voltage levels (30 kV) and frequencies (2 kHz) tested. Later, Niemira (2012), studying the effect of plasma jet processing parameters, such as treatment time, distance, and feed gas for inactivation of pathogens inoculated on almonds, demonstrated lower inactivation rates. The use of an air plasma jet for short times of 20 s in conjunction with a higher distance of 6 cm between the plasma emitter and samples were attributed to the lower reduction of *E. coli* (1.34 log) in this case.

In a study conducted by Basaran et al. (2008) low-pressure cold plasma was demonstrated to be a suitable fungal decontamination method. In this work plasma was generated by using air and sulfur hexafluoride (SF_6) and tested against *Aspergillus parasiticus* on hazelnut, peanut, and pistachio samples. Air plasma treatment for 5 min resulted in a 1 log reduction of *A. parasiticus*, and a further 5 min treatment resulted in an additional 1 log reduction. However, SF_6 plasma application was more effective, resulting in an approximately 5 log decrease in the fungal population following the same treatment duration. When effectiveness of plasma treatment was tested against aflatoxins, 20 min of air plasma treatment resulted in a 50% reduction in total aflatoxins, while only a 20% reduction in total aflatoxins was observed after 20 min of treatment with SF_6 plasma. Higher reduction levels of aflatoxigenic fungi inoculated on hazelnuts due to the air plasma were reported by Dasan et al. (2016). Plasma treatment for 5 min using an atmospheric pressure fluidized-bed plasma jet system (655 W, 25 kHz, 3000 L/h flow rate, dry air) resulted in reductions of 4.50 log CFU/g in *A. flavus* and 4.19 log CFU/g in *A. parasiticus* populations with no change in fungi load recorded during 30 days of storage compared to the control.

Pathogenic bacteria can contaminate seeds and grow during sprouting, which also becomes a food safety concern for most reported sprout-associated outbreaks. The low-pressure plasma treatment was demonstrated to be effective for inactivation of *Aspergillus* spp. and *Penicillum* spp. associated with a range of seeds (Selcuk et al., 2008). Depending on the type of seed surface, plasma gas type (air and SF_6), plasma treatment time (up to 20 min), and the microbial population density, the percentage of seed infection was reduced to below 1%, without lowering the seed quality below the commercial threshold of 85% seed germination. In another study, Mitra et al. (2013) studied the effects of cold atmospheric plasma for decontamination of background microflora of the seeds of chickpea (*Cicer arietinum*). They observed significant reduction in the natural microbial populations (2 log) attached to the chickpea surface after extended plasma treatments of up to 5 min. Cold plasma technology has also been investigated as an alternative technology for the reduction of fungal growth and the various mycotoxins associated with grains. Butscher et al. (2015) tested a low-pressure fluidized-bed plasma reactor for the treatment of wheat grains in order to inactivate artificially deposited *B. amyloliquefaciens* endospores on the surface of produce. With this setup, concentration of *B. amyloliquefaciens* endospores was reduced by 2.15 log units within 30 s of effective treatment time, generated at a higher power input of 900 W. However, an overall experimental duration of almost an hour was required and the surface temperature of the grains was found to be 90°C at 900 W. Nonetheless, according to the energy influx measurements and the solution of the heat equation, thermal inactivation and wheat grain degradation was excluded. In addition, other analyses showed no negative effects of plasma treatment on the flour and baking properties. Later, focusing on the decontamination of grain products, the same research group utilized an atmospheric pressure DBD-generated pulsed plasma treatment for inactivation of *Geobacillus stearothermophilus*. Inoculated wheat grains as well as polypropylene model substrates, which were used to investigate the influence of substrate shape and surface properties on the efficacy of plasma inactivation, were treated in a pulsed argon plasma discharge which applied different combinations of treatment time, pulse voltage, and frequency. While with 10 min of treatment reductions by an ca. 5 log were achieved on polypropylene granules, endospore inactivation on wheat grains was less efficient and reached a maximum reduction of approximately 3 log units after 60 min; however, there was no impact on the nutritional quality parameters of produce (Butscher et al., 2016). It is known that complex produce surface characteristics complicate a homogeneous plasma treatment, impacting the consistency of plasma-mediated microbial inactivation. In the work of Butscher et al. (2016), the reduced inactivation efficacy of plasma, even after an extended treatment time of 60 min, could be attributed to the complexity of the grains surface structure, where microorganisms can be protected by the uneven surface, loose pieces of bran, or hidden deep inside the ventral furrow. Moreover, in case of products with a high surface-to-volume ratio, the concentration of plasma-generated reactive species tends to decrease during the processing due to interaction with the food surface itself, rather than with the microorganisms on that surface (Hertwig et al., 2015). This suggests that an extensive research is needed for further optimization or tailoring of the plasma treatment parameters to achieve an efficient and consistent inactivation of pathogens attached on different types of commodities with different surface characteristics.

3 Microbiological Safety of Animal Origin Foods

3.1 Meat, Fish, and Poultry

The safety of foods obtained from animal sources, particularly meat and meat products, remains a major challenge for the global food industry. The most common pathogens associated with meat products include *B. cereus*, *Campylobacter* spp., *Clostridium perfringens*, *Clostridium botulinum*, *E. coli*, *L. monocytogenes*, *Salmonella* spp.,

S. aureus, *Yersinia enterocolitica*, *Aeromonas*, *Brucella*, *Clostridium difficile*, *Enterobacter*, and *Shigella* (Stoica et al., 2014). A particular problem of high importance is that, while meat products are commonly frozen for preservation, pathogens like enterohemorrhagic *E. coli* (EHEC) can survive in meat products even after 180 days of frozen storage (Ro et al., 2015). To address the growing demands of consumers for high-quality meat products without the use of artificial preservatives, and without compromising the safety, it is necessary to develop and implement novel intervention technologies in the meat industry. In extending the success of cold plasma in decontaminating fresh produce, researchers have fortunately observed good success with meat products as well. The potential of cold plasma technology as a novel decontamination intervention in meat processing sector can be justified from the broad range of microorganisms it can inactivate, as tabulated in Table 2. In this section, we will selectively discuss the implications and importance of the studies summarized in Table 2.

3.1.1 The In-Package Plasma Technologies for Meat Safety

The use of indirect plasma in conjunction with utilization of closed chambers for decontamination of meat produce have been highlighted in recent studies conducted by Rød et al. (2012) and Fröhling et al. (2012). In the study of Rød et al. (2012), sliced, ready-to-eat meat product (bresaola) inoculated with *Listeria innocua* were treated inside sealed bags using a DBD plasma source at 15.5, 31, and 62 W input power for 2–60 s. The discharge operated in the bags filled with 30% oxygen and 70% argon. The authors observed a reduction by 0.8 to 1.6 log CFU/g, irrespective of the treatment times. Furthermore, they also noted highest inactivation of an ca. 1.6 log CFU/g with multiple treatments for 20 s to a time interval of 10 min at operating power of 62 W. Fröhling et al. (2012) employed a microwave plasma source (2.45 GHz, 1.2 kW; operating in air at 20 L/min) for the indirect plasma treatment of pork *Longissimus dorsi* muscle placed inside a glass bottle. They reported the aerobic viable count on the plasma-treated pork to remain between 2 and 3 log CFU/g during a storage period of 20 days at 5°C, while the count increased to 9.69 log CFU/g in control.

Employing a flexible thin-layer DBD plasma operating at 2 W average power from a bipolar square-waveform voltage at 15 kHz, Jayasena et al. (2015) treated fresh pork and beef inside sealed plastic packages. They noted that a 10 min treatment resulted in the reduction of *L. monocytogenes*, *E. coli* O157:H7 and *Salmonella* Typhimurium by 2.04, 2.54, and 2.68 log CFU/g in pork-butt samples and 1.90, 2.57, and 2.58 log CFU/g in beef loin, respectively.

It is important to note that the key advantage of in-package cold plasma treatment is that the bactericidal molecules are generated and contained in the package, allowing extended exposure to pathogenic microbes, while reverting back to the original

TABLE 2 A Summary of the Research on Inactivation of Microorganisms on Meat and Poultry by Cold Plasma Treatment

Plasma Source	Gas	Bacteria	Product	Maximal log Reduction	Reference
Corona discharge plasma jet	Air	*Escherichia coli* O157:H7, *Listeria monocytogenes*	Frozen and unfrozen pork	*E. coli*~1.5 log, *Listeria*~1.0 log in 120 s	Choi et al. (2016)
Atmospheric pressure DBD	Air	*Salmonella enterica*, *Campylobacter jejuni*	Chicken breast and thigh	*S. enterica*~2.5 log, *C. jejuni*~2.4 log in 3 min	Dirks et al. (2012)
Microwave plasma	Air	Total aerobic bacteria	Pork	~2 log	Fröhling et al. (2012)
Dielectric barrier discharge	Nitrogen/oxygen mixture	*Listeria monocytogenes*, *Escherichia coli* O157:H7, *Salmonella typhimurium*	Pork	*Listeria*~2.0 log, *E. coli*~2.5, *Salmonella*~2.6	Jayasena et al. (2015)
Dielectric barrier discharge	Nitrogen/oxygen mixture	*Listeria monocytogenes*, *Escherichia coli* O157:H7, *Salmonella typhimurium*	Beef	*Listeria*~1.9, *E. coli*~2.5, *Salmonella*~2.5	Jayasena et al. (2015)
Dielectric barrier discharge	He, He/O_2 mixture	Total aerobic bacteria	Bacon	~4.5 log in 90 s	Kim et al. (2011)
Dielectric barrier discharge	He, He/(0.3%)O_2 mixture	*E. coli*, *L. monocytogenes*	Pork loin	*E. coli*~0.55 log, *Listeria*~0.59 log in 10 min	Kim et al. (2013a)
Atmospheric pressure plasma jet	N_2 (6 L/min) and O_2 (10 sccm)	*Salmonella* Typhimurium	Chicken Breast	~1.2 log in 10 min	Kim et al. (2013b)
Atmospheric pressure plasma jet	N_2 (6 L/min) and O_2 (10 sccm)	*Salmonella* Typhimurium	Pork loin	~1.5 log in 10 min	Kim et al. (2013b)
RF-driven atmospheric pressure DBD plasma	Ar (20,000 sccm)	*Staphylococcus aureus*	Beef jerky	~1.8 log in 8 min	Kim et al. (2014a)

Continued

TABLE 2 A Summary of the Research on Inactivation of Microorganisms on Meat and Poultry by Cold Plasma Treatment—cont'd

Plasma Source	Gas	Bacteria	Product	Maximal log Reduction	Reference
Atmospheric pressure plasma jet	He, He+O_2, N_2, or N_2+O_2	*Listeria monocytogenes*	Cooked chicken breast and ham	4.73 log on chicken (N_2+O_2) in 2 min; 6.52 log on ham (N_2, N_2+O_2) in 2 min	Lee et al. (2011)
Cold plasma jet (23–38.5 kHz, 6.5–16 kV)	Helium (5 L/min) +oxygen (100 mL/min)	*Listeria innocua*	Chicken meat and skin	Skin ~ 1.0 log Meat > 3.0 log	Noriega et al. (2011)
Dielectric barrier discharge	70% Ar and 30% O_2	*Listeria innocua*	Bresaola (dry-cured beef)	*Listeria* > 1.5 log	Rød et al. (2012)
Pulsed low pressure discharge (0.8 MPa, 20–100 kHz)	Helium	Psychrotrophs, yeast and mold count	Pork (*Longissimus dorsi*)	Psychrotrophs ~ 3 log; Yeast and mold ~ 3 log	Ulbin-Figlewicz et al. (2015)
Pulsed low pressure discharge (0.8 MPa, 20–100 kHz)	Argon	Psychrotrophs, yeast and mold count	Pork (*Longissimus dorsi*)	Psychrotrophs ~ 2 log; Yeast and mold ~ 2.6 log	Ulbin-Figlewicz et al. (2015)
Pulsed low pressure discharge (0.8 MPa, 20–100 kHz)	Nitrogen	Psychrotrophs, mesophiles, total yeast and mold count	Pork (*Longissimus dorsi*)	Psychrotrophs-negligible reduction; Yeast and mold ~ 1.0 log	Ulbin-Figlewicz et al. (2015)
Pulsed low pressure discharge (20 kPa, 20–100 kHz)	Helium	Psychrotrophs, mesophiles, total yeast and mold count, *Listeria monocytogenes*	Pork	Yeast and mold ~ 1.9 log Psychrotrophs ~ 1.6 log *Listeria*—No effect	Ulbin-Figlewicz et al. (2014)

gas within a few hours of storage (Misra et al., 2013). Unlike most conventional food technologies, the antimicrobial treatment inside a sealed package ensures the prevention of post-processing contamination.

3.1.2 The Effects of Plasma Process Parameters

It has been reported that for a plasma pen operating in helium, a higher voltage, excitation frequency, and presence of oxygen in the gas composition for generation of plasma resulted in higher inactivation rates of *L. innocua* on chicken muscle and skin (Noriega et al., 2011). Song et al. (2009) noted that *L. monocytogenes* was effectively reduced or eliminated by large area-type atmospheric pressure plasma applied to ham, but input power, exposure time, and the type of food affected the efficiency of the inactivation. In particular, they reported that a higher input power and longer exposure times resulted in a greater inactivation.

The inactivation efficacy of N_2 is generally found to be higher than He. This is because He in comparison to N_2 yields more active species, specifically N_2^+ and N^+ groups (Lee et al., 2011). Furthermore, when He or N_2 is mixed with O_2, the overall bacterial inactivation efficacy of the plasma increases; c.f. Lee et al. (2011) where inactivation of *L. monocytogenes* on chicken breast and ham has been shown to increase upon the addition of O_2. Up to a 2–3 log reduction in the population of *L. monocytogenes* (KCTC 3596), *E. coli* (KCTC 1682), and *Salmonella* Typhimurium (KCTC 1925) using $He+O_2$ mixture, as compared to 1–2 log cycles with He alone, has also been documented (Kim et al., 2011). The increase in antimicrobiological action in the presence of oxygen arises from the action of singlet oxygen and ozone.

Using a corona discharge plasma jet operating in air (20 kV, 58 kHz), within 2 min of treatment span, an ca. 1.5 log and >1.0 log units reductions in *E. coli* O157:H7 and *L. monocytogenes* respectively, have been observed in pork samples (Choi et al., 2016). When investigating plasma-led inactivation of microorganisms inoculated onto pork loins, Kim et al. (2013a) noted that *E. coli* and *L. monocytogenes* populations significantly decreased on pork loin from 0.5 to 0.55 log CFU/g and from 0.43 to 0.59 log CFU/g, respectively, upon extending the treatment time from 5 to 10 min, respectively.

3.1.3 The Effects of Food Surface

Kim et al. (2014a) employed an RF-driven atmospheric pressure plasma system operating in Ar (flow rate: 20,000 sccm; 200 W power) for the treatment of beef jerky contaminated with *S. aureus* ATCC 12600. They observed an ca. 3–4 log reduction in beef jerky after 10 min of treatment. However, the inactivation rate was much faster when cells were inoculated on a polystyrene surface, indicating that the surface morphology of the product was a deciding factor for process efficacy.

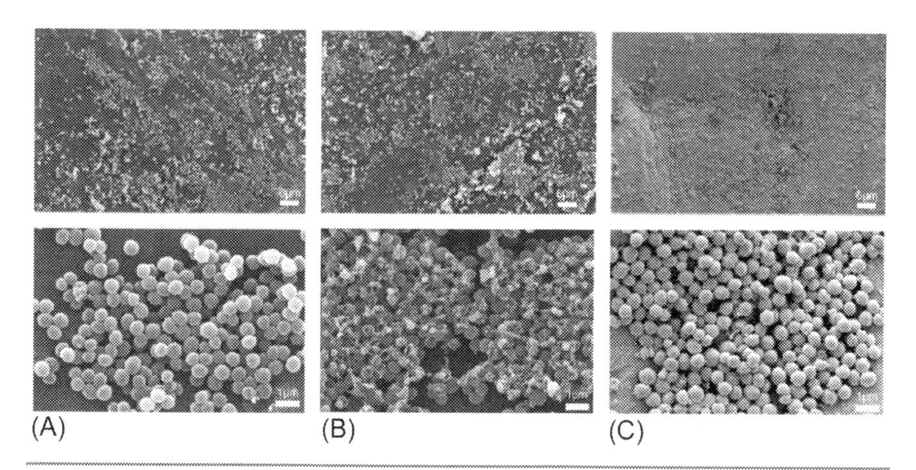

Fig. 2 SEM images of *S. aureus* on the beef jerky; (A) before plasma treatment, (B) after plasma treatment for 5 min, and (C) after heat treatment to 70°C for 5 min. *(From Kim, J.-S., Lee, E.-J., Choi, E.H., Kim, Y.-J., 2014. Inactivation of* Staphylococcus aureus *on the beef jerky by radio-frequency atmospheric pressure plasma discharge treatment. Innovative Food Sci. Emerg. Technol., 22, 124–130. with permission from Elsevier.)*

In an earlier study, Noriega et al. (2011) also suggested the surface topography of the food influenced the antibacterial efficacy of plasma treatment. Via scanning electron microscopy (SEM), Kim et al. (2014a) found that the bacterial cells disintegrated into pieces and many pores were created (see Fig. 2). On contrary, in the bacterial cells subjected to thermal treatments at 70°C for 5 min, the damages were far less.

Moon et al. (2009) reported that the sterilization efficacy of a radio-frequency dielectric barrier discharge plasma against *E. coli* inoculated on the pork surface was comparable with that of a conventional UV sterilizer. Dirks et al. (2012) reported that DBD plasma operating in air decreased the background microflora on a chicken breast by 0.84 and 0.85 log following 15 and 30 s of exposure, respectively. However, the decontamination effect was found to be lower on chicken thigh skin, with 0.33 and 0.21 log reduction in the background microflora after 15 and 30 s, respectively. These differences can be attributed to the surface roughness, which allows bacteria to hide out.

3.1.4 The Effects of Cell Type and Cell Concentration

Similar to the case of plant produce, the inactivation count of bacteria achieved when treating meat depends on the initial population of the cells. For example, it has been reported that starting at 2, 3, and 4 log CFU of *C. jejuni* on chicken breast, the maximum reductions obtained using a DBD air plasma for 3 min were 1.65, 2.45, and 2.45 log, respectively (Dirks et al., 2012). With the same initial concentrations, the reductions achieved on chicken skin were 1.42, 1.87, and 3.11 log, respectively, after 3 min of plasma treatment.

Using a direct He plasma treatment (10 lpm) and He (10 lpm)/oxygen (10 sccm) mixture at an operating power of 125 W, Kim et al. (2011) recorded a decrease in the total aerobic bacteria on bacon from 7.08 to 2.5 log CFU/g in 90 s. Furthermore, they noted that following a storage of the treated bacon for 7 days at 4°C, the total aerobic count remained low. This implied that the cells damaged by cold plasma were not capable of repair mechanisms.

3.1.5 Hygienic Design of Meat-Processing Machinery

In a distinct study, Leipold et al. (2010) for the first time demonstrated the decontamination of a rotary-type cutting knife, typically employed in the meat industry, using a DBD type configuration, while operating in air at atmospheric pressure. Specifically, they leveraged the metallic knife itself as a ground electrode, and were successful in achieving up to 5 log reduction of *L. innocua* within 340 s. Considering that the microbiological inactivation was achieved while the knife was in operation, this method could potentially allow for a reduced risk of cross-contamination between separate batches of meat (Misra et al., 2011).

In another approach for hygienic design of food-contact surfaces in processing equipment, poly(ethylene glycol) (PEG)-like structures have been generated on stainless steel using di(ethylene glycol) vinyl ether (DiEGVE) RF plasma environments (Wang et al., 2003). Such modification has been shown to remarkably decrease the attachment of *L. monocytogenes* and biofilm formation by >90% compared to unmodified stainless steel. The plasma source employed for this investigation was a capacitively coupled parallel-plate cold plasma reactor.

3.1.6 Decontamination of Fish and Fish Products

The treatment of cold-smoked salmon inside a sealed package, using a DBD plasma (operated at an applied voltage: 13 kV at 15 kHz frequency) in Ar and Ar+7% CO_2 mixture has been reported (Chiper et al., 2011). Up to a ~3 log CFU/g reduction in the population of *Photobacterium phosphoreum* in the fish samples within 60–120 s of cold plasma treatment was observed. However, unlike *P. phosphoreum*, neither Ar plasma nor Ar+7% CO_2 plasma treatments was found effective against *L. monocytogenes* or *Lactobacillus sakei* for up to 120 s. In another study, Park and Ha (2015) studied the effects of cold oxygen plasma on the reductions of *Penicillium citrinum* and *Cladosporium cladosporioides* on the surface of dried filefish fillets (*Stephanolepis cirrhifer*). Their plasma source employed relied on the use of high-energy ultraviolet radiation between 180 and 270 nm for the ionization (a process developed by BioZone Scientific). The authors noted a reduction of >1 log CFU/g of both bacteria on the treated fillets after >10 min.

3.2 Egg and Egg Products

Salmonella spp. has been shown to be present in eggs (Berchieri et al., 2001) and poses a major food-safety risk in table eggs and food products using eggs. To address the safety issues of *Salmonella* related outbreaks, several nonthermal approaches for the surface decontamination of eggs have been developed since 1990s including: pulsed light technology, ozone, UV radiation, and electrolyzed water (Bermúdez-Aguirre and Corradini, 2012), all of which have demonstrated only a limited degree of success. Cold plasma technologies have also been explored for decontamination of eggs.

In one of the earliest documented work, Davies and Breslin (2003) reported gas phase air plasma to be ineffective against *Salmonella enteritidis* PT4 inoculated on eggs. However, their study did not present sufficient details of the plasma source and parameters. It is only several years later when a resistive barrier discharge operating at atmospheric pressure in air developed by Ragni et al. (2010) is found effective in reducing *S. enteritidis* and *Salmonella typhimurium* by 4.5 and 3.5 log CFU/eggshell, respectively, within 90 min of treatment. Herein, the level of inactivation was found to vary with the relative humidity of the gas, which in turn rules the plasma chemistry. In the most recent study, Donner and Keener (2011) treated eggs inside sealed plastic packages followed by 24 h storage. They observed an ca. 3 $\log_{10}$ reduction in *S. enteritidis*.

The successful inactivation of *L. monocytogenes* in cooked egg white and egg yolk using an atmospheric pressure plasma jet has also been noted in literature (Lee et al., 2012b). The study documented a >5 log reduction in the count after only 2 min of treatment using He, N_2, either alone or with added O_2. The addition of O_2 was found to enhance the inactivation effects.

The use of a radio-frequency (RF) plasma jet has been shown to bring about synergistic influence on the inhibitory effects of the plant essential oils (clove oil, sweet basil oil, and lime oil) against *E. coli*, *S. typhimurium*, and *S. aureus* residing on chicken eggs (Matan et al., 2013). Specifically, the study found that growth of the three bacteria on eggshell mixed with clove oil (10 μL/mL) or its key chemical constituent (eugenol at 5 μL/mL) is completely inhibited upon exposing to plasma at 40 W. Overall, cold plasma treatments have been found effective in decreasing pathogenic microorganisms in egg and egg products, as evidenced from recent literature. However, much remains to be studied, before conclusive decisions can be made.

3.3 Milk and Milk Products

While pasteurization of milk remains the most effective technique until date to ensure safety of milk against pathogenic bacteria and extending the shelf life of the product, the use of high temperatures often results in undesirable quality changes. Therefore, the dairy and food science community remains interested in alternative milk pasteurization technologies. Recently limited studies have attempted to explore the possibility of using cold plasma for sterilization of milk and cheese. The salient results of these studies are described in this section.

The effectiveness of a corona discharge generated using a 9 kV alternating current (AC) input, in decontaminating milk inoculated with *E. coli* was evaluated by Gurol et al. (2012). Thanks to the multifaceted action of plasma chemical species, the *E. coli* population in whole milk fell from 7.78 to 3.63 log CFU/mL within 20 min of treatment. Irrespective of the fat content, mean reduction values of approximately 50% were recorded after only 3 min of treatment. Plasma application has been found to cause insignificant changes to the pH, color, proteins, fatty acid composition, and volatile compounds in milk (Gurol et al., 2011). Using an encapsulated plasma source operating in air (which is based on the principle of barrier discharge), successful inactivation of *E. coli*, *L. monocytogenes*, and *S.* Typhimurium in milk by approximately 3.8, 4.0, and 3.7 log CFU/mL has been reported (Kim et al., 2015).

Song et al. (2009) have reported the inactivation of a 3-strain cocktail of *L. monocytogenes* in sliced cheese with up to a 6 log reduction after 120 s treatment. The plasma source employed in their study was an atmospheric pressure RF plasma (13.65 MHz, 125 W power) operating in helium gas. This group also revealed a positive impact of increasing the input power and plasma treatment time on the inactivation efficacy. Significant reductions in the population of *E. coli* and *S. aureus* inoculated on cheese slices, following treatment with a DBD plasma operating in He and He/O_2 mixture have also been confirmed (Lee et al., 2012a). However, after 10 min of plasma treatment, damage to the cheese slices was noticed.

When subjecting to encapsulated DBD plasma in air, the populations of *E. coli*, *Salmonella* Typhimurium, and *L. monocytogenes* have been observed to follow 2.67, 3.10, and 1.65 decimal reductions at 60 s, 45 s, and 7 min treatments, respectively (Yong et al., 2015a). This study also revealed further decreases in the bacterial populations during posttreatment storage. Specifically, for a posttreatment storage for 5 min, the populations of *E. coli*, *S.* Typhimurium, and *L. monocytogenes* on a cheese slice ($\sim$5 log CFU/g) were observed to decrease by 1.75, 1.97, and 1.65 log CFU/g, respectively. In a related approach, Yong et al. (2015b) developed a flexible thin-layer barrier discharge with the idea that it could potentially be part of the food package itself. Using this setup the researchers observed that the population of *E. coli* O157:H7, *L. monocytogenes*, and *Salmonella* Typhimurium on sliced cheddar cheese decreased by 3.2, 2.1, and 5.8 log CFU/g, respectively, following 10 min of plasma treatment. The innovative designs of plasma sources and the pioneering studies in decontamination of cheese established by Prof. Jo and his group suggest that there exists ample potential for commercial developments. However, the timelines for such developments remain elusive.

4 Concluding Remarks

Modern-day consumers in both developed and developing countries demand high-quality and convenient food products, be it of plant or animal origin, with natural flavor and taste, and appreciate the fresh appearance of minimally processed foods. However, amid these demands, the safety of food products remains a major challenge for the global food industry. In fact, the challenges are growing due to the emergence of new strains of pathogens, their low infectious doses, increased virulence, resistance to antibiotics, and cross-contamination or recontamination of foods, food-contact surfaces, and water within the food production chain. Cold plasma technology is a relatively new approach, aiming to improve microbiological safety in conjunction with the maintenance of the sensory attributes of the treated foods. A critical analysis of the summary of the studies to date, investigating the decontamination of plant- and animal-based foods, leads to the conclusion that there is a huge potential for the commercial scale exploitation of the plasma technologies. An absence of standardized methods, with regard to bacterial strains, initial microbial load, and type of food for evaluation of the antimicrobial properties of decontamination techniques, complicates the comparison between microbial reductions achieved by utilization of various cold plasma systems. General conclusions about overall cold plasma antimicrobial efficacy cannot be made at this point.

References

Abadias, M., Usall, J., Anguera, M., Solsona, C., Viñas, I., 2008. Microbiological quality of fresh, minimally-processed fruit and vegetables, and sprouts from retail establishments. Int. J. Food Microbiol. 123, 121–129.

Aneja, K.R., Dhiman, R., Aggarwal, N.K., Kumar, V., Kaur, M., 2014. Microbes associated with freshly prepared juices of citrus and carrots. Int. J. Food Sci. 2014, 7.

Baier, M., Foerster, J., Schnabel, U., Knorr, D., Ehlbeck, J., Herppich, W.B., Schlüter, O., 2013. Direct non-thermal plasma treatment for the sanitation of fresh corn salad leaves: evaluation of physical and physiological effects and antimicrobial efficacy. Postharvest Biol. Technol. 84, 81–87.

Baier, M., Görgen, M., Ehlbeck, J., Knorr, D., Herppich, W.B., Schlüter, O., 2014. Non-thermal atmospheric pressure plasma: screening for gentle process conditions and antibacterial efficiency on perishable fresh produce. Innovative Food Sci. Emerg. Technol. 22, 147–157.

Baier, M., Ehlbeck, J., Knorr, D., Herppich, W.B., Schlüter, O., 2015a. Impact of plasma processed air (PPA) on quality parameters of fresh produce. Postharvest Biol. Technol. 100, 120–126.

Baier, M., Janßen, T., Wieler, L.H., Ehlbeck, J., Knorr, D., Schlüter, O., 2015b. Inactivation of Shiga toxin-producing *Escherichia coli* O104:H4 using cold atmospheric pressure plasma. J. Biosci. Bioeng. 120, 275–279.

Basaran, P., Basaran-Akgul, N., Oksuz, L., 2008. Elimination of *Aspergillus parasiticus* from nut surface with low pressure cold plasma (LPCP) treatment. Food Microbiol. 25, 626–632.

Berchieri Jr., A., Wigley, P., Page, K., Murphy, C., Barrow, P., 2001. Further studies on vertical transmission and persistence of *Salmonella enterica* serovar Enteritidis phage type 4 in chickens. Avian Pathol. 30, 297–310.

Berger, C.N., Sodha, S.V., Shaw, R.K., Griffin, P.M., Pink, D., Hand, P., Frankel, G., 2010. Fresh fruit and vegetables as vehicles for the transmission of human pathogens. Environ. Microbiol. 12, 2385–2397.

Bermúdez-Aguirre, D., Corradini, M.G., 2012. Inactivation kinetics of *Salmonella* spp. under thermal and emerging treatments: a review. Food Res. Int. 45, 700–712.

Bermúdez-Aguirre, D., Wemlinger, E., Pedrow, P., Barbosa-Cánovas, G., Garcia-Perez, M., 2013. Effect of atmospheric pressure cold plasma (APCP) on the inactivation of *Escherichia coli* in fresh produce. Food Control 34, 149–157.

Beuchat, L.R., Komitopoulou, E., Beckers, H., Betts, R.P., Bourdichon, F., Fanning, S., Joosten, H.M., Ter Kuile, B.H., 2013. Low-water activity foods: increased concern as vehicles of foodborne pathogens. J. Food Prot. 76, 150–172.

Borges, A., Simões, L.C., Saavedra, M.J., Simões, M., 2014. The action of selected isothiocyanates on bacterial biofilm prevention and control. Int. Biodeterior. Biodegrad. 86 (Pt A), 25–33.

Butscher, D., Schlup, T., Roth, C., Müller-Fischer, N., Gantenbein-Demarchi, C., Rudolf von Rohr, P., 2015. Inactivation of microorganisms on granular materials: reduction of *Bacillus amyloliquefaciens* endospores on wheat grains in a low pressure plasma circulating fluidized bed reactor. J. Food Eng. 159, 48–56.

Butscher, D., Zimmermann, D., Schuppler, M., Rudolf von Rohr, P., 2016. Plasma inactivation of bacterial endospores on wheat grains and polymeric model substrates in a dielectric barrier discharge. Food Control 60, 636–645.

CDC, 2011. Multistate Outbreak of Listeriosis Linked to Whole Cantaloupes from Jensen Farms. Centers for Disease Control and Prevention, Colorado.

Chiper, A.S., Chen, W., Mejlholm, O., Dalgaard, P., Stamate, E., 2011. Atmospheric pressure plasma produced inside a closed package by a dielectric barrier discharge in Ar/CO_2 for bacterial inactivation of biological samples. Plasma Sources Sci. Technol. 20, 10.

Choi, S., Puligundla, P., Mok, C., 2016. Corona discharge plasma jet for inactivation of *Escherichia coli* O157:H7 and *Listeria monocytogenes* on inoculated pork and its impact on meat quality attributes. Ann. Microbiol 66, 685–694.

Critzer, F., Kelly-Wintenberg, K., South, S., Golden, D., 2007. Atmospheric plasma inactivation of foodborne pathogens on fresh produce surfaces. J. Food Prot. 70, 2290.

Dasan, B.G., Mutlu, M., Boyaci, I.H., 2016. Decontamination of *Aspergillus flavus* and *Aspergillus parasiticus* spores on hazelnuts via atmospheric pressure fluidized bed plasma reactor. Int. J. Food Microbiol. 216, 50–59.

Davies, R., Breslin, M., 2003. Investigations into possible alternative decontamination methods for *Salmonella enteritidis* on the surface of table eggs. J. Veterinary Med. Ser. B 50, 38–41.

Deng, S., Ruan, R., Mok, C.K., Huang, G., Lin, X., Chen, P., 2007. Inactivation of *Escherichia coli* on almonds using nonthermal plasma. J. Food Sci. 72, M62–M66.

Dirks, B.P., Dobrynin, D., Fridman, G., Mukhin, Y., Fridman, A., Quinlan, J.J., 2012. Treatment of raw poultry with nonthermal dielectric barrier discharge plasma to reduce *Campylobacter jejuni* and *Salmonella enterica*. J. Food Prot. 75, 22–28.

Donner, A., Keener, K.M., 2011. Investigation of in-package ionization. J. Purdue Undergrad. Res. 1, 10–15.

EFSA, 2013. Scientific report of EFSA and ECDC. The European Union Summary Report on trends and sources of zoonoses, zoonotic agents and food-borne outbreaks in 2011. EFSA J. 11, 3129.

Espie, S., Marsili, L., MacGregor, S.J., Anderson, J.G., 2001. Investigation of dissolved ozone production using plasma discharge in liquid. Pulsed Power Plasma Sci., Dig. Tech. Pap. 1, 616–619.

Fernandez, A., Noriega, E., Thompson, A., 2013. Inactivation of *Salmonella enterica* serovar Typhimurium on fresh produce by cold atmospheric gas plasma technology. Food Microbiol. 33, 24–29.

Fröhling, A., Durek, J., Schnabel, U., Ehlbeck, J., Bolling, J., Schlüter, O., 2012. Indirect plasma treatment of fresh pork: decontamination efficiency and effects on quality attributes. Innovative Food Sci. Emerg. Technol. 16, 381–390.

Giaouris, E., Heir, E., Hébraud, M., Chorianopoulos, N., Langsrud, S., Møretrø, T., Habimana, O., Desvaux, M., Renier, S., Nychas, G.-J., 2014. Attachment and biofilm formation by foodborne bacteria in meat processing environments: causes, implications, role of bacterial interactions and control by alternative novel methods. Meat Sci. 97, 298–309.

Gómez-López, V.M., Ragaert, P., Debevere, J., Devlieghere, F., 2007. Pulsed light for food decontamination: a review. Trends Food Sci. Technol. 18, 464–473.

Goodburn, C., Wallace, C.A., 2013. The microbiological efficacy of decontamination methodologies for fresh produce: a review. Food Control 32, 418–427.

Gu, G., Cevallos-Cevallos, J.M., van Bruggen, A.H.C., 2013. Ingress of *Salmonella enterica* Typhimurium into tomato leaves through hydathodes. PLoS ONE. 8, e53470.

Gurol, C., Ekinci, F.Y., Aslan, N., Guerzoni, E., Vannini, L., Korachi, M., 2011. Non thermal plasma as an alternative tool for milk processing. Curr. Opin. Biotechnol. 22 (Suppl. 1), S100.

Gurol, C., Ekinci, F.Y., Aslan, N., Korachi, M., 2012. Low temperature plasma for decontamination of *E. coli* in milk. Int. J. Food Microbiol. 157, 1–5.

Harris, L., Palumbo, M., Beuchat, L., Danyluk, M., 2015. Outbreaks of foodborne illness associated with the consumption of tree nuts, peanuts, and sesame seeds [table and references]. In: Outbreaks From Tree Nuts, Peanuts, and Sesame Seeds. UC Davis, Davis, CA.

Hertwig, C., Reineke, K., Ehlbeck, J., Knorr, D., Schlüter, O., 2015. Decontamination of whole black pepper using different cold atmospheric pressure plasma applications. Food Control 55, 221–229.

Hygreeva, D., Pandey, M.C., Radhakrishna, K., 2014. Potential applications of plant based derivatives as fat replacers, antioxidants and antimicrobials in fresh and processed meat products. Meat Sci. 98, 47–57.

Jahid, I.K., Han, N., Ha, S.-D., 2014. Inactivation kinetics of cold oxygen plasma depend on incubation conditions of Aeromonas hydrophila biofilm on lettuce. Food Res. Int. 55, 181–189.

Jahid, I.K., Han, N., Zhang, C.-Y., Ha, S.-D., 2015. Mixed culture biofilms of *Salmonella* Typhimurium and cultivable indigenous microorganisms on lettuce show enhanced resistance of their sessile cells to cold oxygen plasma. Food Microbiol. 46, 383–394.

Jayasena, D.D., Kim, H.J., Yong, H.I., Park, S., Kim, K., Choe, W., Jo, C., 2015. Flexible thin-layer dielectric barrier discharge plasma treatment of pork butt and beef loin: effects on pathogen inactivation and meat-quality attributes. Food Microbiol. 46, 51–57.

Kim, S.-H., Wei, C.-I., 2012. Biofilms. In: Gómez-López, V.M. (Ed.), Decontamination of Fresh and Minimally Processed Produce. Wiley-Blackwell, USA, pp. 59–75.

Kim, B., Yun, H., Jung, S., Jung, Y., Jung, H., Choe, W., Jo, C., 2011. Effect of atmospheric pressure plasma on inactivation of pathogens inoculated onto bacon using two different gas compositions. Food Microbiol. 28, 9–13.

Kim, H.-J., Yong, H.I., Park, S., Choe, W., Jo, C., 2013a. Effects of dielectric barrier discharge plasma on pathogen inactivation and the physicochemical and sensory characteristics of pork loin. Curr. Appl. Phys. 13, 1420–1425.

Kim, H.-J., Yong, H.I., Park, S., Kim, K., Bae, Y.S., Choe, W., Oh, M.H., Jo, C., 2013b. Effect of inactivating *Salmonella* Typhimurium in raw chicken breast and pork loin using an atmospheric pressure plasma jet. J. Anim. Sci. Technol. 55, 545–549.

Kim, J.-S., Lee, E.-J., Choi, E.H., Kim, Y.-J., 2014a. Inactivation of *Staphylococcus aureus* on the beef jerky by radio-frequency atmospheric pressure plasma discharge treatment. Innovative Food Sci. Emerg. Technol. 22, 124–130.

Kim, J.E., Lee, D.U., Min, S.C., 2014b. Microbial decontamination of red pepper powder by cold plasma. Food Microbiol. 38, 128–136.

Kim, H.-J., Yong, H.I., Park, S., Kim, K., Choe, W., Jo, C., 2015. Microbial safety and quality attributes of milk following treatment with atmospheric pressure encapsulated dielectric barrier discharge plasma. Food Control 47, 451–456.

Klockow, P.A., Keener, K.M., 2009. Safety and quality assessment of packaged spinach treated with a novel ozone-generation system. LWT Food Sci. Technol. 42, 1047–1053.

Lacombe, A., Niemira, B.A., Gurtler, J.B., Fan, X., Sites, J., Boyd, G., Chen, H., 2015. Atmospheric cold plasma inactivation of aerobic microorganisms on blueberries and effects on quality attributes. Food Microbiol. 46, 479–484.

Lee, H.J., Jung, H., Choe, W., Ham, J.S., Lee, J.H., Jo, C., 2011. Inactivation of *Listeria monocytogenes* on agar and processed meat surfaces by atmospheric pressure plasma jets. Food Microbiol. 28, 1468–1471.

Lee, H.-J., Jung, S., Jung, H.-S., Park, S.-H., Choe, W.-H., Ham, J.-S., Jo, C., 2012a. Evaluation of a dielectric barrier discharge plasma system for inactivating pathogens on cheese slices. J. Anim. Sci. Technol. 54, 191–198.

Lee, H.-J., Song, H.-P., Jung, H.-S., Choe, W.-H., Ham, J.-S., Lee, J.-H., Jo, C.-R., 2012b. Effect of atmospheric pressure plasma jet on inactivation of Listeria monocytogenes, quality, and genotoxicity of cooked egg white and yolk. Korean J. Food Sci. Anim. Resour. 32, 561–570.

Leipold, F., Kusano, Y., Hansen, F., Jacobsen, T., 2010. Decontamination of a rotating cutting tool during operation by means of atmospheric pressure plasmas. Food Control 21, 1194–1198.

Ma, R., Wang, G., Tian, Y., Wang, K., Zhang, J., Fang, J., 2015. Non-thermal plasma-activated water inactivation of food-borne pathogen on fresh produce. J. Hazard. Mater. 300, 643–651.

Martín, B., Perich, A., Gómez, D., Yangüela, J., Rodríguez, A., Garriga, M., Aymerich, T., 2014. Diversity and distribution of *Listeria monocytogenes* in meat processing plants. Food Microbiol. 44, 119–127.

Matan, N., Nisoa, M., Matan, N., 2013. Antibacterial activity of essential oils and their main components enhanced by atmospheric RF plasma. Food Control 39, 97–99.

Misra, N.N., Tiwari, B.K., Raghavarao, K.S.M.S., Cullen, P.J., 2011. Nonthermal plasma inactivation of food-borne pathogens. Food Eng. Rev. 3, 159–170.

Misra, N.N., Ziuzina, D., Cullen, P.J., Keener, K.M., 2013. Characterization of a novel atmospheric air cold plasma system for treatment of packaged biomaterials. Trans. ASABE 56, 1011–1016.

Misra, N.N., Patil, S., Moiseev, T., Bourke, P., Mosnier, J.P., Keener, K.M., Cullen, P.J., 2014. In-package atmospheric pressure cold plasma treatment of strawberries. J. Food Eng. 125, 131–138.

Mitra, A., Li, Y.-F., Klämpfl, T.G., Shimizu, T., Jeon, J., Morfill, G.E., Zimmermann, J.L., 2013. Inactivation of surface-borne microorganisms and increased germination of seed specimen by cold atmospheric plasma. Food Bioprocess Technol. 7, 645–653.

Montenegro, J., Ruan, R., Ma, H., Chen, P., 2002. Inactivation of *E-coli* O157:H7 using a pulsed nonthermal plasma system. J. Food Sci. 67, 646–648.

Moon, S.Y., Kim, D.B., Gweon, B., Choe, W., Song, H.P., Jo, C., 2009. Feasibility study of the sterilization of pork and human skin surfaces by atmospheric pressure plasmas. Thin Solid Films 517, 4272–4275.

Niemira, B.A., 2012. Cold plasma reduction of *Salmonella* and *Escherichia coli* O157:H7 on almonds using ambient pressure gases. J. Food Sci. 77, M171–M175.

Niemira, B.A., Sites, J., 2008. Cold plasma inactivates *Salmonella Stanley* and *Escherichia coli* O157: H7 inoculated on golden delicious apples. J. Food Prot. 71, 1357–1365.

Noriega, E., Shama, G., Laca, A., Diaz, M., Kong, M.G., 2011. Cold atmospheric gas plasma disinfection of chicken meat and chicken skin contaminated with *Listeria innocua*. Food Microbiol. 28, 1293–1300.

Olaimat, A.N., Holley, R.A., 2012. Factors influencing the microbial safety of fresh produce: a review. Food Microbiol. 32, 1–19.

Ölmez, H., Temur, S.D., 2010. Effects of different sanitizing treatments on biofilms and attachment of *Escherichia coli* and *Listeria monocytogenes* on green leaf lettuce. LWT Food Sci. Technol. 43, 964–970.

Palmers, S., Grauwet, T., Celus, M., Wibowo, S., Kebede, B.T., Hendrickx, M.E., Van Loey, A., 2015. A kinetic study of furan formation during storage of shelf-stable fruit juices. J. Food Eng. 165, 74–81.

Park, S.Y., Ha, S.-D., 2015. Application of cold oxygen plasma for the reduction of *Cladosporium cladosporioides* and *Penicillium citrinumon* the surface of dried filefish (*Stephanolepis cirrhifer*) fillets. Int. J. Food Sci. Technol. 50, 966–973.

Pasquali, F., Stratakos, A.C., Koidis, A., Berardinelli, A., Cevoli, C., Ragni, L., Mancusi, R., Manfreda, G., Trevisani, M., 2016. Atmospheric cold plasma process for vegetable leaf decontamination: a feasibility study on radicchio (red chicory, *Cichorium intybus* L.). Food Control 60, 552–559.

Patil, S., Bourke, P., Frias, J.M., Tiwari, B.K., Cullen, P.J., 2009. Inactivation of *Escherichia coli* in orange juice using ozone. Innovative Food Sci. Emerg. Technol. 10, 551–557.

Patil, S., Valdramidis, V.P., Cullen, P.J., Frias, J., Bourke, P., 2010. Inactivation of *Escherichia coli* by ozone treatment of apple juice at different pH levels. Food Microbiol. 27, 835–840.

Perni, S., Liu, D.W., Shama, G., Kong, M.G., 2008a. Cold atmospheric plasma decontamination of the pericarps of fruit. J. Food Prot. 71, 302–308.

Perni, S., Shama, G., Kong, M.G., 2008b. Cold atmospheric plasma disinfection of cut fruit surfaces contaminated with migrating microorganisms. J. Food Prot. 71, 1619–1625.

Pleadin, J., Vulić, A., Perši, N., Škrivanko, M., Capek, B., Cvetnić, Ž., 2015. Annual and regional variations of aflatoxin B1 levels seen in grains and feed coming from Croatian dairy farms over a 5-year period. Food Control 47, 221–225.

Ragni, L., Berardinelli, A., Vannini, L., Montanari, C., Sirri, F., Guerzoni, M.E., Guarnieri, A., 2010. Non-thermal atmospheric gas plasma device for surface decontamination of shell eggs. J. Food Eng. 100, 125–132.

Rangel, J.M., Sparling, P.H., Crowe, C., Griffin, P.M., Swerdlow, D.L., 2005. Epidemiology of *Escherichia coli* O157:H7 outbreaks, United States, 1982–2002. Emerg. Infect. Dis. 11, 603–609.

Raybaudi-Massilia, R.M., Mosqueda-Melgar, J., Soliva-Fortuny, R., Martín-Belloso, O., 2009. Control of pathogenic and spoilage microorganisms in fresh-cut fruits and fruit juices by traditional and alternative natural antimicrobials. Compr. Rev. Food Sci. Food Saf. 8, 157–180.

Ro, E.Y., Ko, Y.M., Yoon, K.S., 2015. Survival of pathogenic enterohemorrhagic *Escherichia coli* (EHEC) and control with calcium oxide in frozen meat products. Food Microbiol. 49, 203–210.

Rød, S.K., Hansen, F., Leipold, F., Knøchel, S., 2012. Cold atmospheric pressure plasma treatment of ready-to-eat meat: inactivation of *Listeria innocua* and changes in product quality. Food Microbiol. 30, 233–238.

Scholtz, V., Pazlarova, J., Souskova, H., Khun, J., Julak, J., 2015. Nonthermal plasma—a tool for decontamination and disinfection. Biotechnol. Adv. 33, 1108–1119.

Selcuk, M., Oksuz, L., Basaran, P., 2008. Decontamination of grains and legumes infected with *Aspergillus* spp. and *Penicillum* spp. by cold plasma treatment. Bioresour. Technol. 99, 5104–5109.

Sharma, G., Rao, S., Bansal, A., Dang, S., Gupta, S., Gabrani, R., 2014. Pseudomonas aeruginosa biofilm: potential therapeutic targets. Biologicals 42, 1–7.

Shi, X.-M., Zhang, G.-J., Wu, X.-L., Li, Y.-X., Ma, Y., Shao, X.-J., 2011. Effect of low-temperature plasma on microorganism inactivation and quality of freshly squeezed orange juice. IEEE Trans. Plasma Sci. 39, 1591–1597.

Solomon, E.B., Sharma, M., 2009. Microbial attachment and limitations of decontamination methodologies. In: Sapers, G.M., Solomon, E.B., Matthews, K.R. (Eds.), Produce Contamination Problem: Causes and Solutions. Elsevier Inc., San Diego, CA, pp. 21–45.

Song, H.P., Kim, B., Choe, J.H., Jung, S., Moon, S.Y., Choe, W., Jo, C., 2009. Evaluation of atmospheric pressure plasma to improve the safety of sliced cheese and ham inoculated by 3-strain cocktail *Listeria monocytogenes*. Food Microbiol. 26, 432–436.

Srey, S., Park, S.Y., Jahid, I.K., Ha, S.-D., 2014. Reduction effect of the selected chemical and physical treatments to reduce *L. monocytogenes* biofilms formed on lettuce and cabbage. Food Res. Int. 62, 484–491.

Stoica, M., Stoean, S., Alexe, P., 2014. Overview of biological hazards associated with the consumption of the meat products. J. Agroaliment. Process. Technol. 20, 192–197.

Surowsky, B., Frohling, A., Gottschalk, N., Schluter, O., Knorr, D., 2014. Impact of cold plasma on *Citrobacter freundii* in apple juice: inactivation kinetics and mechanisms. Int. J. Food Microbiol. 174, 63–71.

Ulbin-Figlewicz, N., Jarmoluk, A., Marycz, K., 2014. Antimicrobial activity of low-pressure plasma treatment against selected foodborne bacteria and meat microbiota. Ann. Microbiol. 65, 1537–1546.

Ulbin-Figlewicz, N., Brychcy, E., Jarmoluk, A., 2015. Effect of low-pressure cold plasma on surface microflora of meat and quality attributes. J. Food Sci. Technol. 52, 1228–1232.

Van Boxstael, S., Habib, I., Jacxsens, L., De Vocht, M., Baert, L., Van De Perre, E., Rajkovic, A., Lopez-Galvez, F., Sampers, I., Spanoghe, P., De Meulenaer, B., Uyttendaele, M., 2013. Food safety issues in fresh produce: bacterial pathogens, viruses and pesticide residues indicated as major concerns by stakeholders in the fresh produce chain. Food Control 32, 190–197.

Wang, Y., Somers, E., Manolache, S., Denes, F., Wong, A., 2003. Cold plasma synthesis of poly(ethylene glycol)-like layers on stainless-steel surfaces to reduce attachment and biofilm formation by *Listeria monocytogenes*. J. Food Sci. 68, 2772–2779.

Wang, H., Zhou, B., Feng, H., 2012a. Surface characteristics of fresh produce and their impact on attachment and removal of human pathogens on produce surfaces. In: Gómez-López, V.M. (Ed.), Decontamination of Fresh and Minimally Processed Produce. Wiley-Blackwell, USA, pp. 43–57.

Wang, R.X., Nian, W.F., Wu, H.Y., Feng, H.Q., Zhang, K., Zhang, J., Zhu, W.D., Becker, K.H., Fang, J., 2012b. Atmospheric-pressure cold plasma treatment of contaminated fresh fruit and vegetable slices: inactivation and physiochemical properties evaluation. Eur. Phys. J. D 66, 1–7.

Warning, A., Datta, A.K., 2013. Interdisciplinary engineering approaches to study how pathogenic bacteria interact with fresh produce. J. Food Eng. 114, 426–448.

Yong, H.I., Kim, H.-J., Park, S., Alahakoon, A.U., Kim, K., Choe, W., Jo, C., 2015a. Evaluation of pathogen inactivation on sliced cheese induced by encapsulated atmospheric pressure dielectric barrier discharge plasma. Food Microbiol. 46, 46–50.

Yong, H.I., Kim, H.-J., Park, S., Kim, K., Choe, W., Yoo, S.J., Jo, C., 2015b. Pathogen inactivation and quality changes in sliced cheddar cheese treated using flexible thin-layer dielectric barrier discharge plasma. Food Res. Int. 69, 57–63.

Yu, H., Perni, S., Shi, J.J., Wang, D.Z., Kong, M.G., Shama, G., 2006. Effects of cell surface loading and phase of growth in cold atmospheric gas plasma inactivation of *Escherichia coli* K12. J. Appl. Microbiol. 101, 1323–1330.

Zhang, M., Oh, J.K., Cisneros-Zevallos, L., Akbulut, M., 2013. Bactericidal effects of nonthermal low-pressure oxygen plasma on S. typhimurium LT2 attached to fresh produce surfaces. J. Food Eng. 119, 425–432.

Zhou, G.H., Xu, X.L., Liu, Y., 2010. Preservation technologies for fresh meat—a review. Meat Sci. 86, 119–128.

Ziuzina, D., Patil, S., Cullen, P.J., Keener, K.M., Bourke, P., 2014. Atmospheric cold plasma inactivation of *Escherichia coli, Salmonella enterica* serovar Typhimurium and *Listeria monocytogenes* inoculated on fresh produce. Food Microbiol. 42, 109–116.

Ziuzina, D., Han, L., Cullen, P.J., Bourke, P., 2015. Cold plasma inactivation of internalised bacteria and biofilms for *Salmonella enterica* serovar Typhimurium, *Listeria monocytogenes* and *Escherichia coli*. Int. J. Food Microbiol. 210, 53–61.

Chapter 10

Quality of Cold Plasma Treated Plant Foods

N.N. Misra
GTECH, Research & Development, General Mills India Pvt Ltd, Mumbai, India

1 Introduction

Plant foods are often subjected to minimal or thermal processing prior to consumption. The ultimate aim of any processing, whether conventional or novel, including cold plasma, is to impart edibility, palatability, and prolong shelf life, while maintaining nutritional and organoleptic food qualities. Food quality is a multifaceted concept for which consumers use both quality attributes and quality cues to form their assessment of perceived quality (Oude Ophuis and Van Trijp, 1995). Quality cues are observable product characteristics that can be intrinsic (eg, appearance, color, shape, size, structure) or extrinsic (eg, price, brand, nutritional information, production information, country of origin). The general quality parameters for foods have been summarized in Fig. 1, which broadly classifies these into physical, chemical, and microbiological.

In the context of the fresh produce industry, an integration of handling steps from farm to retail is critical to the final quality. Apart from the nutritional and sensory

Cold Plasma in Food and Agriculture. http://dx.doi.org/10.1016/B978-0-12-801365-6.00010-X

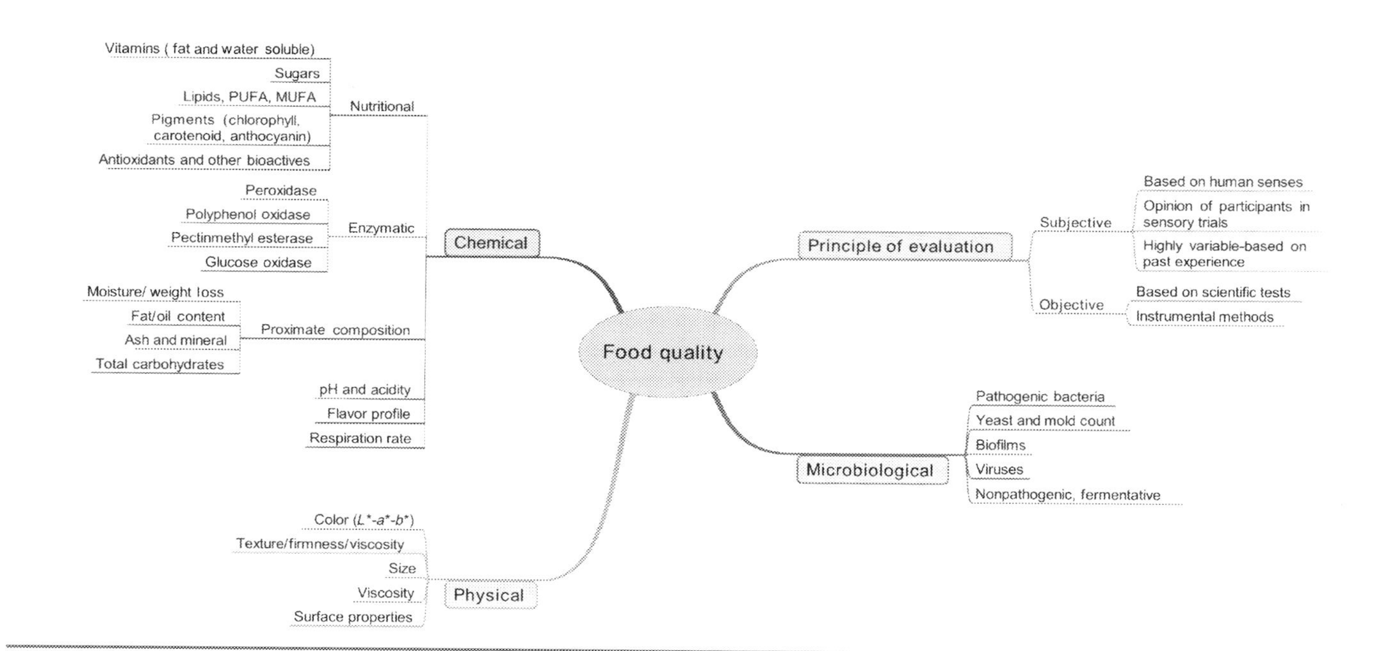

Fig. 1 An overview of some commonly used quality criteria for foods.

quality, the important physicochemical parameters that are often employed for judging the quality of fresh fruits and vegetables, and their products, include color, texture/firmness, weight loss, pH, acidity, and respiration rate. It may be recalled that the chemical species of plasma responsible for microbiological inactivation are primarily active radicals, and also tend to react with the food components via physical as well as chemical routes. The optimization of the plasma processes for food decontamination, without losing quality, is also challenging due to the fact that foods have natural variability. When treating plant foods, properties such as surface-to-volume ratio, the gas permeability of the cuticle, the porosity, and the susceptibility to oxidation are some important factors that should be considered. However, during cold plasma treatment of foods, several process parameters can be adjusted to optimize the process efficacy (eg, applied voltage/power, frequency, gas/gas mixture, and treatment time) for achieving desired levels of microbial inactivation, while retaining the quality of foods.

This chapter aims at providing a thorough overview of the reports of how cold plasma affects the quality traits of fruits, vegetables, and their products. The quality parameters discussed include: color, texture, microstructure, physiological activity, flavor, chemical constituents, and sensory characteristics. Where available, an elucidation of the possible mechanisms underlying the observed changes is also provided.

2 Physical Quality

2.1 Color

The changes in color of plasma-treated foods is decided by several factors: the plasma treatment conditions; the product characteristics—cut fruit versus intact, solid, or liquid, pigments responsible for color (carotenoid/anthocyanin/chlorophyll, etc.); and storage conditions and duration of storage. The color of plasma-treated fresh produce often changes during storage as a result of the partial inactivation of enzymes and microorganisms, which can initiate undesirable chemical reactions. A review of the literature pertinent to the color of plasma-treated produce is provided hereunder. The effects of plasma on plant pigments are reviewed later within this chapter.

Insignificant changes in color of carrots and tomatoes following cold argon plasma treatment using a plasma needle array (3.95–12.83 kV, 60 Hz, 0.5–10 min exposure time) has been reported by Bermúdez-Aguirre et al. (2013). However, under the same conditions, a noticeable change in the color of lettuce has been reported for treatment times beyond 7 min. No change in the surface color of peppercorns, either with remote or direct plasma treatments, using a radio-frequency jet, has been observed by Hertwig et al. (2015). Regarding treatment with ionized air from a microwave plasma source, the loss of color in tomatoes and carrots has been observed (Baier et al., 2015). Furthermore, the browning in the carrot also increases during storage. Following plasma treatment, a loss of color in cucumber and carrot slices has been recorded (Wang et al., 2012).

In cold plasma-treated red chicory leaves, an increase in redness and decrease in chroma during storage has been reported (Pasquali et al., 2016). For dielectric barrier discharge (DBD) plasma treatments in air, cut kiwi fruits have been observed to exhibit a lesser degree of darkening during storage, compared to untreated samples (Ramazzina et al., 2015). An increase in the darkness (conversely a decrease in lightness value L^*) has also been observed in blueberries following treatment with dry air plasma. However, in this case, the darkening has been attributed to the melting of the cutaneous wax in the berries (Lacombe et al., 2015). In the context of the gas plasma treatment of cut apples using a DBD, plasma exposure for 30 min followed by 4 h of storage has been observed to decrease the browning with respect to control (Tappi et al., 2014). The decrease in the production of dark pigments has been hypothesized to be an outcome of the inhibition of enzymatic browning phenomena. When using a plasma jet, one should note that the change in surface lightness could also be due to the loss of surface moisture (Wang et al., 2012).

In the context of fruit juices, argon cold plasma has been observed to turn pomegranate juice darker, with an increase in blueness (Bursać Kovačević et al., 2016). These changes have been explained on the basis of possible chemical changes in the constituting pigments; that is, polymerization of the phenolic compounds. In a recent study, an insignificant decrease in the color of air plasma-treated (70 kV, 50 Hz) orange juice inside a sealed package has been reported for treatment times up to 60 s (Almeida et al., 2015).

2.2 Firmness

Changes in the texture of fruits and vegetables are commonly related to transformations in cell wall polymers (Oey et al., 2008). A majority of the studies report acceptable retention of firmness in plasma-treated fruits and vegetables. For example, acceptable firmness retention for in-package air plasma-treated strawberries (voltage 60 kV, 50 Hz) after 24 h of storage at 10°C has also been recorded (Misra et al., 2014c). In a related study exploring cold plasma treatments of strawberry in modified atmosphere packaging (MAP), the firmness retention was found to be better in a high-oxygen environment (65% O_2, 16% N_2, 19% CO_2) than a high-nitrogen environment (90% N_2, 10% O_2) (Misra et al., 2014b). A nitrogen-component-rich plasma results in a considerable loss of firmness. Insignificant changes in the elastic modulus of apples, cucumbers, tomatoes, and carrots have been reported after cooled microwave air plasma treatment for up to 10 min (2.45 GHz, 1.2 kW power, 20 L/min flow rate) (Baier et al., 2015). However, ionized air treatment has been shown to retain firmness better than ozone treatment in cucumbers (Li et al., 2012). When washing strawberries with plasma-activated water, no deviations in the firmness have been observed even after 4 days of storage (Ma et al., 2015).

A loss of firmness in air plasma-exposed blueberries for treatment times exceeding 60 s has been noticed in recent literature (Lacombe et al., 2015). The softening of the blueberries was attributed to mechanical damage due to the high air-flow rates of the plasma jet, in addition to thermal effects. In a storage study, a good retention of firmness in plasma-treated cherry tomatoes inside sealed plastic packages has been reported for up to 13 days (Misra et al., 2014a). When treating a cut apple with DBD plasma in air, unlike other studies, an increase in the firmness has been reported by Tappi et al. (2014). However, following storage for 24 h, "crunchiness" of the plasma-treated apple slices was confirmed to be lower. A later publication by the same group revealed that DBD air plasma treatment had an insignificant effect on the firmness of celery pieces, even during storage (Berardinelli et al., 2016). Although, not quantified, most studies in literature have confirmed that plasma treatment causes damage to the outermost cells of the produce, especially in cut fruits and vegetables. This is not necessarily reflected in the macroscopic texture and firmness of the produce.

2.3 Surface Features and Microstructure

Following cold plasma treatments with argon gas plasma jet (27.12 MHz, gas flow: 20 sccm), damage to the microstructure of lettuce has been reported (Grzegorzewski et al., 2011). Specifically, wide areas with thick platelets and small-sized granular structures in untreated lettuce leaves (Fig. 2A) have been observed to disappear with

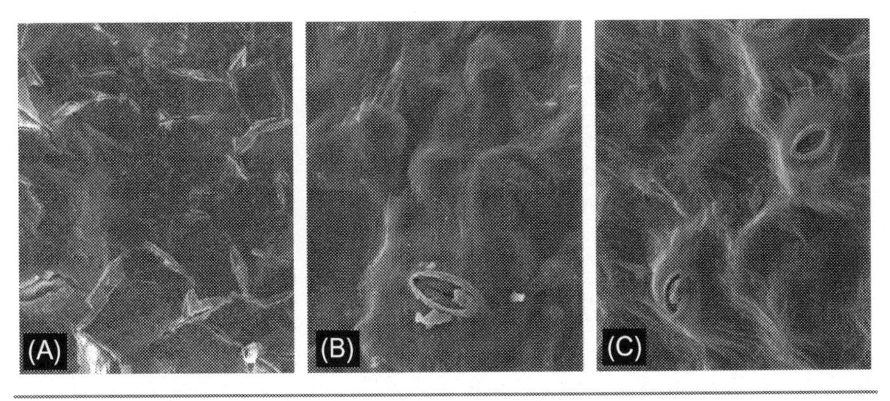

Fig. 2 SEM micrographs of lamb's lettuce (*V. locusta*) leaf adaxial surfaces at 100× magnification: (A) untreated leaf; (B) surface after exposure to argon plasma jet at 20 W for 20 s, and (C) 20 W for 60 s. *(From Grzegorzewski, F., Ehlbeck, J., Schlüter, O., Kroh, L.W., Rohn, S., 2011. Treating lamb's lettuce with a cold plasma—influence of atmospheric pressure Ar plasma immanent species on the phenolic profile of* Valerianella locusta. *LWT Food Sci. Technol. 44, 2285–2289, with permission from Elsevier.)*

an increase in exposure time (Fig. 2B). Desiccation and stiffening of the surface after 60 s of treatment has also been observed (Fig. 2C).

The erosion and rupture of the epidermal tissue layers of lettuce have been attributed to the action of energetic Ar^+ ions and/or reactive oxygen species (ROS) impinging the surface. In another study involving cold plasma treatment of red chicory using a DBD, surface erosion of the leaves caused by oxidation of cell components has been hypothesized (Pasquali et al., 2016). Contrary to this, Bermúdez-Aguirre et al. (2013) observed no changes in the microstructure of lettuce following cold argon plasma treatments with a needle array. For treatments exceeding 60 s, Niemira and team have reported notable damage in air plasma-treated blueberries, particularly a rupturing and bruising of the skin and wilting of the sepals (Lacombe et al., 2015). Diffusion of gases across the fruit boundary and a loss of water vapors from the fruit occur either through aqueous/waxy layers of the epidermis or through gaseous pores (Kader and Saltveit, 2003). Therefore, a destruction of the waxy layer will eventually result in an increased respiration rate and weight loss during storage.

In the context of cold plasma treatment of seeds, the change in the surface features, both chemical and physical, result in a change in the wettability. With radio-frequency plasma in air, a decrease in the apparent contact angle of lentils (*Lens culinaris*), bean (*Phaseolus vulgaris*), and wheat (*Triticum*, species C9) surfaces has been observed (Bormashenko et al., 2012). The modification of food properties using cold plasma could have very interesting consequences with potential applications in the cereal products industry. An increase in the hydrophilicity, an increase in water absorption rate, and a decrease in the cooking time of basmati rice, following radio-frequency cold plasma treatment has been reported (Thirumdas et al., 2015). After exposure to cold plasma, scanning electron microscopy has revealed formation of fissures and depressions on the rice surface. These surface changes are believed to provide routes for faster water absorption and therefore, result in a reduced cooking time.

3 Physiological Activity and Respiration Rate

The respiration of fresh produce involves the oxidation of organic molecules of the cells (such as starch, sugar and organic acids) to simpler molecules (CO_2 and H_2O), and production of energy (ATP), heat and other molecules. The respiration rate of fresh produce is often a good index of the storage life of horticultural products: the lower the rate, the longer the life. The respiration rate is largely related to the physiological activity.

Baier et al. (2013) evaluated the antimicrobial action and associated physiological effects of cold plasma processing on lamb's lettuce. They observed that cold plasma can result in an inhibition of photosynthetic activity at increased power settings. These effects were attributed to the thermal damage due to the nature of plasma source, and to the stress induced by active particles in plasma. Insignificant impact of ionized air treatments (from a microwave plasma source) on the photosynthesis efficiency of apples has been reported, at least for up to 48 h posttreatment (Baier et al., 2015). Cucumbers treated under the same conditions have been found to behave differently; a decrease in activity followed by a recovery, reaching similar levels as that of control samples, has been observed. This pattern of physiological stress followed by recovery has been explained on the basis of superficial cell injury, followed by stiffening, due to moisture loss during storage.

In a study evaluating the effects of cold plasma decontamination on cut apples, the possibility of an alteration of the cellular respiratory pathway was suggested (Tappi et al., 2014). The study revealed a complex response, considering a decrease in O_2 consumption in plasma-treated apples, but not necessarily a decrease in CO_2 evolution. Tappi et al. (2014) suspected the alteration in the cellular respiratory pathway as the cause for changes in respiration rates of cold plasma-treated produce. Insignificant differences among control and in-package air plasma-treated cherry tomatoes have been reported (Misra et al., 2014a). In fact, the respiration rates of tomatoes treated for 60 s and 180 s were found to be lower than that of the control. At the end of a 13-day storage period, the respiration rates appeared to converge to similar values. In another study involving in-package plasma treatment of strawberries, the respiration rate of the treated produce was noted to decrease following a lag period (Misra et al., 2014c). This delay has been attributed to the decreased microbial count. A subsequent study by the same group confirmed that the instantaneous rise in the respiration rate, followed by a considerable drop, also occurs when the plasma is induced in a high-oxygen or high-nitrogen mixture (Misra et al., 2014b). The respiratory gas dynamics of strawberries treated in plasma generated in different gas mixtures is summarized in Fig. 3.

4 Chemical Quality

4.1 Change in pH

Acidification of liquids exposed to the air plasma is consistently reported in literature. The change in pH and acidity of plasma-treated produce is closely related to the dynamics of plasma chemistry. Via liquid chromatography on samples of the water cathode, Chen et al. (2008) have confirmed that the concentration of NO_3^-, which originates from the acid HNO_3 in the discharge, increased with the plasma exposure resulting in acidification of the liquid. When fresh fruits and vegetables are treated with plasma, the chemical species in gas phase react with liquid water present on the food surface as a thin film to form the acids.

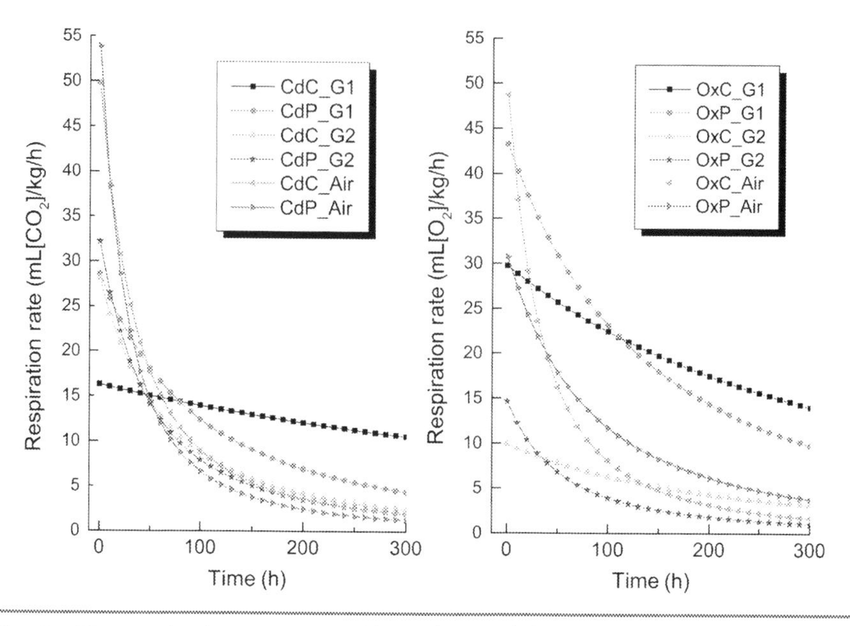

Fig. 3 Changes in the gas composition of control (C) and 5 min in-package DBD plasma-treated (P) strawberries in air, G1 (65% O_2+16% N_2+19% CO_2) and G2 (90% N_2+10% O_2) gas mixtures; Cd refers to carbon dioxide and Ox refers to oxygen (Misra et al., 2014b,c).

Contrary to most reports and theories, no significant change has been observed in the pH of strawberries treated with plasma-activated water (PAW, 98% Ar+2% O_2) (Ma et al., 2015). This is most likely due to the absence of nitrogen, which is primarily responsible for a pH drop. No significant change in the pH of DBD air plasma-treated orange juice (at 60 kHz, 30 kV) has been reported (Shi et al., 2011). Here, the buffering action of the juice is suspected to counteract the plasma-liquid chemistry. An insignificant change in the pH of in-package DBD air plasma-treated (at 50 Hz, 40 kV) cherry tomatoes (relative to control) after a storage period of 13 days has also been reported (Misra et al., 2014a). All of these reports allow to conclude that the effects of plasma on the pH of complex food matrices is counteracted by several factors including: buffering action, physiological activity of the living tissues, and the possibility of the liquid emanating from the damaged tissues on the surface washing off the acids on the surface.

4.2 Proteins and Enzyme Activity

A rapid inactivation of peroxidase (POD) in crude tomato extract was first demonstrated by Pankaj et al. (2013). The authors reported that the treatment time and voltage, both factors had significant effects on the inactivation, which followed sigmoidal decay kinetics. Surowsky et al. (2013) reported the inactivation of POD as well as polyphenol oxidase (PPO) in model solutions using a cold plasma jet with different gases. The circular dichroism spectroscopy confirmed a decrease in the amount of α-helices in both enzymes, which strongly correlated with the loss of activity. The changes in fluorescence emission in enzymes with plasma treatments also confirmed the structural modifications. Via FTIR spectroscopy of DBD plasma-treated strawberries, the possibility of changes in the constituent proteins of the strawberry itself has also been put forth (Misra et al., 2015b). Tappi et al. (2014) observed that considerable inactivation of PPO was achieved in cut apples upon their treatment with air plasma for 20 and 30 min, but not 10 min (see Fig. 4).

The inactivation of horseradish peroxidase (HRP) induced by cold plasma discharge at the Ar gas-solution interface has been demonstrated (Ke and Huang, 2013). It was observed through spectroscopic methods that the hydrogen peroxide and UV light attacked the heme group to yield fluorescent products, while hydroxyl radicals contributed to the destruction of HRP and increased the rate of degradation. Contrary to the effect of plasma on most enzymes, lipase activity has been observed to increase after plasma treatment for 1 min (Li et al., 2011). This effect has been

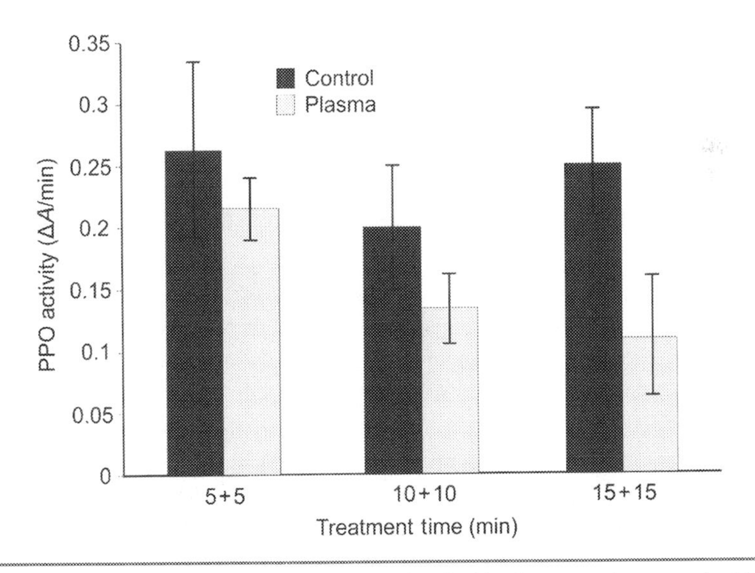

Fig. 4 Polyphenol oxidase (PPO) activity (ΔA/min) of fresh-cut apples treated with DBD plasma compared to untreated control samples. The times ($x+x$) indicate specified treatment duration (in min) for each side of the cut surface. *(Data from Tappi, S., Berardinelli, A., Ragni, L., Dalla Rosa, M., Guarnieri, A., Rocculi, P., 2014. Atmospheric gas plasma treatment of fresh-cut apples. Innovative Food Sci. Emerg. Technol. 21, 114–122.)*

attributed to the action of plasma directly on the enzyme, rather than any long-lived chemical species in the buffer. A tryptophan-mediated reaction was confirmed using fluorescence microscopy.

Deng et al. (2007) suggested that the main agents of a helium/oxygen mixture plasma responsible for the activity loss of proteins are excited atomic oxygen and excited nitrogen species. In 2000, a study evaluating the reaction of amino acids with ozone concluded that the susceptibility to oxidation under aqueous ozonation conditions are, in order of reactivity: Met > Trp > Tyr > His (Kotiaho et al., 2000). Considering that nitric oxide treatments lead to an inactivation of enzymes in fruits (Deng et al., 2013), and the generation of considerable amounts of nitrogen oxides (NOx) in air plasma (Moiseev et al., 2014), the plasma-led inactivation of enzymes can partly be summed as the action of not only ROS like ozone or singlet oxygen, but also NOx. A change in the secondary structure of proteins following exposure to plasma is now well-known (Segat et al., 2014; Surowsky et al., 2013).

Misra et al. (2015a) reported that the secondary structure of gluten becomes more stable in weak wheat flour subjected to an atmospheric pressure cold plasma treatment in air. Via infrared spectroscopy, they revealed that an increase in DBD plasma treatment beyond 60 kV for 5 min increased the parallel and antiparallel β-sheets, indicating the loss of an orderly structure of the proteins. As a consequence, an improvement in the dough strength and optimum mixing time in strong as well as weak wheat flours was also observed in mixographic studies.

4.3 Antioxidant Activity

Several reports in literature have validated that cold plasma treatment does not significantly affect the antioxidant activity in foods. Recent studies on the cold plasma treatment of red chicory (*Cichorium intybus* L.) and fresh-cut kiwifruit using a DBD operating in air revealed no noticeable changes in the antioxidant activities of extracts for treatments up to 30 and 20 min, respectively (Pasquali et al., 2016; Ramazzina et al., 2015). The antioxidant activities in these studies have been assessed on the basis of 3-ethyl-benzothiazoline-6-sulfonic acid (ABTS) radical scavenging activity, oxygen radical absorbance capacity (ORAC), 2,2-diphenyl-1-picrylhydrazyl (DPPH) scavenging activity, and ferric reducing ability of plasma (FRAP) assay. An insignificant effect of corona discharge plasma operating in air on antioxidant activity of dried laver, an edible red algae seaweed, has been confirmed (Kim et al., 2015). Using a microwave plasma source operating in N_2, (N_2+O_2), He, and (He$+O_2$), Song et al. (2015) observed insignificant changes in the antioxidant activity of plasma-treated lettuce.

The fact that the antioxidant activity of foods remains unaffected by plasma treatments is actually counterintuitive. This is because one can expect the antioxidants in food to react with free radicals. However, the extent of such a reaction will be governed by the concentration of the plasma species, recombination rates, the antioxidant versus radical reaction rates, and the diffusivity of the plasma radicals into the food matrix.

4.4 Starch

Being a semicrystalline system, the starch granule consists of crystalline and amorphous regions. Whether the crystalline structure of starch can be described as either A, B, or C-type depends on the packing of the amylopectin side chain into double helices (Zhang et al., 2015). The effective modification of starch on exposure to oxygen glow plasma, and the multiscale characterization of the oxidized starches, is reported in literature. Using X-ray diffraction, Zhang et al. (2013) have observed that oxygen plasma treatment results in a higher degree of destruction to the amorphous materials (molecular and supramolecular structures) in corn and potato starch, and relatively insignificant changes in the crystalline structure. Oxygen plasma has been shown to act on the C6 location of wheat starch, through modification of hydroxyl group to carbonyl group (Khorram et al., 2015). Relatively insignificant effects of low-pressure plasma treatment on the crystalline structure of rice have been reported (Thirumdas et al., 2015). An alteration of the starch structures at multiple scales—molecular, mesoscopic, and macroscopic—inevitably reflects in its functional properties related to gelatinization, thickening, and gelling (Zhang et al., 2015). Through an extensive set of experiments and building upon the principles of gelatinization, Zhang et al. (2015) proposed a structure-functionality relationship for (N_2 or He) glow plasma-treated potato starch (see Fig. 5). According to this model, the functionality changes have been explained as follows:

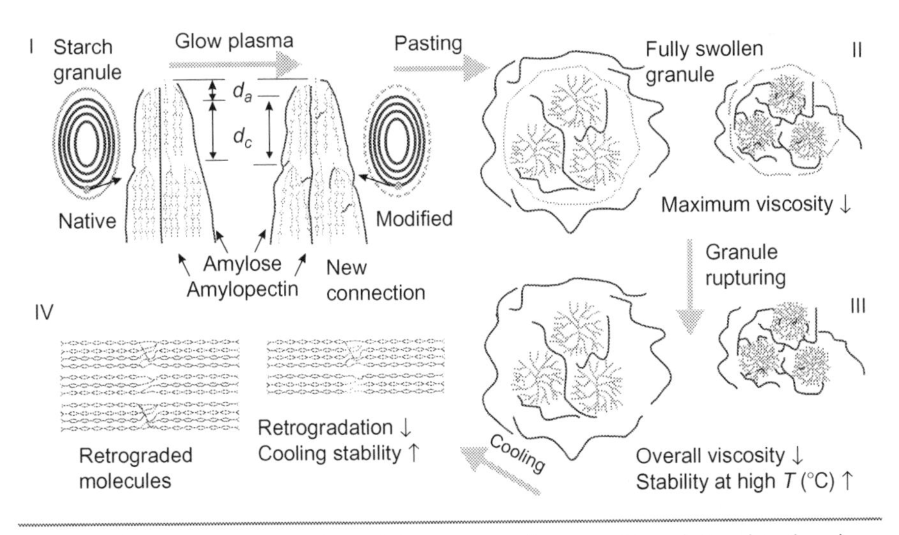

Fig. 5 Schematic representation of the structure-functionality relationship for glow-plasma treated starch. *(Based on Zhang, B., Chen, L., Li, X., Li, L., Zhang, H., 2015. Understanding the multi-scale structure and functional properties of starch modulated by glow-plasma: a structure-functionality relationship. Food Hydrocoll. 50, 228–236, with permission from Elsevier.)*

(I) The polymerization/crosslinking induced by glow plasma makes starch molecules more branched and networked;
(II) The crosslinking hinders the stretchability of the molecules during granule swelling, which results in a decrease in the maximum viscosity of the starch paste;
(III) The limited stretch of the chain, even after the granule rupturing, results in overall viscosity lowering, and an enhanced high-temperature paste stability;
(IV) As a net effect, the possible rearrangement during cooling is limited, which causes weaker retrogradation and improved paste-cooling stability.

4.5 Lipids

The action of radicals on lipids is well-known to induce oxidation and formation of primary and secondary oxidation products, which are responsible for off-odors. However, in the context of plasma, while this issue was pointed out very early (Misra et al., 2011), not much experimental work has been carried out. The accelerated oxidation of lipids using a plasma source has been demonstrated only recently (Vandamme et al., 2015), although, not in the context of plant lipids, but for fish oils. Also, cold plasma has been leveraged for its rapid esterification of waste frying oils to produce biodiesel (Cubas et al., 2015). Fortunately, fruits and vegetable being poor sources of fats and oils, the plasma chemical species-led oxidation does not appear to be a significant issue of concern in plant foods. However, lipid oxidation could be problematic when treating grains and flours.

4.6 Ascorbic Acid

In one of the foremost studies, Shi et al. (2011) reported that following DBD plasma exposure, vitamin C content decreased. However, the authors also stated that compared to the control, the reduction was insignificant. Using a microwave plasma source operating in N_2, N_2+O_2, He, and He$+O_2$, Song et al. (2015) observed insignificant changes in the ascorbic acid content of plasma-treated lettuce, irrespective of the applied power and storage temperature.

Up to a 4% loss of vitamin C has been noticed in a study involving plasma treatment of cut carrots, cucumbers, and pears (Wang et al., 2012). DBD plasma treatments at 60 and 80 kV applied voltage have been found to decrease the ascorbic acid content in whole strawberries (Misra et al., 2015b). The observed loss of ascorbic acid during processing has been attributed to the action of ozone and other oxidizing species of plasma (Misra et al., 2015b; Wang et al., 2012). It has been speculated that ascorbic acid degrades either by direct attack of ozone, described by the Criegee mechanism, to produce ozonide or by indirect reaction because of secondary oxidants, such as singlet oxygen and excited molecular oxygen (Enami et al., 2008; Kanofsky and Sima, 1991).

4.7 Pigments and Phenolic Constituents

Chlorophyll is the green pigment found in the leafy parts of plants. When treating leafy vegetables and green fruits, a certain degree of oxidation of pigments due to free radicals has been noted by researchers. In the case of fresh-cut kiwi fruits, a considerable decrease in the chlorophyll content during storage occurs. However, the extent of pigment degradation in DBD plasma-treated cut kiwi fruits has been observed to be lower compared to untreated samples (Ramazzina et al., 2015). The decrease in degradation is suspected to result from the (partial) inactivation of the chlorophyllase enzyme in the matrix.

Carotenoids are important for the orange-yellow and red appearance of fruits and vegetables. Information about the fate of carotenoids during plasma processing has not received much attention. A decrease in the carotenoids content of cut kiwi fruit upon subjecting to DBD plasma treatment in air was reported in 2015 (Ramazzina et al., 2015). The oxidation of carotenoids on the surface of carrots has also been hypothesized (Wang et al., 2012); however, this needs experimental validation.

Anthocyanins are polyphenolic flavonoids, often responsible for the red to blue color of many fruits and vegetables, and are localized in the cell vacuole. In one of the foremost studies, pure flavonoid compounds and those in lamb's lettuce have been observed to follow a time-dependent degradation upon oxygen plasma treatment (Grzegorzewski et al., 2011). Later, it was observed that cold plasma treatments at 60 and 80 kV for up to 5 min had no noticeable effects on the anthocyanin content of whole strawberries (Misra et al., 2015b). However, in the same study the phenolic acids have been noted to remain unchanged after plasma treatment. In a 2016 study, an increase in anthocyanin content in pomegranate juice by 21–35% upon treatment with an Ar plasma jet has been reported (Bursać Kovačević et al., 2016). The increase has been hypothesized to result from an increase in the extraction efficacy, chemical structure modification, and possible copigmentation of anthocyanins after plasma treatments. This group has also shown that pasteurization (80°C for 2 min) and Ar cold plasma treatment (jet configuration, 2.5 kV at 25 kHz frequency) result in an increase in total phenolic content of pomegranate juice by 29.55% and 33.03%, respectively (Herceg et al., 2016). An increase in anthocyanin content by 34% and phenolic acid content by 15% in sour cherry Marasca (*Prunus cerasus* var. Marasca) juice has also been reported (Elez Garofulić et al., 2015). The plasma treatment time and sample volume have been identified as the major factors influencing the anthocyanins and phenolic acids in sour cherry juice. A significant reduction in the total anthocyanin content of blueberries (as cyanidin-3-glucoside equivalent) after 90 s of dry air plasma exposure has also been reported (Lacombe et al., 2015). Considering that the posttreatment temperature in this study exceeded 38°C, the degradation was attributed to thermal effects; nevertheless, the possible action of oxidizing species (eg, ozone and other radicals) has also been admitted.

In a distinct study, cold plasma in nitrogen from a DBD operating at 30 kV has been explored for possible benefits during the withering of green tea leaves (Kuloba et al., 2014). Contrary to the initial hypothesis, the research group observed that the polyphenol content decreased with increasing withering time, which was indeed faster, when withering was carried out in a N_2 plasma environment.

4.8 Volatile Constituents

The effect of cold plasma on the volatile constituents is an interesting aspect which has not received much attention. The ability of plasma to destroy volatile compounds was reported by Chen et al. (2010), who observed up to an 80% decrease in the levels of dimethylamine in 5 min following treatments with a corona system at 40 kV. Ma and Lan (2015) were the first to report the effects of plasma treatment on flavor constituents of a food system. They evaluated the effects of plasma treatment (at 10 kV for 5 min, 30°C) and heat processing (85°C, 30 min) on the volatile constituents of tomato juice. After plasma treatment, an increase in *trans*-2-hexenal, *o*-xylene, L-threitol, and glycerol in tomato juice was noticed by the authors (see Fig. 6).

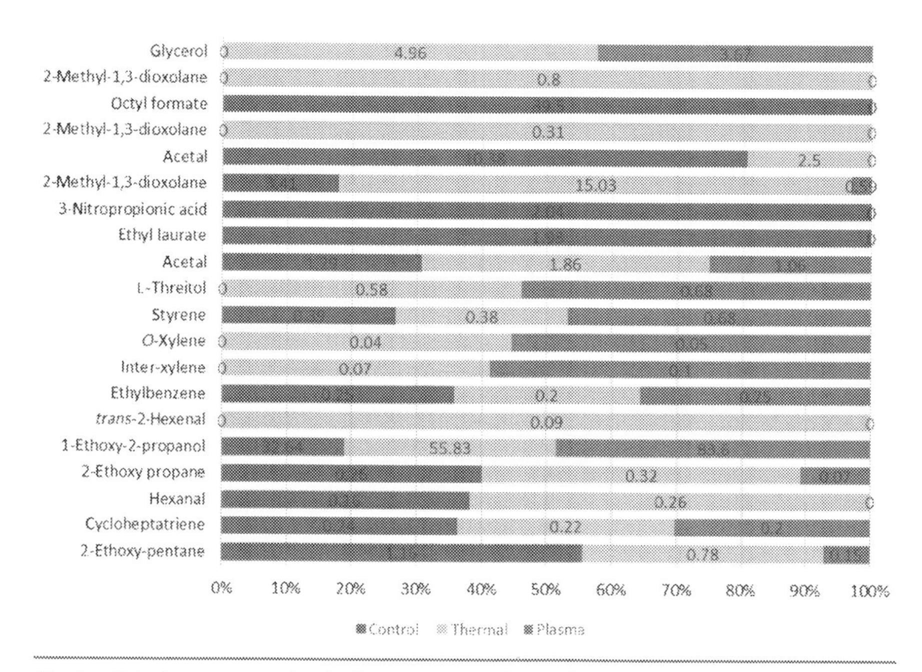

Fig. 6 Relative content of volatiles in control, thermally processed, and plasma-treated tomato juice. *(Data from Ma, T.J., Lan, W.S., 2015. Effects of non-thermal plasma sterilization on volatile components of tomato juice. Int. J. Environ. Sci. Technol. 12, 3767–3772.)*

An increase in the contents of *trans*-2-hexenal and *n*-hexanal in plasma-treated tomato juice compared to heat processing was also observed.

In another study, a loss of hexyl acetate, aldehyde, 2-hexenal, γ-decanolactone (5-hexyl-dihydro-3H-furan-2-one), and mesifuran (2,5-dimethyl-4-hydroxy-2H-furan-3-one) in cold plasma-treated strawberries and a significant increase in linalool have been noticed (Misra, 2014). These treatments were carried out at 60 and 80 kV for 1 and 5 min. Such a loss of volatile constituents is undesirable. It is suspected that the plasma species, viz. ozone, nitric oxides, and hydroxyl radicals, generated by the reaction of ozone with water (Hong et al., 2013), react with the fruit volatiles, leading to their destruction. In addition, the inactivation of enzymes could also be associated with flavor changes, considering their involvement in flavor biosynthesis and metabolism pathways. Flavor changes in plant foods resulting from effects on enzyme activity have been confirmed for high pressure (Oey et al., 2008) and pulsed electric-field processing (Aguiló-Aguayo et al., 2009). Although the inactivation of few enzymes under the influence of plasma has been reported (discussed earlier), a confirmation of enzyme-mediated volatile dynamics is still pending. Another aspect worth exploring is the change in volatile profile during storage, where the possibility of recovery of the perceived aroma remains a question.

5 Sensory Attributes

For evaluating any novel food technology, it is necessary to carry out sensory analysis, in addition to the pure chemical and mechanical studies. A significant loss of freshness, color, odor, texture, and overall acceptability in plasma-treated red chicory leaves during storage at 4°C after 3 days has been reported (Pasquali et al., 2016). Such negative effects on organoleptic attributes have been attributed to plausible anaerobic conditions in vegetable tissues.

For microwave plasma-treated apple slices, no perceivable changes in the texture and appearance have been documented (Schnabel et al., 2015) during sensory studies. However, a "sour" or "chemical" type of odor has been associated with plasma-treated apple slices during the descriptive sensory analysis.

Products prepared out of plasma-decontaminated dried seaweeds were reported to have no change in the sensory attributes, namely in appearance, color, flavor, taste, and texture (Kim et al., 2015). In treating lettuce with a microwave plasma source, no significant changes have been recorded in the sensory attributes at 400 and 900 W of input power (Song et al., 2015). Specifically, the study evaluated the overall appearance and the flavor of lettuce. Looking at the spectrum of studies that have been conducted so far, the lack of sensory studies over the course of a storage period is clear. Moreover, a descriptive sensory analysis could help in identifying if any changes due to plasma in the flavor constituents and lipids, etc., are perceivable or not.

6 Conclusions

Until now, research has largely focused on evaluation of the general appearance of cold plasma-treated foods, their organoleptic characteristics, and physicochemical properties. The stability of sensitive food components, such as vitamins and other bioactive constituents, needs more focus before firm conclusions about the benefits of plasma technology can be drawn. Optimization studies need to be undertaken to reduce or remove the negative effect on quality, as evidenced during the storage. It is well recognized that unravelling the mechanisms behind the changes in the chemical constituents of plasma-treated foods of plant origin is difficult. This is because of the inherent nature of nonequilibrium plasma, which involves a multitude of physical and chemical phenomena occurring at various scales in time and space. Many unknowns are still unresolved, which range from the interaction mechanisms by which ionized gases modify plant tissues, their organization, and the constituting biomacromolecules, to the in-depth advanced physical and chemical processes of the plasma sources used for such applications.

References

Aguiló-Aguayo, I., Oms-Oliu, G., Soliva-Fortuny, R., Martín-Belloso, O., 2009. Flavour retention and related enzyme activities during storage of strawberry juices processed by high-intensity pulsed electric fields or heat. Food Chem. 116, 59–65.

Almeida, F.D.L., Cavalcante, R.S., Cullen, P.J., Frias, J.M., Bourke, P., Fernandes, F.A.N., Rodrigues, S., 2015. Effects of atmospheric cold plasma and ozone on prebiotic orange juice. Innovative Food Sci. Emerg. Technol. 32, 127–135.

Baier, M., Foerster, J., Schnabel, U., Knorr, D., Ehlbeck, J., Herppich, W.B., Schlüter, O., 2013. Direct non-thermal plasma treatment for the sanitation of fresh corn salad leaves: evaluation of physical and physiological effects and antimicrobial efficacy. Postharvest Biol. Technol. 84, 81–87.

Baier, M., Ehlbeck, J., Knorr, D., Herppich, W.B., Schlüter, O., 2015. Impact of plasma processed air (PPA) on quality parameters of fresh produce. Postharvest Biol. Technol. 100, 120–126.

Berardinelli, A., Pasquali, F., Cevoli, C., Trevisani, M., Ragni, L., Mancusi, R., Manfreda, G., 2016. Sanitisation of fresh-cut celery and radicchio by gas plasma treatments in water medium. Postharvest Biol. Technol. 111, 297–304.

Bermúdez-Aguirre, D., Wemlinger, E., Pedrow, P., Barbosa-Cánovas, G., Garcia-Perez, M., 2013. Effect of atmospheric pressure cold plasma (APCP) on the inactivation of *Escherichia coli* in fresh produce. Food Control 34, 149–157.

Bormashenko, E., Grynyov, R., Bormashenko, Y., Drori, E., 2012. Cold radiofrequency plasma treatment modifies wettability and germination speed of plant seeds. Sci. Rep. 2, 741.

Bursać Kovačević, D., Putnik, P., Dragović-Uzelac, V., Pedisić, S., Režek Jambrak, A., Herceg, Z., 2016. Effects of cold atmospheric gas phase plasma on anthocyanins and color in pomegranate juice. Food Chem. 190, 317–323.

Chen, Q., Saito, K., Shirai, H., 2008. Atmospheric pressure plasma using electrolyte solution as cathode. In: The International Interdisciplinary-Symposium on Gaseous and Liquid Plasmas, Sendai, Japan.

Chen, J., Yang, J., Pan, H., Su, Q., Liu, Y., Shi, Y., 2010. Abatement of malodorants from pesticide factory in dielectric barrier discharges. J. Hazard. Mater. 177, 908–913.

Cubas, A.L., Machado, M.M., Pinto, C.R., Moecke, E.H., Dutra, A.R., 2015. Biodiesel production using fatty acids from food industry waste using corona discharge plasma technology. Waste Manag. 47 (Pt A), 149–154.

Deng, X., Shi, J., Kong, M., 2007. Protein destruction by a helium atmospheric pressure glow discharge: capability and mechanisms. J. Appl. Phys. 101, 074701.

Deng, L., Pan, X., Chen, L., Shen, L., Sheng, J., 2013. Effects of preharvest nitric oxide treatment on ethylene biosynthesis and soluble sugars metabolism in 'Golden Delicious' apples. Postharvest Biol. Technol. 84, 9–15.

Elez Garofulić, I., Režek Jambrak, A., Milošević, S., Dragović-Uzelac, V., Zorić, Z., Herceg, Z., 2015. The effect of gas phase plasma treatment on the anthocyanin and phenolic acid content of sour cherry Marasca (*Prunus cerasus* var. Marasca) juice. LWT Food Sci. Technol. 62, 894–900.

Enami, S., Hoffmann, M.R., Colussi, A.J., 2008. Acidity enhances the formation of a persistent ozonide at aqueous ascorbate/ozone gas interfaces. Proc. Natl. Acad. Sci. 105, 7365–7369.

Grzegorzewski, F., Ehlbeck, J., Schlüter, O., Kroh, L.W., Rohn, S., 2011. Treating lamb's lettuce with a cold plasma—influence of atmospheric pressure Ar plasma immanent species on the phenolic profile of *Valerianella locusta*. LWT Food Sci. Technol. 44, 2285–2289.

Herceg, Z., Kovačević, D.B., Kljusurić, J.G., Jambrak, A.R., Zorić, Z., Dragović-Uzelac, V., 2016. Gas phase plasma impact on phenolic compounds in pomegranate juice. Food Chem. 190, 665–672.

Hertwig, C., Reineke, K., Ehlbeck, J., Knorr, D., Schlüter, O., 2015. Decontamination of whole black pepper using different cold atmospheric pressure plasma applications. Food Control 55, 221–229.

Hong, S.-M., Min, Z.W., Mok, C., Kwon, H.-y., Kim, T.-k., Kim, D.-h., 2013. Aqueous degradation of imidacloprid and fenothiocarb using contact glow discharge electrolysis: degradation behavior and kinetics. Food Sci. Biotechnol. 22, 1773–1778.

Kader, A.A., Saltveit, M.E., 2003. Respiration and gas exchange. In: Bartz, J.A., Brecht, J.K. (Eds.), Postharvest Physiology and Pathology of Vegetables. Marcel Dekker, Inc., New York, pp. 229–246.

Kanofsky, J.R., Sima, P., 1991. Singlet oxygen production from the reactions of ozone with biological molecules. J. Biol. Chem. 266, 9039–9042.

Ke, Z., Huang, Q., 2013. Inactivation and heme degradation of horseradish peroxidase induced by discharge plasma. Plasma Process. Polym. 10, 731–739.

Khorram, S., Zakerhamidi, M.S., Karimzadeh, Z., 2015. Polarity functions' characterization and the mechanism of starch modification by DC glow discharge plasma. Carbohydr. Polym. 127, 72–78.

Kim, J.-W., Puligundla, P., Mok, C., 2015. Microbial decontamination of dried laver using corona discharge plasma jet (CDPJ). J. Food Eng. 161, 24–32.

Kotiaho, T., Eberlin, M.N., Vainiotalo, P., Kostiainen, R., 2000. Electrospray mass and tandem mass spectrometry identification of ozone oxidation products of amino acids and small peptides. J. Am. Soc. Mass Spectrom. 11, 526–535.

Kuloba, P.W., Gumbe, L.O., Okoth, M.W., Obanda, M., Ng'ang'a, F.M., 2014. An investigation into low-temperature nitrogen plasma environment effect on the content of polyphenols during withering in made Kenyan tea. Int. J. Food Sci. Technol. 49, 1020–1026.

Lacombe, A., Niemira, B.A., Gurtler, J.B., Fan, X., Sites, J., Boyd, G., Chen, H., 2015. Atmospheric cold plasma inactivation of aerobic microorganisms on blueberries and effects on quality attributes. Food Microbiol. 46, 479–484.

Li, H.-P., Wang, L.-Y., Li, G., Jin, L.-H., Le, P.-S., Zhao, H.-X., Xing, X.-H., Bao, C.-Y., 2011. Manipulation of lipase activity by the helium radio-frequency, atmospheric-pressure glow discharge plasma jet. Plasma Process. Polym. 8, 224–229.

Li, J., Yan, S., Wang, Q., Li, Y., Wang, Q., 2012. Effects of ionized air treatments on postharvest physiology and quality of fresh cucumber. J. Food Process. Preserv. 38, 271–277.

Ma, T.J., Lan, W.S., 2015. Effects of non-thermal plasma sterilization on volatile components of tomato juice. Int. J. Environ. Sci. Technol. 12, 3767–3772.

Ma, R., Wang, G., Tian, Y., Wang, K., Zhang, J., Fang, J., 2015. Non-thermal plasma-activated water inactivation of food-borne pathogen on fresh produce. J. Hazard. Mater. 300, 643–651.

Misra, N.N., 2014. Effect of Nonthermal Plasma on Quality of Fresh Produce (Ph.D. Thesis). Dublin Institute of Technology, Dublin, Ireland.

Misra, N.N., Tiwari, B.K., Raghavarao, K.S.M.S., Cullen, P.J., 2011. Nonthermal plasma inactivation of food-borne pathogens. Food Eng. Rev. 3, 159–170.

Misra, N.N., Keener, K.M., Bourke, P., Mosnier, J.P., Cullen, P.J., 2014a. In-package atmospheric pressure cold plasma treatment of cherry tomatoes. J. Biosci. Bioeng. 118, 177–182.

Misra, N.N., Moiseev, T., Patil, S., Pankaj, S.K., Bourke, P., Mosnier, J.P., Keener, K.M., Cullen, P.J., 2014b. Cold plasma in modified atmospheres for post-harvest treatment of strawberries. Food Bioprocess Technol. 7, 3045–3054.

Misra, N.N., Patil, S., Moiseev, T., Bourke, P., Mosnier, J.P., Keener, K.M., Cullen, P.J., 2014c. In-package atmospheric pressure cold plasma treatment of strawberries. J. Food Eng. 125, 131–138.

Misra, N.N., Kaur, S., Tiwari, B.K., Kaur, A., Singh, N., Cullen, P.J., 2015a. Atmospheric pressure cold plasma (ACP) treatment of wheat flour. Food Hydrocoll. 44, 115–121.

Misra, N.N., Pankaj, S.K., Frias, J.M., Keener, K.M., Cullen, P.J., 2015b. The effects of nonthermal plasma on chemical quality of strawberries. Postharvest Biol. Technol. 110, 197–202.

Moiseev, T., Misra, N.N., Patil, S., Cullen, P.J., Bourke, P., Keener, K.M., Mosnier, J.P., 2014. Post-discharge gas composition of a large-gap DBD in humid air by UV–Vis absorption spectroscopy. Plasma Sources Sci. Technol. 23, 065033.

Oey, I., Lille, M., Van Loey, A., Hendrickx, M., 2008. Effect of high-pressure processing on colour, texture and flavour of fruit- and vegetable-based food products: a review. Trends Food Sci. Technol. 19, 320–328.

Oude Ophuis, P.A.M., Van Trijp, H.C.M., 1995. Perceived quality: a market driven and consumer oriented approach. Food Qual. Prefer. 6, 177–183.

Pankaj, S.K., Misra, N.N., Cullen, P.J., 2013. Kinetics of tomato peroxidase inactivation by atmospheric pressure cold plasma based on dielectric barrier discharge. Innovative Food Sci. Emerg. Technol. 19, 153–157.

Pasquali, F., Stratakos, A.C., Koidis, A., Berardinelli, A., Cevoli, C., Ragni, L., Mancusi, R., Manfreda, G., Trevisani, M., 2016. Atmospheric cold plasma process for vegetable leaf decontamination: a feasibility study on radicchio (red chicory, *Cichorium intybus* L.). Food Control 60, 552–559.

Ramazzina, I., Berardinelli, A., Rizzi, F., Tappi, S., Ragni, L., Sacchetti, G., Rocculi, P., 2015. Effect of cold plasma treatment on physico-chemical parameters and antioxidant activity of minimally processed kiwifruit. Postharvest Biol. Technol. 107, 55–65.

Schnabel, U., Niquet, R., Schlüter, O., Gniffke, H., Ehlbeck, J., 2015. Decontamination and sensory properties of microbiologically contaminated fresh fruits and vegetables by microwave plasma processed air (PPA). J. Food Process. Preserv. 39, 653–662.

Segat, A., Misra, N.N., Fabbro, A., Buchini, F., Lippe, G., Cullen, P.J., Innocente, N., 2014. Effects of ozone processing on chemical, structural and functional properties of whey protein isolate. Food Res. Int. 66, 365–372.

Shi, X.-M., Zhang, G.-J., Wu, X.-L., Li, Y.-X., Ma, Y., Shao, X.-J., 2011. Effect of low-temperature plasma on microorganism inactivation and quality of freshly squeezed orange juice. IEEE Trans. Plasma Sci. 39, 1591–1597.

Song, A.Y., Oh, Y.J., Kim, J.E., Song, K.B., Oh, D.H., Min, S.C., 2015. Cold plasma treatment for microbial safety and preservation of fresh lettuce. Food Sci. Biotechnol. 24, 1717–1724.

Surowsky, B., Fischer, A., Schlueter, O., Knorr, D., 2013. Cold plasma effects on enzyme activity in a model food system. Innovative Food Sci. Emerg. Technol. 19, 146–152.

Tappi, S., Berardinelli, A., Ragni, L., Dalla Rosa, M., Guarnieri, A., Rocculi, P., 2014. Atmospheric gas plasma treatment of fresh-cut apples. Innovative Food Sci. Emerg. Technol. 21, 114–122.

Thirumdas, R., Deshmukh, R.R., Annapure, U.S., 2015. Effect of low temperature plasma processing on physicochemical properties and cooking quality of basmati rice. Innovative Food Sci. Emerg. Technol. 31, 83–90.

Vandamme, J., Nikiforov, A., Dujardin, K., Leys, C., De Cooman, L., Van Durme, J., 2015. Critical evaluation of non-thermal plasma as an innovative accelerated lipid oxidation technique in fish oil. Food Res. Int. 72, 115–125.

Wang, R.X., Nian, W.F., Wu, H.Y., Feng, H.Q., Zhang, K., Zhang, J., Zhu, W.D., Becker, K.H., Fang, J., 2012. Atmospheric-pressure cold plasma treatment of contaminated fresh fruit and vegetable slices: inactivation and physiochemical properties evaluation. Eur. Phys. J. D 66, 1–7.

Zhang, B., Xiong, S., Li, X., Li, L., Xie, F., Chen, L., 2013. Effect of oxygen glow plasma on supramolecular and molecular structures of starch and related mechanism. Food Hydrocoll. 37, 69–76.

Zhang, B., Chen, L., Li, X., Li, L., Zhang, H., 2015. Understanding the multi-scale structure and functional properties of starch modulated by glow-plasma: a structure-functionality relationship. Food Hydrocoll. 50, 228–236.

Chapter 11

Quality of Cold Plasma Treated Foods of Animal Origin

H.-J. Kim*, D.D. Jayasena†, H.I. Yong‡, C. Jo‡
*National Institute of Crop Science, RDA, Suwon, Republic of Korea, †Uva Wellassa University, Badulla, Sri Lanka, ‡Seoul National University, Seoul, Republic of Korea

1 Introduction

Diets have changed significantly over the last 50 years, with a global trend toward higher intake of foods and food products of animal origin such as meat, milk, eggs, and fish. This category of foods is an excellent source of high-quality protein and micronutrients essential for normal development and good health (Smith et al., 2013). In spite of the high consumption of these foods and their above-mentioned benefits, serious concerns regarding the safety of foods of animal origin and their products continue to rise, both in developed and developing countries.

The most serious safety issues resulting in consumer health problems and economical losses in most countries are associated with foodborne diseases. For instance, in the United States, 76 million cases of foodborne diseases occur annually, resulting in 325,000 hospitalizations and 5000 deaths (Mead et al., 1999; Tauxe et al.,

Cold Plasma in Food and Agriculture. http://dx.doi.org/10.1016/B978-0-12-801365-6.00011-1

2010); this leads to high medical costs and loss in productivity worth 6.6–37.1 billion USD (Tauxe et al., 2010). Furthermore, it is reported that the incidence of foodborne illness caused by *Listeria monocytogenes* has increased in many European countries (EFSA, 2007a; Jensen et al., 2010; Mook et al., 2011). Contaminated ready-to-eat meat and seafood have been recognized as high-risk products (EFSA, 2007b; USDA-FSIS, 2003, 2010). The major foodborne pathogens identified include *Salmonella, Escherichia coli* O157:H7, shigatoxigenic *E. coli, L. monocytogenes*, and *Shigella* spp. (Sivapalasingam et al., 2004). In addition, the Gram-positive *L. monocytogenes* and *Staphylococcus aureus* are pathogens of concern in ready-to-eat meat, dairy, and poultry products in which the growth of these organisms may occur during storage (Lozano et al., 2009; Sofos, 2008). Therefore, implementation of an effective protection system against biochemical and microbial deterioration that may occur during preparation, storage, and distribution is necessary to ensure the safety of products for consumption (Fratianni et al., 2010; Lucera et al., 2012).

Several technologies, including conventional thermal treatments, have been used to control microbial deterioration in foods and food products of animal origin. For instance, thermal sterilization is traditionally employed to effectively inactivate pathogens. However, there are several issues related to the quality of thermally processed foods due to their negative impacts on nutritive value, functional properties, and sensory qualities (Jayasena et al., 2015; Kim et al., 2013a). This has led to the exploration of novel nonthermal technologies including chemical treatment, ozonation, UV treatment, ultrasound, irradiation, and high-pressure processing. However, these technologies also have several disadvantages, such as a high cost of application, their requirement for specialized equipment, their extended processing times, the generation of undesirable residues, and low efficiencies (Kim et al., 2011, 2013a; Misra et al., 2011; Yun et al., 2010). Plasma treatment is an emerging nonthermal technology with high potential for application to foods, including those of animal origin.

Various types of plasma sources have been tested for their antimicrobial efficacy in foods of animal origin, such as pork, chicken, eggs, fish, cheese, ham, beef jerky, and bacon (Table 1). The results obtained indicate that plasma treatment is a potential antimicrobial technique for the inactivation of foodborne pathogens in the aforementioned foods. However, an understanding of the effect of plasma treatment on the quality of foods of animal origin is still very limited and further investigations are highly desirable. In this chapter, a summary of the quality aspects of cold plasma treated foods of animal origin and their products is provided.

TABLE 1 Applications of Nonthermal Plasma Treatment as a Sterilization Method for Foods and Food Products of Animal Origin

Treated Product	Microorganisms Tested	Plasma Source/ Carrier Gas	References
Beef loin and pork butt	*Escherichia coli* O157:H7 *Listeria monocytogenes* *Salmonella* Typhimurium	Flexible thin-layer dielectric barrier discharge (DBD) plasma/atmospheric air	Jayasena et al. (2015)
	Total aerobic microbial flora, yeast and mold, and psychrotrophic microorganisms	Low-pressure cold plasma/helium and argon	Ulbin-Figlewicz et al. (2014)
Pork	*Escherichia coli*	Atmospheric pressure radio frequency glow discharge plasma/ helium	Moon et al. (2009)
Pork loin	*Escherichia coli* *Listeria monocytogenes*	Atmospheric pressure DBD plasma/helium or the mixture of helium and oxygen	Kim et al. (2013a)
	Total aerobic microbial flora	Indirect microwave plasma/air	Fröhling et al. (2012)
	Total aerobic microbial flora and yeast and mold	Low-pressure cold plasma/helium, argon and nitrogen	Ulbin-Figlewicz et al. (2013)
Pork loin and chicken breast	*Salmonella* Typhimurium	Atmospheric pressure plasma (APP) jet/ nitrogen or mixture of nitrogen and oxygen	Kim et al. (2013b)
Chicken meat and skin	*Listeria innocua*	Pen-type APP/helium and oxygen	Noriega et al. (2011)
Chicken breast and thigh	*Salmonella enterica* and *Campylobacter jejuni*	Atmospheric pressure DBD plasma/ atmospheric air	Dirks et al. (2012)
Chicken breasts	*Escherichia coli*	APP jet/nitrogen or mixture of nitrogen and oxygen	Yong et al. (2014)
Beef jerky	*Staphylococcus aureus*	Radio frequency atmospheric pressure plasma/argon	Kim et al. (2014)
Bresola	*Listeria innocua*	Atmospheric pressure DBD plasma/argon and oxygen mixture	Rød et al. (2012)

Continued

TABLE 1 Applications of Nonthermal Plasma Treatment as a Sterilization Method for Foods and Food Products of Animal Origin—cont'd

Treated Product	Microorganisms Tested	Plasma Source/ Carrier Gas	References
Bacon	*Escherichia coli* *Listeria monocytogenes* *Salmonella* Typhimurium	APP/helium	Kim et al. (2011)
Ham	3-Strain *Listeria monocytogenes*	APP/helium	Song et al. (2009)
Chicken hams	*Listeria innocua*	APP jet/helium, mixture of helium and oxygen, nitrogen, and mixture of nitrogen and oxygen	Lee et al. (2011b)
	Campylobacter jejuni	Radio frequency APP/ argon	Kim et al. (2013c)
Eggs	*Salmonella* Enteritidis *Salmonella* Typhimurium	Resistive barrier discharge plasma/ atmospheric air	Ragni et al. (2010)
Eggs shell	*Salmonella* Typhimurium	Low-pressure discharge plasma/ nitrogen and oxygen	Mok and Song (2013)
Milk	*Escherichia coli* O157:H7 *Listeria monocytogenes* *Salmonella* Typhimurium	Atmospheric pressure DBD plasma/ atmospheric air	Kim et al. (2015)
Whole, semi-skimmed, and skimmed milk	*Escherichia coli*	Atmospheric corona discharge plasma/ atmospheric air	Gurol et al. (2012)
Sliced cheese (라인이 구분되어야 할 것 같습니다) Sliced cheese (라인이 구분되어야 할 것 같습니다)	*Escherichia coli* *Listeria monocytogenes* *Salmonella* Typhimurium *Escherichia coli*	Atmospheric pressure DBD plasma/ atmospheric air Atmospheric pressure DBD Plasma/helium and oxygen	Yong et al. (2015a) Yong et al. (2015b) Lee et al. (2012a,b)
Smoked salmon	*Listeria monocytogenes*	APP/helium and oxygen	Lee et al. (2011a)
Cold-smoked salmon	*Lactobacillus sakei* *Photobacterium phosphoreum*	DBD plasma/argon and carbon dioxide	Chiper et al. (2011)
Dried filefish fillets	*Penicillium citrinum* *Cladosporium cladosporioides*	Atmospheric pressure plasma/oxygen	Park and Ha (2014)

2 Meat and Meat Products

2.1 Physical Quality

Several studies have reported inconsistent data on the effect of cold plasma on physical meat quality parameters, such as texture and color. Consumers generally use surface color as an indicator of the freshness and quality of meat at the time of purchase (Nam et al., 2011). In general, consumers prefer a bright red color of fresh meat to a tan or brown discoloration (Hood and Riordan, 1973). Appropriate precautions should be implemented to maintain the color of meat, as postmortem handling and processing may change its appearance significantly. The inherent factor most responsible for raw meat color is myoglobin, which consists of a globular protein, heme iron, and porphyrin ring. The free-binding site of heme iron is available for binding with oxygen, carbon monoxide, nitric oxide, or other small molecules. Specific binding formations at the sixth position and state of iron (ferrous or ferric) are largely responsible for the various meat colors observed, mainly red but also purple, brown, and others (Faustmann and Cassens, 1990; Mancini and Hunt, 2005).

Jayasena et al. (2015) recently treated fresh pork butt and beef loin with a flexible thin-layer dielectric barrier discharge (DBD) plasma system and demonstrated that this treatment did not significantly change the L^*-values (lightness) of pork butt and beef loin samples or the b^*-value of pork butt samples. Furthermore, b^*-value of DBD plasma-treated pork butt samples was not significantly changed; however, the b^*-value of beef samples exposed to 10 min of DBD plasma treatment was significantly higher than that of untreated samples. In contrast, increasing the exposure time to the same plasma treatment lowered a^*-values (redness) of pork and beef samples (Jayasena et al., 2015). This is consistent with the findings of Fröhling et al. (2012). The low a^*-value and high b^*-value suggest a greener surface color, although this could not be clearly detected by the naked eye. Fröhling et al. (2012) reported a green coloring effect in fresh pork due to indirect plasma treatment. Green coloring of meat may occur when myoglobin reacts with hydrogen peroxide, resulting in the formation of choleglobin, or when sulfmyoglobin forms in presence of hydrogen sulfide (H_2S) and oxygen. Hydrogen peroxide generated during indirect plasma treatment may react with myoglobin, resulting in a green appearance of meat (Faustmann and Cassens, 1990; Fröhling et al., 2012). Furthermore, the higher b^*-value of fresh pork may be attributed to the oxidation of deoxymyoglobin or oxymyoglobin during plasma treatment, resulting in the formation of oxidized myoglobin or metmyoglobin of a brownish color (Fröhling et al., 2012; Jayasena et al., 2015; Mancini and Hunt, 2005). This oxidation process may be accelerated by increasing the plasma exposure times, resulting in a higher concentration of metmyoglobin (Fröhling et al., 2012), which may increase the b^*-value of meat.

In contrast to the above results, Kim et al. (2013a) found that DBD plasma treatment, using helium as the input gas, significantly decreased the L^*-value of pork loin. This is consistent with the findings of Cheng et al. (2010) and Kim et al. (2011). However, the a^*- and b^*-values of pork loins subjected to this treatment were not notably different from those of untreated meat samples. Meat moisture content is known to correlate with lightness; therefore, the decrease observed in the L^*-value may have been due to the evaporation of moisture from the surface of meat during plasma treatment (Kim et al., 2011, 2013a).

However, not all studies have demonstrated that a discoloration of fresh meats occurs after plasma treatment. When pork samples were treated by low-pressure plasma with nitrogen and helium as the input gases, no significant difference in the L^*-, a^*-, and b^*-values was found between treated samples and the control (Ulbin-Figlewicz et al., 2013). Dirks et al. (2012) also reported that no visible changes in color occurred when chicken breast or chicken thigh skin were treated with DBD plasma, even for the maximum time period of 3 min. Likewise, Moon et al. (2009) reported that no visible changes were detected in the color of pork samples following plasma treatment (Fig. 1).

Previous studies evaluating color changes in processed meat products after plasma treatment found significant changes in L^*-, a^*-, and b^*-values, or in the total color difference (ΔE) in beef jerky following plasma treatment (Kim et al., 2014). No significant differences in L^*- and b^*-values were observed following a DBD plasma treatment of ready-to-eat bresaola inside sealed linear low-density polyethylene bags; however, the a^*-value was found to decrease (Rød et al., 2012). Upon visual examination, the plasma-treated bresaola samples were found to have lost the bright

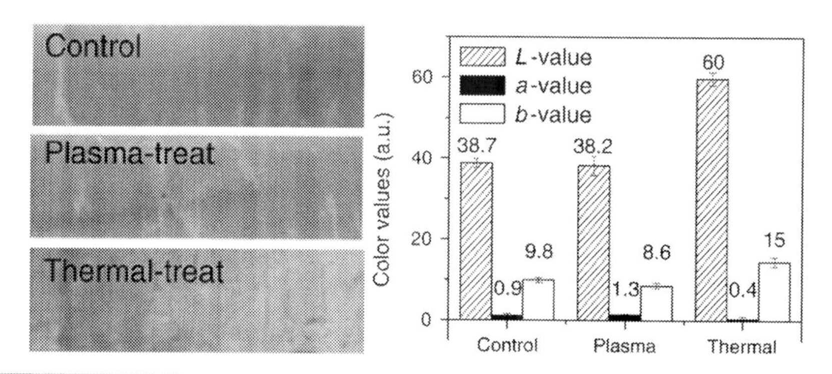

Fig. 1 Surface color of pork after atmospheric pressure radio frequency glow discharge and thermal treatment. When the pork sample contacted with heat, a color changed from pinkish red (top) to brownish gray (bottom). No thermal damage in color was observed from plasma treatment (middle). *(Based on Moon, S.Y., Kim, D.B., Gweon, B., Choe, W., Song, H.P., Jo, C., 2009. Feasibility study of the sterilization of pork and human skin surfaces by atmospheric pressure plasmas. Thin Solid Films 517, 4272–4275 with permission from Elsevier.)*

red color, and appeared reddish brown. Similarly, significant reductions (approximately 40% and 70%) in the a^*-values of all DBD plasma-treated bresaola samples were detected after 1 day and 14 days of storage, respectively, compared to those of the control (Rød et al., 2012). However, the a^*-values of all plasma-treated samples were comparable, irrespective of applied power (15.5 and 62 W) and treatment time (5 and 20 s). In contrast, Kim et al. (2011) reported that the L^*-value of plasma-treated bacon was decreased at higher input power and exposure time. This result was more obvious when the helium/oxygen mixture was used as the input gas, compared to when only helium. In the same study, the a^*-value of plasma-treated bacon was observed to have increased, irrespective of the input gas used.

Cold plasma technology has recently been applied for color enhancement of meat products. Plasma-liquid interaction results in the acidification of the liquid, combined with the generation of reactive oxygen species (ROS) and nitrogen species such as nitrate (NO_3^-) and nitrite (NO_2^-) (Oehmigen et al., 2010). Based on this principle, Jung et al. (2015a) hypothesized that water containing nitrite formed by DBD plasma treatment may be used as a nitrite source for cured meat products. Nitrite has been used as a curing agent for imparting the characteristic pinkish color to cured meat, and for the inhibition of pathogens such as *Clostridium botulinum*. Nitrite added to meat batter may be oxidized by sequestering oxygen and interacting with the sixth coordination site of myoglobin to form nitric oxide-myoglobin, which is responsible for the pink color observed in cured meats (Honikel, 2008; Sebranek and Bacus, 2007). Jung et al. (2015b) reported that the color of sausages manufactured with plasma-treated water were similar to those of sausages cured with conventional sodium nitrite. More recently, a similar method using direct gas-phase plasma treatment instead of water during manufacturing steps, such as mixing, chopping, and cutting, was developed to produce nitrite sources in a chamber, which resulted in the pink color characteristic of cured meat products (Lim et al., 2015). The authors proposed that this method, if successfully implemented in the meat-processing industry, should eliminate the need to prepare nitrites from synthetic or natural sources.

Texture is an important attribute of the eating quality of meat. The findings of Jayasena et al. (2015) revealed that texture parameters of pork butt and beef loin, including hardness, springiness, cohesiveness, gumminess, and chewiness were not affected by a flexible thin-layer DBD plasma treatment, irrespective of the meat type. Similarly, no significant change was reported in the shear force value of plasma-treated beef jerky compared to that of the control (Kim et al., 2014).

2.2 Chemical Quality

Plasma can generate free radicals and ROS, including ozone, which may compromise the functions of fatty acids, especially unsaturated fatty acids, thereby inducing lipid oxidation and producing lipid oxidation byproducts such as malondialdehyde

(MDA) and hexanal (Jayasena et al., 2015; Kim et al., 2013a; Liu et al., 2008; Montie et al., 2000). The level of lipid oxidation in meat samples can be quantified by measuring the 2-thiobarbituric acid reactive substances (TBARS) value. It has previously been reported that the detectable threshold for off-flavor in meat is 0.5–2.0 mg MDA/kg (Chang et al., 1961; Gray et al., 1996). In general, TBARS values increase with increase in plasma power, treatment time, and duration of storage. In addition, the changes in TBARS values could depend on the plasma type, carrier gas, or characteristics of the plasma-treated sample: eg, fat content and fatty acid composition (Rød et al., 2012). On a relevant note, as high oxygen concentrations lead to some degree of lipid oxidation, the level of oxygen in the packaging material should be taken into account when meat is plasma treated.

The limited data available indicates that cold plasma treatment of meat increases the lipid oxidation, assessed on the basis of TBARS values. However, these studies report that the TBARS values obtained are below the threshold of 0.5–2.0 mg MDA/kg; as a result, the off-flavors are likely to be undetectable. Jayasena et al. (2015) reported that the TBARS values of fresh pork butt and beef loin samples show a slight increase with increasing treatment time, particularly after exposure for 10 min, during flexible thin-layer DBD plasma treatment (Fig. 2). Similarly, TBARS values for the DBD plasma-treated pork loin samples have been detected to be higher than those for the

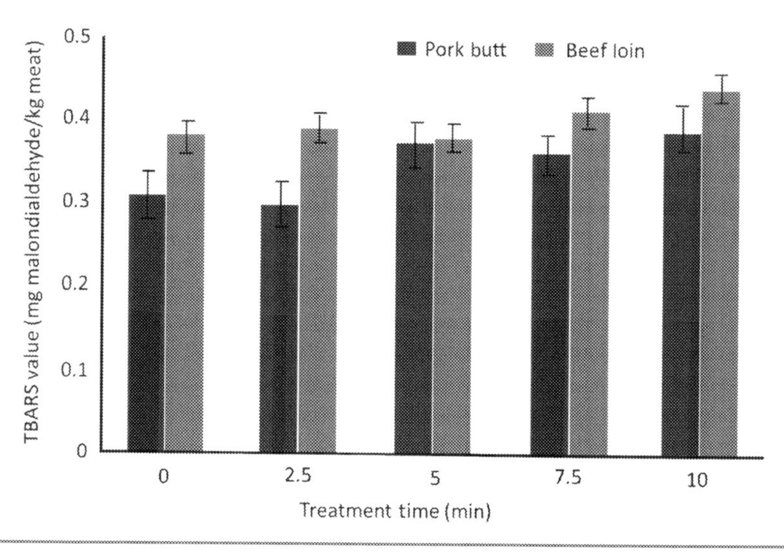

Fig. 2 2-Thiobarbituric acid reactive-substance values (mg malondialdehyde/kg meat) of pork butt and beef loin samples treated with flexible thin-layer dielectric barrier discharge plasma. *(Data from Jayasena, D.D., Kim, H.J., Yong, H.I., Park, S., Kim, K., Jo, C., 2015. Flexible thin-layer dielectric barrier discharge plasma treatment of pork butt and beef loin: effects on pathogens inactivation and meat-quality attributes. Food Microbiol. 46, 51–57, with permission from Elsevier.)*

untreated samples (Kim et al., 2013a). In particular, higher TBARS values have been observed when a combination of helium and oxygen is used as the carrier gas, than when helium is used alone. Kim et al. (2011) have also reported an increase in the TBARS value of plasma-treated bacon compared to control, after 7 days of refrigerated storage.

When treating bresaola with DBD plasma in a polyethylene bag, significantly higher TBARS values have been observed in treated samples (~0.25–0.4 mg MDA/kg) relative to control (~0.1–0.15 mg MDA/kg) stored in original commercial packages containing 0.02% O_2 (Rød et al., 2012). However, off-flavors could remain undetected in either treated or control samples as the threshold of off-flavor in meat is above 0.5 mg MDA/kg (Rød et al., 2012). It may be noted that the low TBARS value could be due to the very lean nature of bresaola, and that greater lipid oxidation is likely to occur in more fatty products.

The pH of meat products is known to affect the technological processing ability, as well as most sensory traits (Monin, 1998). The pH of fresh pork loins has been reported to decrease significantly following treatments with indirect plasma (Fröhling et al., 2012) and DBD plasma, reaching up to 5.3 (Kim et al., 2013a). The decrease of pH was only 0.1 which may not influence the quality of the meat (Kim et al., 2013a). During indirect plasma treatment, 0.6% NO, 1.8% NO_2, 0.07% CO_2, 0.04% HNO_3, 0.08% HNO_2, and 0.03% H_2O have been recorded using FT-IR spectroscopy. The accumulation of such acidogenic molecules on the meat surface is often held responsible for the pH drop in the product (Fröhling et al., 2012; Kim et al., 2013a). When the plasma is induced in a noble gas, the pH may not change significantly. For example, Kim et al. (2011) reported no notable changes in the pH of bacon after treatment with helium plasma.

Interestingly, no significant changes in the fatty acid composition of beef jerky have been reported after 5 min of cold plasma treatment (Kim et al., 2014). The water activity of plasma-treated chicken ham has also been found comparable to untreated samples for up to 6 min of treatment time. However, exposure times greater than 8 min significantly decrease the water activity of chicken ham (Kim et al., 2013c).

2.3 Sensory Properties

Sensory evaluation is the analysis of product attributes, as perceived by the human senses (Lyon et al., 2001). Consumers are often interested in sensory quality parameters such as appearance, aroma/odor, taste, texture, and sound. Studies conducted so far reveal that cold plasma can have negative impacts on several sensory parameters of meat. For instance, Kim et al. (2013a) reported that sensory parameters such as appearance, color, odor, and overall acceptability in DBD-treated fresh pork loin had significantly lower scores than in nontreated samples. Lipid oxidation byproducts such as alkanes, alkenes, aldehydes, alcohols, ketones, and acids (Jayasena and Jo, 2014) generated during plasma treatment may produce off-odors which are sensorially described as a fishy, metallic, rancid, and oxidized flavor (Kochhar, 1996). Thus, lower sensory quality may be due to the high fat content in meat and meat products, which leads to increased production of lipid oxidation byproducts during plasma treatment (Jayasena et al., 2015; Lee et al., 2012a,b). However, no significant sensory differences have been reported between the plasma-treated and control samples after cooking (Kim et al., 2013a).

Jayasena et al. (2015) recently reported that cooked pork butt and beef loin samples treated with DBD plasma show comparable scores for appearance, color, off-odor, and overall acceptability, but not taste, compared with controls (see Fig. 3). DBD plasma treatment negatively affected consumer preference for the taste of both meat samples. Unacceptable sensorial perceptions in the beef loin were reported only after 10 min of plasma treatment (Jayasena et al., 2015). Literature reveals that

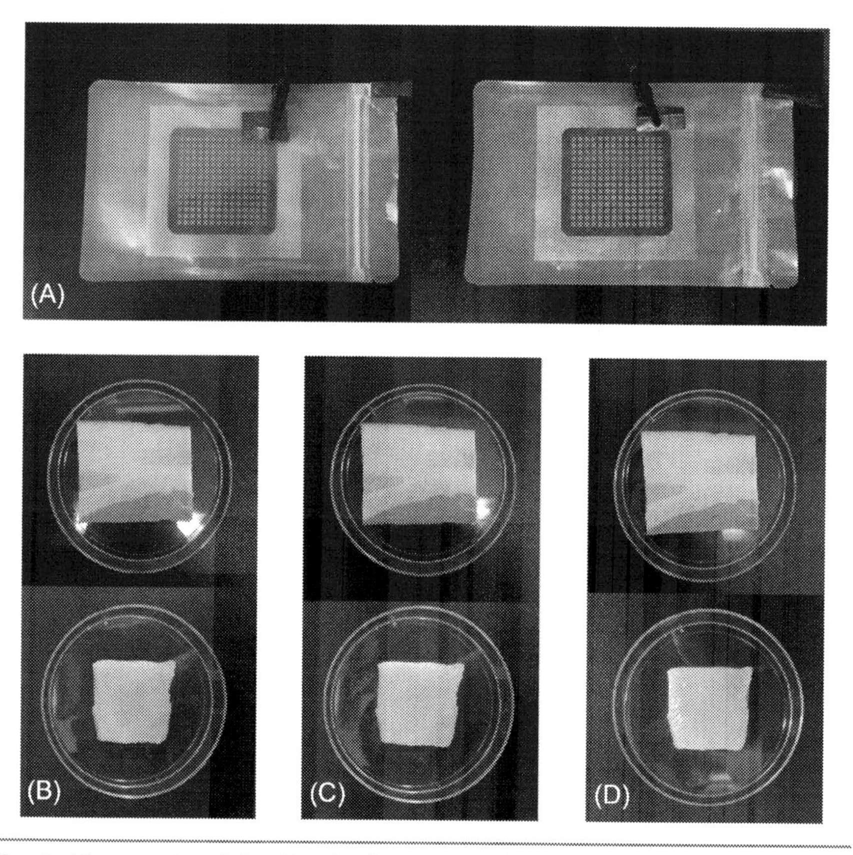

Fig. 3 Photographs of flexible thin-layer DBD plasma (A); surface appearance of bacon and chicken breast after the DBD plasma treatment for 0 (B), 5 (C), and 10 min (D); DBD plasma treatment conditions were identical to those described by Jayasena et al. (2015).

sensory-evaluation data on cold plasma treated foods are scant. Therefore, there is a need for further evaluation of the sensory qualities of cold plasma-treated meat and meat products.

3 Egg and Egg Products

Eggshell quality is a primary concern to the egg industry, because disinfection of the eggshell surface is vital for the prevention of egg spoilage and disease caused by consumption of spoiled eggs or egg products (Fuhrmann et al., 2010). Previous studies reported that the quality of table eggs, including eggshell color, weight, and yolk index, were not affected by plasma treatment (Ragni et al., 2010). Microscopic analysis revealed no differences between the plasma-treated and control eggs in terms of scores for both the cuticle and the inner surface of the shell membrane (Ragni et al., 2010).

Lee et al. (2012b) found that the L^*-values of cooked egg white and yolk subjected to atmospheric pressure plasma treatment were significantly lower than those of the control. However, there were no notable differences between the a^*- and b^*-values of nonthermal plasma-treated cooked egg white and yolk and those of the control samples.

The pH of egg white is an important internal quality parameter for egg freshness. Fresh eggs usually have a pH value ranging from 7.6 to 7.9 and a cloudy appearance due to the presence of carbon dioxide (Dutta et al., 2003). In general, plasma treatments for extended exposure time results in a lowering of pH. Korachi et al. (2010) observed a decrease in pH from 7.5 to 1.2 after 20 min of plasma treatment in aqueous media. Unlike aqueous medium, in most cases no significant differences between the pH of plasma-treated and control samples for fresh egg white (Ragni et al., 2010), cooked egg white, and yolk (Lee et al., 2012b) have been observed. In addition, the TBARS value of the plasma-treated cooked egg samples has been found to remain unaffected after 0 and 7 days, when treated with helium or helium and oxygen plasma (Lee et al., 2012b).

A recent study indicates that cold plasma does not adversely affect the sensory parameters of cooked egg white samples (Lee et al., 2012b). However, significant reductions in flavor, taste, and overall acceptability have been observed in plasma-treated cooked egg yolk.

4 Milk and Dairy Products

No significant changes in the color of milk were observed by Gurol et al. (2012), with the exposure of milk samples to cold plasma (9 kV) for up to 15 min. However, a slight change in milk color occurs after 20 min. Conversely, in another report the

L^*- and b^*-values of commercial whole milk have been reported to increase, whereas the a^*-value decreased after DBD plasma treatment (Kim et al., 2015). However, the researchers did state that these changes remain undetected to naked eye.

Yong et al. (2015b) studied the decontamination of sliced cheddar cheese using a DBD. Post plasma treatments, the researchers observed a significant decrease in the L^*-value and an increase in the a^*-value. But, the team also found that the b^*-value, total color difference (ΔE), sensory appearance, and color scores show insignificant differences relative to the control. In a similar study conducted by Lee et al. (2012a) on cheese slices, a significant decrease in L^*-values and an increase in the b^*-value has been observed in DBD plasma-treated cheese slices compared to control. Food browning is typically a consequence of reactions between amino groups (proteins, peptides, amino acids, and amines) and reducing sugars, oxidized lipids, vitamin C, and quinones. Reactions involving quinones are termed enzymatic browning reactions, whereas those involving other groups are considered nonenzymatic browning (Zamora and Hidalgo, 2005). The Maillard reaction, a nonenzymatic browning reaction, could occur between amino acids or milk proteins and milk lactose. Konteles et al. (2009) have drawn analogy from gamma-irradiation led discoloration, and suggested the aforementioned reactions to be responsible for any similar effects during plasma treatment.

Kim et al. (2015) reported that a 10 min treatment of whole milk with encapsulated DBD plasma decreased the pH value from 6.9 to 6.6. Treatment with cold plasma has been previously reported to result in an increase in the acidity of liquids (Bruggeman et al., 2008). This may be attributed to the multistep reactions of the plasma-generated reactive species, including NO_x, O, and O_3, with water at the gas-water interface (the quasisteady gas cavity surface, as well as on the surfaces of microdroplets of liquid inside the gas cavity) (Liu et al., 2010).

Cold plasma treatment for 5 and 10 min has been found to increase lipid oxidation in cheese samples (Yong et al., 2015b). Ozone generated during flexible thin-layer DBD plasma treatment may act as an accelerator of lipid oxidation (Roehm et al., 1971; Yong et al., 2015b). High levels of lipid oxidation byproducts, formed due to the high fat concentration in cheese slices, may induce off-odor in plasma-treated samples (Lee et al., 2012a). Unlike cheese, the fatty acid composition of whole milk has been reported to remain unchanged following cold plasma treatment (Kim et al., 2015).

The pasteurization of milk by cold plasma could have positive or negative effects on the sensory attributes of the end product (Cruz et al., 2012). Kim et al. (2015) reported that a preliminary sensory test revealed slight changes in the flavor and taste of milk subjected to encapsulated DBD plasma as compared with nontreated milk. When sliced cheddar cheese is treated with flexible thin-layer DBD plasma for 5 and 10 min, a significant reduction in flavor and overall acceptability, as well as increase in off-odor, has been reported (Yong et al., 2015b); cf. Fig. 4. Plasma

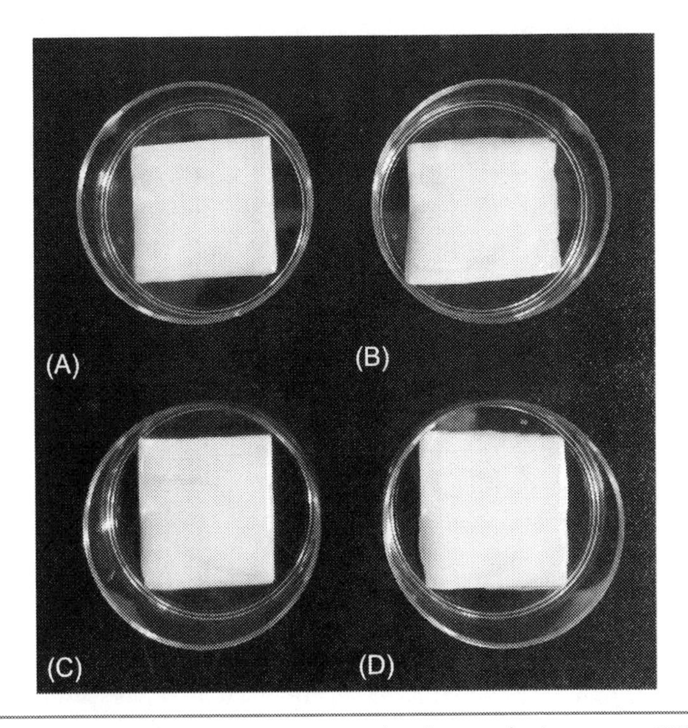

Fig. 4 Sliced cheddar cheese after an encapsulated atmospheric pressure DBD plasma treatment for 0 (A), 5 (B), 10 (C), and 15 min (D); DBD plasma treatment conditions were identical to those described by Yong et al. (2015a).

treatment using helium and helium with oxygen has been reported to cause significant reductions in desirable flavor, odor, and acceptability of cheese slices after 1, 5, and 10 min of treatment (Lee et al., 2012a). The acceptability has been found to be lower for a helium and oxygen mixture, than for helium alone. Therefore, the formation of ozone and its reaction with the fatty acids is most likely the factor responsible for this.

5 Fish Products

To date, very few studies have attempted to use plasma technology to enhance the quality and safety of fish and fish products. Lee et al. (2011a) showed that cold plasma treatment for 2 min with helium gas mixed with oxygen inhibited the growth of *L. monocytogenes* on smoked salmon by ca. 1.0 log CFU/g. In another study, Chiper et al. (2011) demonstrated that the concentrations of *Photobacterium phosphoreum*—a bacterium associated with seafood spoilage—in cold-smoked salmon were significantly reduced by cold plasma treatments (~3 log CFU/g in 60–120 s).

This inactivation effect was comparable between different gas compositions used (Ar and Ar+7% CO_2). However, cold plasma treatments did not elicit significant inactivation of *L. monocytogenes* or *Lactobacillus sakei* in cold-smoked salmon, irrespective of the gas used. In addition, Park and Ha (2014) reported that the counts of *Cladosporium cladosporioides* and *Penicillium citrinum* on dried filefish fillets significantly decreased with the increasing cold oxygen plasma treatment times (3–20 min). However, fillets subjected to 20 min of oxygen plasma treatment displayed increased TBARS values, which may partially contribute to the generation of a putrid and/or metallic smell, and a decreased overall sensory acceptance (Park and Ha, 2014). In contrast, fillets treated with 10 min of cold oxygen plasma received satisfactory TBARS and consumer acceptance. Based on these findings, the authors concluded that a 10-min cold oxygen plasma treatment could be effective in reducing >90% and inactivating the mold without causing deleterious changes to the physicochemical and sensory qualities of the fillets.

6 Conclusion

Currently available data reveals that cold plasma technology could be utilized as a novel antimicrobial intervention for the inactivation of pathogens and improvement of safety of foods of animal origin. However, studies performed to date also show certain limitations of cold plasma treatment in foods of animal origin, such as the acceleration of lipid oxidation and a negative impact on sensory characteristics. Further investigations are required to elucidate the effects of cold plasma on a broad range of quality aspects, including the sensory characteristics of various foods of animal origin, and the retention of these aspects during storage. Furthermore, the impact of cold plasma on the nutritional value of meat, eggs, fish, and their products requires comprehensive investigations. Additionally, the analysis of the marketability/consumer acceptance of cold plasma treated foods of animal origin is crucial, because the potential value of novel food technologies can only be realized if their application does not negatively affect the consumers' purchase decision-making.

Acknowledgments

This work was supported by R&D Program (Plasma Farming, Project No. EN1425-1) through the National Fusion Research Institute of Korea (NFRI) funded by the Government funds and Institute of Green Bio Science and Technology, Seoul National University.

References

Bruggeman, P., Ribežl, E., Maslani, A., Degroote, J., Malesevic, A., Rego, R., Vierendeels, J., Leys, C., 2008. Characteristics of atmospheric pressure air discharge with a liquid cathode and a metal anode. Plasma Sources Sci. Technol. 17, 025012.

Chang, P.-Y., Younathan, M.T., Watts, B.M., 1961. Lipid oxidation in pre-cooked beef preserved by refrigeration, freezing, and irradiation. Food Technol. 15, 168–171.

Cheng, S.Y., Yuen, C.W.M., Kan, C.W., Cheuk, K.K.L., Daoud, W.A., Lam, P.L., Tsoi, W.Y.I., 2010. Influence of atmospheric pressure plasma treatment on various fibrous materials: performance properties and surface adhesion analysis. Vacuum 84 (12), 1466–1470.

Chiper, A.S., Chen, W., Mejlholm, O., Dalgaard, P., Stamate, E., 2011. Atmospheric pressure plasma produced inside a closed package by a dielectric barrier discharge in Ar/CO_2 for bacterial inactivation of biological samples. Plasma Sources Sci. Technol. 20, 025008.

Cruz, A.G., Cadena, R.S., Faria, J.A.F., Molini, H.M.A., Dantas, C., Ferreria, M.M.C., Deliza, R., 2012. PARAFAC: adjustment for modelling consumer study covering probiotic and conventional yogurt. Food Res. Int. 45, 211–215.

Dirks, B.P., Dobrynin, D., Fridman, G., Mukhin, Y., Fridman, A., Quinlan, J.J., 2012. Treatment of raw poultry with nonthermal dielectric barrier discharge plasma to reduce *Campylobacter jejuni* and *Salmonella enterica*. J. Food Prot. 75 (1), 22–28.

Dutta, R., Hines, E.L., Gardner, J.W., Udrea, D.D., Boilot, P., 2003. Non-destructive egg freshness determination: an electronic nose based approach. Meas. Sci. Technol. 14, 190–198.

European Food Safety Authority, 2007a. Request for updating the former SCVPH opinion on *Listeria monocytogenes* risk related to ready-to-eat foods and scientific advice on different levels of *Listeria monocytogenes* in ready-to-eat foods and the related risk for human illness. Scientific opinion of the panel on biological hazards. EFSA J. 599, 1–42.

European Food Safety Authority, 2007b. Review of the community summary report on trends and sources of zoonoses, zoonotic agents and antimicrobial resistance in the European Union in 2005. Scientific opinion of the Scientific Panel on Biological Hazards (BIOHAZ) and Animal Health and Welfare (AHAW). EFSA J. 600, 1–32.

Faustmann, C., Cassens, R.G., 1990. The biochemical basis for discoloration in fresh meat: a review. J. Muscle Foods 1, 217–243.

Fratianni, F., Martino, L.D., Melone, A., Feo, V.D., Coppola, R., Nazzaro, F., 2010. Preservation of chicken breast meat treated with thyme and balm essential oils. J. Food Sci. 75 (8), 528–535.

Fröhling, A., Durek, J., Schnabel, U., Ehlbeck, J., Bolling, J., Schlüter, O., 2012. Indirect plasma treatment of fresh pork: decontamination efficiency and effects on quality attributes. Innov. Food Sci. Emerg. Technol. 16, 381–390.

Fuhrmann, H., Rupp, N., Buchner, A., Braun, P., 2010. The effect of gaseous ozone treatment on egg components. J. Sci. Food Agric. 90, 593–598.

Gaunt, L.F., Beggs, C.B., Georghiou, G.E., 2006. Bactericidal action of the reactive species produced by gas-discharge nonthermal plasma at atmospheric pressure: a review. IEEE Trans. Plasma Sci. 34, 1257–1269.

Gray, J.I., Gomaa, E.A., Buckley, D.J., 1996. Oxidative quality and shelf life of meats. Meat Sci. 43, S111–S123.

Gurol, C., Ekinci, F.Y., Aslan, N., Korachi, M., 2012. Low temperature plasma for decontamination of *E. coli* in milk. Int. J. Food Microbiol. 157, 1–5.

Honikel, K.O., 2008. The use and control of nitrate and nitrite for the processing of meat products. Meat Sci. 78, 68–76.

Hood, D.E., Riordan, E.B., 1973. Discoloration in pre-packaged beef: measurement by reflectance spectrophotometry and shopper discrimination. Int. J. Food Sci. Technol. 8, 333–343.

Jayasena, D.D., Jo, C., 2014. Potential application of essential oils as natural antioxidants in meat and meat products: a review. Food Rev. Int. 30, 71–79.

Jayasena, D.D., Kim, H.J., Yong, H.I., Park, S., Kim, K., Jo, C., 2015. Flexible thin-layer dielectric barrier discharge plasma treatment of pork butt and beef loin: effects on pathogens inactivation and meat-quality attributes. Food Microbiol. 46, 51–57.

Jensen, A.K., Ethelberg, S., Smith, B., Nielsen, E.M., Larsson, J., Mølbak, K., Christensen, J.J., Kemp, M., 2010. Substantial increase in listeriosis, Denmark 2009. Euro Surveillance 15 (12), pii:19522. Available at: http://www.eurosurveillance.org/ViewArticle.aspx?ArticleId=19522 (accessed 22.07.15).

Jung, S., Kim, H.J., Park, S., Yong, H.I., Choe, J.H., Jeon, H.J., Choe, W., Jo, C., 2015a. Color developing capacity of plasma-treated water as a source of nitrite for meat curing. Korean J. Food Sci. An. 35, 703–706.

Jung, S., Kim, H.J., Park, S., Yong, H.I., Choe, J.H., Jeon, H.J., Choe, W., Jo, C., 2015b. The use of atmospheric pressure plasma-treated water as a source of nitrite for emulsion-type sausage. Meat Sci. 108, 132–137.

Kaushik, N.K., Kim, Y.H., Han, Y.G., Choi, E.H., 2013. Effect of jet plasma on T98G human brain cancer cells. Curr. Appl. Phys. 13, 176–180.

Kim, B., Yun, H., Jung, S., Jung, Y., Jung, H., Choe, W., Jo, C., 2011. Effects of atmospheric pressure plasma on inactivation of pathogens inoculated onto bacon using two different gas compositions. Food Microbiol. 28, 9–13.

Kim, H.J., Yong, H.I., Park, S., Choe, W., Jo, C., 2013a. Effects of dielectric barrier discharge plasma on pathogen inactivation and physicochemical and sensory characteristics of pork loin. Curr. Appl. Phys. 13, 1420–1425.

Kim, H.J., Yong, H.I., Park, S., Kim, K., Bae, Y.S., Choe, W., Oh, M.H., Jo, C., 2013b. Effect of inactivating *Salmonella Typhimurium* in raw chicken breast and pork loin using an atmospheric pressure plasma jet. J. Anim. Sci. Technol. 55, 545–549.

Kim, J.S., Lee, E.J., Cho, E.A., Kim, Y.J., 2013c. Inactivation of *Campylobacter jejuni* using radio-frequency atmospheric pressure plasma on agar plates and chicken hams. Korean J. Food Sci. An. 33, 317–324.

Kim, J.S., Lee, E.J., Choi, E.H., Kim, Y.J., 2014. Inactivation of *Staphylococcus aureus* on the beef jerky by radio-frequency atmospheric pressure plasma discharge treatment. Innov. Food Sci. Emerg. Technol. 22, 124–130.

Kim, H.J., Yong, H.I., Park, S., Kim, K., Choe, W., Jo, C., 2015. Microbial safety and quality attributes of milk following treatment with atmospheric pressure encapsulated dielectric barrier discharge plasma. Food Control 47, 451–456.

Kochhar, S.P., 1996. Oxidation pathways to the formulation of off-flavours. In: Food Taints and Off-flavours. 2nd edn. Balckie Academic & Professional, London, pp. 168–225.

Konteles, S., Sinanoglou, V.J., Batrinou, A., Sflomos, K., 2009. Effects of γ-irradiation on *Listeria monocytogenes* population, color, texture and sensory properties of Feta cheese during cold storage. Food Microbiol. 26, 157–165.

Korachi, M., Aslan, N., 2011. The effect of atmospheric pressure plasma corona discharge on pH, lipid content and DNA of bacterial cells. Plasma Sci. Technol. 13, 99.

Korachi, M., Gurol, C., Aslan, N., 2010. Atmospheric plasma discharge sterilization effects on whole cell fatty acid profiles of *Escherichia coli* and *Staphylococcus aureus*. J. Electrost. 68, 508–512.

Lee, H.B., Noh, Y.E., Yang, H.J., Min, S.C., 2011a. Inhibition of foodborne pathogens on polystyrene, sausage casings and smoked salmon using nonthermal plasma treatments. Korean J. Food Sci. Technol. 43, 513–517.

Lee, H.J., Jung, H., Choe, W., Ham, J.S., Lee, J.H., Jo, C., 2011b. Inactivation of *Listeria monocytogenes* on agar and processed meat surfaces by atmospheric pressure plasma jets. Food Microbiol. 28, 1468–1471.

Lee, H.J., Jung, S., Jung, H., Park, S., Choe, W., Ham, J.S., Jo, C., 2012a. Evaluation of a dielectric barrier discharge plasma system for inactivating pathogens on cheese slices. J. Anim. Sci. Technol. 54, 191–198.

Lee, H.J., Song, H.P., Jung, H., Choe, W., Ham, J.S., Lee, J.H., Jo, C., 2012b. Effect of atmospheric pressure plasma jet on inactivation of *Listeria monocytogenes*, quality, and genotoxicity of cooked egg white and yolk. Korean J. Food Sci. An. 32, 561–570.

Lim, Y.B., Park, S., Kim, H.R., Yong, H.I., Kim, S.H., Lee, H.J., 2015. Plasma treatment process for processed meat and plasma treatment apparatus for processed meat. Korea Patent Submission Number 10-2015-0029641. Submission date at March 3, 2015.

Liu, H., Chen, J., Yang, L., Zhou, Y., 2008. Long-distance oxygen plasma sterilization: effects and mechanisms. Appl. Surf. Sci. 254, 1815–1821.

Liu, F., Sun, P., Bai, N., Tian, Y., Zhou, H., Wei, S., Zhou, Y., Zhang, J., Zhu, W., Becker, K., Fang, J., 2010. Inactivation of bacteria in an aqueous environment by a direct-current, cold-atmospheric pressure air plasma microjet. Plasma Process. Polym. 7, 231–236.

Lozano, C., LóPEZ, M., Gómez-Sanz, E., Ruiz-Larrea, F., Torres, C., Zarazaga, M., 2009. Detection of methicillin-resistant *Staphylococcus aureus* ST398 in food samples of animal origin in Spain. J. Antimicrob. Chemother. 64, 1325–1326.

Lucera, A., Costa, C., Conte, A., Del Nobile, M.A., 2012. Food applications of natural antimicrobial compounds. Front. Microbiol. 3 (287), 1–13.

Lyon, B.G., Lyon, C.E., Meullenet, J.F., Lee, Y.S., 2001. Meat quality: sensory and instrumental evaluations. In: Owens, C.M., Alvarado, C.Z., Sams, A.R. (Eds.), Poultry Meat Processing. CRC Press, Boca Raton, FL, pp. 125–156.

Mancini, R.A., Hunt, M., 2005. Current research in meat color. Meat Sci. 71, 100–121.

Mead, P., Slutsker, L., Dietz, V., McCaig, L.F., Bresee, J.S., Shapiro, C., Griffin, P.M., Tauxe, R.V., 1999. Food related illness and death in the United States. Emerg. Infect. Dis. 5, 607–625.

Misra, N.N., Tiwari, B.K., Raghavarao, K.S.M.S., Cullen, P.J., 2011. Nonthermal plasma inactivation of food-borne pathogens. Food Eng. Rev. 3, 159–170.

Mok, C., Song, D.M., 2013. Low-pressure discharge plasma inactivation of *Salmonella* Typhimurium and sanitation of egg. Food Eng. Prog. 17, 245–250.

Monin, G., 1998. Recent methods for predicting quality of whole meat. Meat Sci. 49, S231–S243.

Montie, T.C., Kelly-Wintenberg, K., Roth, J.R., 2000. An overview of research using the one atmospheric uniform glow discharge plasma (OAUGDP) for sterilization of surfaces and materials. IEEE Trans. Plasma Sci. 28, 41–50.

Mook, P., O'Brien, S.J., Gillespie, I.A., 2011. Concurrent conditions and human listeriosis, England, 1999–2009. Emerg. Infect. Dis. 17, 38–43.

Moon, S.Y., Kim, D.B., Gweon, B., Choe, W., Song, H.P., Jo, C., 2009. Feasibility study of the sterilization of pork and human skin surfaces by atmospheric pressure plasmas. Thin Solid Films 517, 4272–4275.

Nam, K.C., Seo, K.S., Jo, C., Ahn, D.U., 2011. Electrostatic spraying of antioxidants on the oxidative quality of ground beef. J. Anim. Sci. 89, 826–832.

Noriega, E., Shama, G., Laca, A., Díaz, M., Kong, M.G., 2011. Cold atmospheric gas plasma disinfection of chicken meat and chicken skin contaminated with *Listeria innocua*. Food Microbiol. 28, 1293–1300.

Oehmigen, K., Hahnel, M., Brandenburg, R., Wilke, C., Weltmann, K.D., von Woedtke, T., 2010. The role of acidification for antimicrobial activity of atmospheric pressure plasma in liquids. Plasma Process. Polym. 7, 250–257.

Park, S.Y., Ha, S.D., 2014. Application of cold oxygen plasma for the reduction of *Cladosporium cladosporioides* and *Penicillium citrinum* on the surface of dried filefish (*Stephanolepis cirrhifer*) fillets. Int. J. Food Sci. Technol. 50, 966–973.

Ragni, L., Berardinelli, A., Vannini, L., Montanari, C., Sirri, F., Guerzoni, M.E., Guarnieri, A., 2010. Non-thermal atmospheric gas plasma device for surface decontamination of shell eggs. J. Food Eng. 100, 125–132.

Rød, S.K., Hansen, F., Leipold, F., Knøchel, S., 2012. Cold atmospheric pressure plasma treatment of ready-to-eat meat: inactivation of *Listeria innocua* and changes in product quality. Food Microbiol. 30, 233–238.

Roehm, J.N., Hadley, J.G., Menzel, D.B., 1971. Oxidation of unsaturated fatty acids by ozone and nitrogen dioxide. Arch. Environ. Health 23, 142–148.

Sebranek, J.G., Bacus, J.N., 2007. Cured meat products without direct addition of nitrate or nitrite: what are the issues? Meat Sci. 77, 136–147.

Sivapalasingam, S., Friedman, C.R., Cohen, L., Tauxe, R.V., 2004. Fresh produce: a growing cause of outbreaks of foodborne illness in the United States, 1973 through 1997. J. Food Prot. 67, 2342–2353.

Smith, J., Sones, K., Grace, D., MacMillan, S., Tarawali, S., Herrero, M., 2013. Beyond milk, meat, and eggs: role of livestock in food and nutrition security. Anim. Front. 3, 6–13.

Sofos, J.N., 2008. Challenges to meat safety in the 21st century. Meat Sci. 78, 3–13.

Song, H.P., Kim, B., Choe, J.H., Jung, S., Moon, S.Y., Choe, W., Jo, C., 2009. Evaluation of atmospheric pressure plasma on improve the safety of sliced cheese and ham inoculated by 3-strain cocktail *Listeria monocytogenes*. Food Microbiol. 26, 432–436.

Tauxe, R.V., Doyle, M.P., Kuchenmuller, T., Schlundt, J., Stein, C.E., 2010. Evolving public health approaches to the global challenge of foodborne infections. Int. J. Food Microbiol. 139, S16–S28.

Threlfall, E.J., Wain, J., Peters, T., Lane, C., De Pinna, E., Little, C.L., Wales, A.D., Davies, R.H., 2014. Egg-borne infections of humans with *Salmonella*: not only an *S. enteritidis* problem. Worlds Poult. Sci. J. 70, 15–26.

U.S. Food Safety and Inspection Service, 2003. FSIS risk assessment for *Listeria monocytogenes* in deli meats. Available at:http://www.fsis.usda.gov/OPPDE/rdad/FRPubs/97-013F/ListeriaReport.pdf (accessed 20.07.15).

U.S. Food Safety and Inspection Service, 2010. FSIS comparative risk assessment for *Listeria monocytogenes* in ready-to-eat meat and poultry deli meats. Available at:http://www.fsis.usda.gov/shared/PDF/Comparative_RA_Lm_Report_May2010.pdf?redirecthttp=true (accessed 20.07.15).

Ulbin-Figlewicz, N., Brychcy, E., Jarmoluk, A., 2013. Effect of low-pressure cold plasma on surface microflora of meat and quality attributes. J. Food Sci. Technol. 52, 1228–1232.

Ulbin-Figlewicz, N., Zimoch-Korzycka, A., Jarmoluk, A., 2014. Antibacterial activity and physical properties of edible chitosan films exposed to low-pressure plasma. Food Bioprocess Technol. 7, 3646–3654.

Wan, J., Coventry, J., Swiergon, P., Sanguansri, P., Versteeg, C., 2009. Advances in innovative processing technologies for microbial inactivation and enhancement of food safety e pulsed electric field and low-temperature plasma. Trends Food Sci. Technol. 20, 414–424.

Xiao, J.F., Zhang, Y.N., Wu, S.G., Zhang, H.J., Yue, H.Y., Qi, G.H., 2014. Manganese supplementation enhances the synthesis of glycosaminoglycan in eggshell membrane: a strategy to improve eggshell quality in laying hens. Poult. Sci. 93, 380–388.

Yong, H.I., Kim, H.J., Park, S., Choe, W., Oh, M.H., Jo, C., 2014. Evaluation of the treatment of both sides of raw chicken breasts with an atmospheric pressure plasma jet for the inactivation of *Escherichia coli*. Foodborne Pathog. Dis. 11, 652–657.

Yong, H.I., Kim, H.J., Park, S., Alahakoon, A.U., Kim, K., Choe, W., Jo, C., 2015a. Evaluation of pathogen inactivation on sliced cheese induced by encapsulated atmospheric pressure dielectric barrier discharge plasma. Food Microbiol. 46, 46–50.

Yong, H.I., Kim, H.J., Park, S., Kim, K., Choe, W., Yoo, S.J., Jo, C., 2015b. Pathogen inactivation and quality changes in sliced cheddar cheese treated using flexible thin-layer dielectric barrier discharge plasma. Food Res. Int. 69, 57–63.

Yun, H., Kim, B., Jung, S., Kruk, Z.A., Kim, D.B., Choe, W., Jo, C., 2010. Inactivation of *Listeria monocytogenes* inoculated on disposable plastic tray, aluminum foil, and paper cup by atmospheric pressure plasma. Food Control 21, 1182–1186.

Zamora, R., Hidalgo, F.J., 2005. Coordinate contribution of lipid oxidation and Maillard reaction to the nonenzymatic food browning. Crit. Rev. Food Sci. Nutr. 45, 49–59.

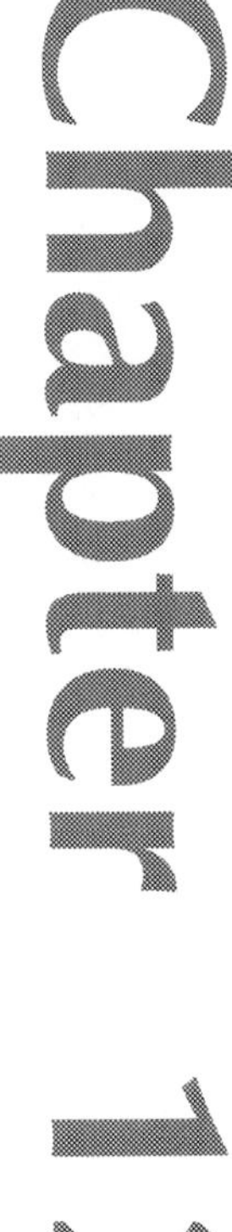

Cold Plasma Applications in Food Packaging

S.K. Pankaj*, S. Thomas†
*Dublin Institute of Technology, Dublin, Ireland, †Mahatma Gandhi University, Kerala, India

1 Introduction

Food packaging plays an important role in the food industry by maintaining the benefits of processing until the time of consumption. Food packaging plays numerous roles like containment, protection, convenience, preservation, communication, traceability, and temper indications. However, packaging must balance food protection with other issues, including energy and material costs, heightened social and environmental consciousness, and strict regulations on pollutants and the disposal of municipal solid waste (Marsh and Bugusu, 2007). Materials for food packaging mainly comprise of glass, paper, metals like aluminum, tinplate, and tin-free steels, and plastics like polyolefins, polyesters, polystyrene, and polyamides. Among these packaging materials, plastics have the major market share due to their low cost, light weight, fabrication flexibility, sealability, appearance, and resistance to breakage,

Cold Plasma in Food and Agriculture. http://dx.doi.org/10.1016/B978-0-12-801365-6.00012-3

tear and puncture. Most plastic materials are characterized by chemically nonpolar inert surfaces, making them nonreceptive to bonding, printing inks, coatings and adhesives; this necessitates the use of various surface treatments to improve these properties. Classical physicochemical methods for modifying the polymer surfaces include ultraviolet light, gamma rays, ion beam techniques, laser treatments, and the use of cold plasma.

The fundamentals of cold plasma have been covered in the initial chapters of this book. It may be recalled that thermal plasmas are characterized by high temperatures. On account of this, cold plasmas are the most suitable for polymer modifications. As discussed in Chapter 4, the generation of cold plasma can vary depending on the energy sources, electrode types and shapes, pressure variations, treatment parameters, gas compositions, and the complex plasma chemistry. For polymer surface modification purposes, corona and dielectric barrier discharges (DBD) are very common. Cold plasma treatments of packaging material surfaces can be used for various purposes, including:

(i) surface pretreatments for cleaning, activation or passivation;
(ii) surface functionalization for introducing specific functional groups on the film surface;
(iii) surface deposition for introducing selected coating materials on the film surface;
(iv) surface etching for creating specific surface roughness and pattern; and
(v) surface sterilization.

In order to practically leverage the cold plasma technologies for food packaging, it is important to study and understand the effects of cold plasma on the polymer materials. This forms the subject of discussion in the following section.

2 Effects of Cold Plasma on Polymers

Within the context of polymer processing and surface treatments, it is highly desirable that any relevant process or technology should exclusively impart the required modifications, without inducing any undesirable changes to the other properties of the polymer. It is worth mentioning at this point that although it is possible to finely control the plasma surface chemistry on a single system, it is almost impossible to get the same results at another system, owing to the considerable variability between reactor systems, design, operating procedures and other uncontrolled variables (Whittle et al., 2013). This leads to varied results of plasma treatments on similar surfaces by different researchers. Keeping this in mind, general effects of cold plasma on polymers have been discussed throughout this chapter.

2.1 Surface Modifications

2.1.1 Physical Changes

The main physical change observed after plasma treatment is the increase in the surface roughness and topography of the treated materials. By observing Fig. 1, one can appreciate the increase in surface roughness, moving from A to D. The increase in the surface roughness is due to the etching and sputtering effect of plasma. The etching of the treated materials is mostly attributed to the bombardment of the energetic particles in the plasma, such as electrons, ions, radicals, neutrals, and excited atoms or molecules, with the surface. Two types of etching mechanisms by plasmas are most accepted and include physical etching, which leads to the removal or reaggregation of low molecular weight fragments, and chemical etching, which results in the breaking of chemical bonds, chain scission, oxidation or chemical degradation of the treated materials. Although the etching effect by plasma is found in all types of discharge systems, such as RF, microwave, corona, and DBD, with plasma-induced in gases like CO_2, O_2, Ar, N_2, and air, the degree of etching is largely a function of the particle

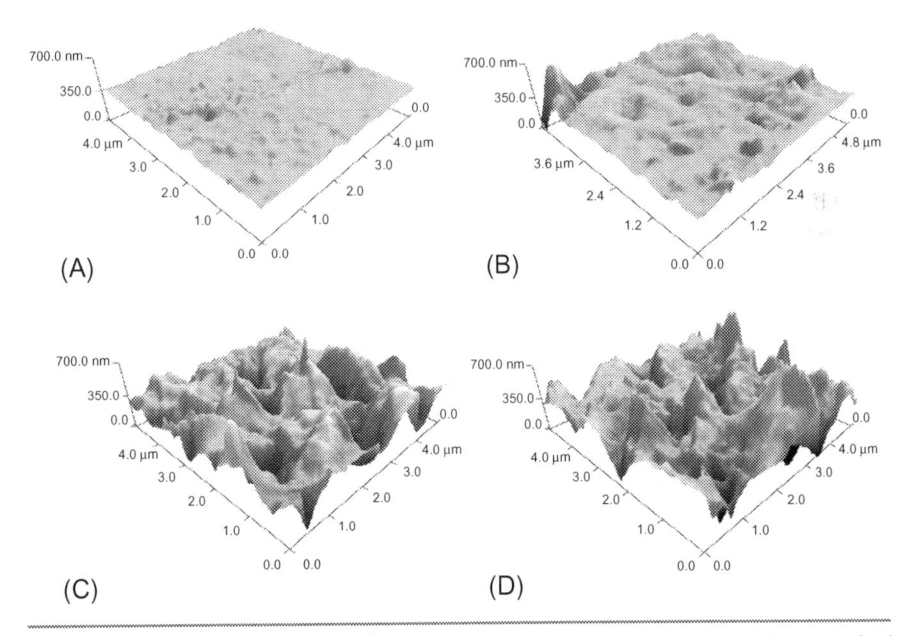

Fig. 1 AFM images of control and plasma-treated zein films. (A) Control, (B) 60 kV/5 min, (C) 70 kV/5 min, and (D) 80 kV/5 min. *(From Pankaj, S.K., Bueno-Ferrer, C., Misra, N.N., Bourke, P., Cullen, P.J., 2014b. Zein film: effects of dielectric barrier discharge atmospheric cold plasma. J. Appl. Polym. Sci. 131, with permission from John Wiley and Sons.)*

energies in the discharge combined with the power and homogeneity of the plasma treatment. In certain cases, plasma treatment may not lead to a significant increase in the average roughness, but it might increase the height of the peaks and valleys, changing the topography of the treated surface. The major application of this increase in surface roughness by cold plasma treatments lies in the enhanced anchoring effects of the polymer surface with increased adhesiveness, wettability, and bond strength.

2.1.2 Chemical Changes

Apart from the physical changes observed on the plasma-treated surface due to the bombardment of reactive species, they can also chemically attach with the polymer chain and change the secondary bonding or overall surface chemical composition. Many times new chemical groups are formed on the treated surface imparting the desired modifications. For example, in the studies conducted by Walton et al. (2010), SF_6 plasma treatment of polyethylene incorporated fluorine-containing species on the polymer surface, which significantly increased the surface hydrophobicity of treated polyethylene.

The change in the surface chemical composition after plasma treatment is mainly dependent on the plasma gas and type of reactive species generated (Misra et al., 2014d). As an example, for air plasmas, nitrogen and oxygen are the main reactive species generating molecules, and a significant increase in the surface oxygen component can be observed with air plasma treatment. The incorporation of these reactive species mostly increases the content or formation of new polar functional groups, like hydroxyl, carboxyl, and carboxylate, on the treated surface. As shown in Fig. 2, the air plasma treatment of corn starch film leads to an increase in oxygen containing groups, and even the formation of new polar groups at higher voltage levels. It is also important to note at this point that in certain cases, contrary to intended change, plasma species might be just chemisorbed on the treated surface, instead of covalent bonding, which can desorb with time or further treatments.

2.1.3 Contact Angle

The changes in the contact angle and surface free energy are dependent on the changes in the surface chemical composition of the polymer surface after plasma treatment. In most cases, a significant increase in the surface free energy of the treated polymers is observed, which can be attributed to the increase in polar groups on the polymer surface after plasma treatment. Pankaj et al. (2015) observed a decrease in the water contact angle from 54° to 21° after DBD plasma treatment of cornstarch film at 80 kV for 5 min. The surface free energy was significantly higher after plasma treatment due to an increase in the polar component of the surface free energy (Table 1). The increase in the surface free energy was found to be more rapid at low treatment times than longer treatment times. Moisture content in plasmas is

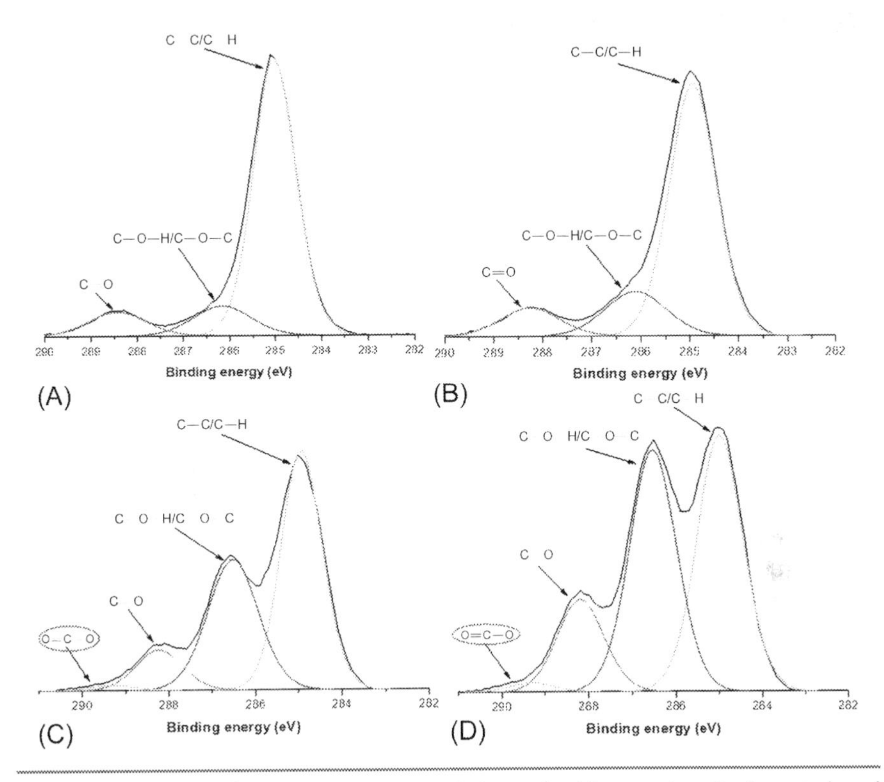

Fig. 2 X-ray photoelectron spectroscopy C 1s peaks (deconvoluted) of control and plasma-treated corn starch films. (A) Control, (B) 60 kV/5 min, (C) 70 kV/5 min, and (D) 80 kV/5 min. *(From Pankaj, S.K., Bueno-Ferrer, C., Misra, N.N., O'Neill, L., Tiwari, B.K., Bourke, P., Cullen, P.J., 2015. Dielectric barrier discharge atmospheric air plasma treatment of high amylose corn starch films. LWT—Food Sci. Technol. 63, 1076–1082, with permission from Elsevier.)*

TABLE 1 The Contact Angle (A) and Surface Free Energy (B) of Plasma-Treated Corn Starch Films

	Control	60 kV/5 min	70 kV/5 min	80 kV/5 min
(A) Contact angle (°)				
Water	54.3±1.3	25.5±0.6	22.6±0.6	21.1±3.5
Ethylene glycol	28.6±2.8	8.8±2.2	3.0±1.0	4.1±0.8
(B) Surface free energy (mN/m)				
Total (γ^{tot})	49.1	62.3	64.2	64.2
Dispersive (γ^{d})	32.8	34.7	37.9	36.5
Polar (γ^{p})	16.3	27.6	26.3	27.7

(Adapted from Pankaj, S.K., Bueno-Ferrer, C., Misra, N.N., O'Neill, L., Tiwari, B.K., Bourke, P., Cullen, P.J., 2015. Dielectric barrier discharge atmospheric air plasma treatment of high amylose corn starch films. LWT—Food Sci. Technol. 63, 1076–1082, with permission from Elsevier.)

usually avoided, although Van Deynse et al. (2014) have shown that adding water vapor to the working gas during plasma treatment could give better surface wettability.

However, as is often admitted in the literature, the increase in the surface free energy may not be permanent over extended periods of time, and upon storage, reversal of plasma-induced changes could be observed. This phenomenon is commonly referred as "aging" or "hydrophobic recovery."

2.2 Crystal Structure

The change in the crystal structure after cold plasma treatments has previously been studied by many researchers. However, the variability in the energy source, discharge systems, plasma gases, and in turn the reactive species and voltage levels, discharge gaps etc., make it difficult to draw any definitive conclusions. Nevertheless, it is widely recognized that plasma treatment leads to degradation and chain scissions of treated polymers, which causes formation of chains with radical tails with different mobility, which thereby results in the rearrangement of the polymer chains and a change in polymer crystallinity (Ataeefard et al., 2009). It is also known that amorphous regions of polymers are more prone to an etching effect with plasma and hence, initial crystallinity of treated polymer often has a significant influence on the final outcome.

2.3 Barrier Properties

With regard to the application of polymers in food packaging, it is highly desirable that an in-package food-processing method or polymer-modification technology does not adversely affect the barrier properties of the packaging material. Most of the studies available in the literature suggest that plasma treatment with gases like air, oxygen, and carbon dioxide have no or little impact on the water vapor and oxygen permeability of the polymers. However, studies were also done where the intended result was to reduce the water vapor permeability after plasma treatment. This can be achieved by deposition of the hydrophobic coating on the polymer surface through plasma treatments. Tenn et al. (2012) have shown that carbon tetrafluoride and tetramethylsilane plasma treatment of poly(ethyl-co-vinyl alcohol) films increase the hydrophobicity of treated films and decrease the water permeability up to 28% due to the formation of fluorine-containing groups and SiO_xC_y compounds on the film surface. This reflects on potential new applications of plasma treatment for creating polymers with tailored water vapor barrier properties, with respect to food-packaging materials.

3 Applications

Cold plasma has a wide range of applications, including remediation of gaseous pollutants, treatments of volatile organic compounds, greenhouse gas abatement, lasers, excimer lamps, display panels, etc. However, from a food-packaging perspective, applications of cold plasma are classically focused on surface modifications. Recently, however, the applications have extended into the novel area of in-package plasma decontamination of food products (Connolly et al., 2013). A detailed description of these applications follows hereunder.

3.1 Surface Activation

Surface activation and functionalization of polymeric materials refers to structural or chemical changes in the surface layer of the polymers to facilitate the attachment of specific functional groups on the surface to achieve desired functionality. Surface activation is typically applied for altering or improving adhesion and printing properties of surfaces. Via cold plasma, surface activation results in the generation of polar groups or cross-linked molecules on the treated surface, which increases the total surface free energy of the polymer. It may also cause etching due to the bombardment of energetic particles, such as electrons, ions, radicals, neutrals, and excited atoms/molecules on the surface; this may improve adhesion for inert, biocompatible materials, which otherwise might not stick together. Plasma species interact with chemical groups on the polymer surface, creating new functional groups like hydroxyl (—OH), carbonyl (—CO), carboxyl (—COOH), amino (—NH_2), and amide (NHCO) that modify the polymer properties from hydrophobic to hydrophilic surfaces, increasing the adhesion, wettability, and biocompatibility, which may be difficult to achieve by conventional activation methods.

3.1.1 Adhesion

Cold plasma allows the modification of surface bonding properties without affecting the bulk properties of the treated polymers (Dixon and Meenan, 2012). Adhesion refers to the tendency of different surfaces to cling together due to the physical and chemical intermolecular interactions at the interface. Plasma treatments can be used to modify the total surface free energy of the polymers for increasing adhesion or antiadhesion, depending on the specific requirement. With respect to food packaging, modification in adhesive properties after plasma treatment can be exploited for creating polymer composites with desired barrier properties against water vapor or oxygen.

3.1.2 Printing

As discussed previously, cold plasma treatments can modify the contact angle and corresponding surface free energy of the polymer surface, which increases the printability of the plasma-treated surfaces. The printability of the polymers can be measured by a number of techniques, including those that measure the physical surface changes resulting from plasma treatments. Although standardized procedures like DIN 53364 (ISO 8296) are available, dynamic measurement of contact angle is often used to investigate wettability, which is also closely related with ink adhesion. The effects of plasma treatment with regard to printability can be visually appreciated from Fig. 3, which shows the drastic enhancement in the ink spreadability of a polyethylene web surface after only 5 s of plasma treatment. Plasma treatment has also been known to be used for the creation of abrasion-resistant imprints and precise color matching and high pixel accuracy for decoration to glass bottles or jars, along with homogeneous coatings with a high scratch resistance without any collateral negative effects (Pankaj et al., 2014b).

3.2 Surface Deposition

Surface deposition by plasma treatment refers to the deposition of a thin layer of desired compounds to a thickness of usually less than one micrometer onto a polymer

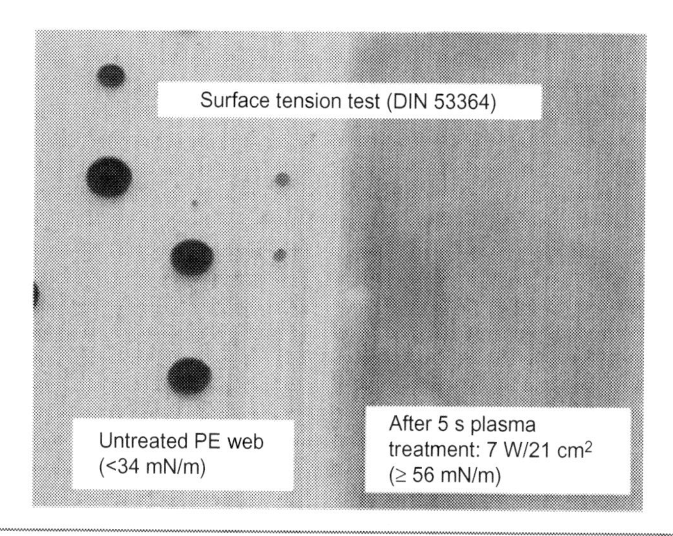

Fig. 3 The effect of fused hollow cathode plasma exposure on ink spreadability of a polyethylene web surface. *(From Bárdos, L., Baránková, H., 2010. Cold atmospheric plasma: sources, processes, and applications. Thin Solid Films 518, 6705–6713, with permission from Elsevier.)*

surface for imparting specific functionalities. Plasma deposition in the field of food packaging has several potential areas of applications (Denes, 2004), enumerated as follows:

(i) deposition of bioactive or antimicrobial compounds in edible and active packaging systems to achieve particular benefits such as shelf-life extension or nutritional quality enhancement;
(ii) plasma-generated nanoparticle systems;
(iii) immobilization of biopolymers like enzymes and oligomers on the surface;
(iv) deposition of an antifouling layer on food-processing surfaces to avoid biofilm formation; and
(v) plasma deposition of barrier layers, to improve barrier properties toward gases (oxygen, carbon dioxide, water vapor) and chemical solvents.

Among these, the deposition of bioactive compounds and barrier layers are of significant importance to food packaging and are discussed in detail in the following sections.

3.2.1 *Bioactive and Antimicrobial Compounds*

The immobilization of bioactive functional compounds like lysozyme, nisin, vanillin, sodium benzoate, glucose oxidase, bovine lactoferrin, lactoferricin, chitosan, nanosilver, trichlosan, or antimicrobial peptides into the packaging material by plasma treatment has been extensively studied within the emerging field of antimicrobial and active packaging (Lei et al., 2014; Pankaj et al., 2014b). Some examples of plasma-mediated coatings on polymer surfaces and/or the deposition of bioactive compounds are summarized in Table 2.

TABLE 2 Examples of Coating on Plasma-Treated Surface or Plasma Deposition of Bioactive Compounds

Bioactive Compound	Substrate	Plasma Treatment	Reference
Vanillin	Apple surface	Corona treatment (12 kV, Ar: 11.8 sccm, 20 min)	Fernández-Gutierrez et al. (2010)
Nanosilver	Polyethylene	RF plasma (30 W, Ar: 30 sccm, diethylene glycol-dimethylether: 0.25 sccm, 13.3 Pa, 20 min)	Del Nobile et al. (2004)
Chitosan	Ethylene copolymer	Corona treatment (48 dyne)	Joerger et al. (2009)
Chitosan/Silver	Ethylene copolymer	Corona treatment (48 dyne)	Joerger et al. (2009)
Glucose oxidase	Biaxially oriented polypropylene	DBD plasma (1 W/m^2, 0.5 min, 2 mm)	Vartiainen et al. (2005)

3.2.2 Barrier Layers

Food packaging plays an important role in acting as a barrier between the food and the external environment. The major environmental factors which could have significant impacts on the food shelf life and quality include oxygen, water vapor and other aromatic compounds. An improvement in barrier properties of packaging materials for food contact applications has been one of the most studied applications in plasma-treated polymers.

As discussed in Section 2.3, one of the most used methods to decrease the water vapor permeability of polymers consists of depositing a thin layer of SiO_x on traditional polymers, like polyethylene terephthalate (PET) foils, or novel bio-based polymers like chitosan films, through plasma-enhanced chemical vapor deposition (CVD). This type of layer is colorless; therefore, it could easily be deposited onto transparent packaging materials, while retaining visibility of the packaged food. Biodegradable polymers like poly(lactic acid) (PLA) have drawn considerable attention due to the adverse environmental impact of traditional polymers. A particular limitation of PLA is its poor barrier property, which can be managed by cold plasma treatment. A decrease of up to 60% in the water vapor transmission rate has been observed after cross-linking PLA with tetramethoxysilane (TMOS) with a possible mechanism shown in Fig. 4. Atmospheric pressure plasma-enhanced CVD of thin functional films has shown the potential to circumvent the limitations imposed by

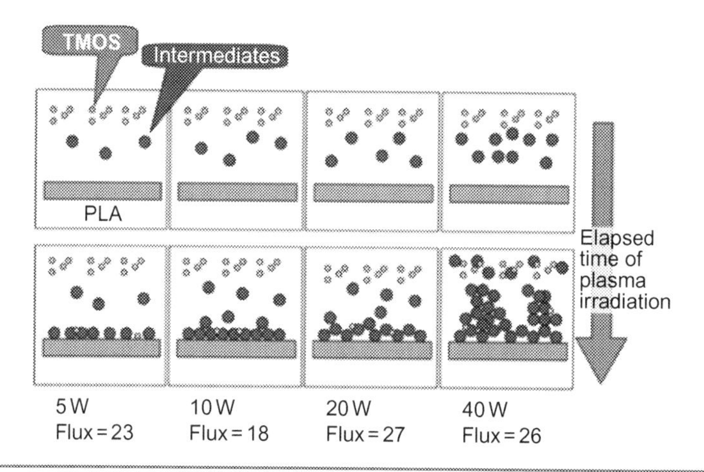

Fig. 4 A possible mechanism of surface coating of poly(lactic acid) (PLA) with tetramethoxysilane (TMOS). *(From Uemura, Y., Maetsuru, Y.-s., Fujita, T., Yoshida, M., Hatate, Y., Yamada, K., 2006. The effect of coatings formed by low temperature tetramethoxysilane plasma treatment on water-vapor permeability of poly(L-lactic acid) film. Korean J. Chem. Eng. 23, 144–147, with permission from Springer.)*

conventional low-pressure deposition methods due to the low cost; there is also no need for vacuum equipment or complex transfer chambers (Starostin et al., 2015). This area of application needs further study to create high-barrier polymers for gases and volatiles too.

3.3 Surface Sterilization

One of the classical roles of packaging is to provide protection to the contained food products. After the food-processing stage, the packaging should protect the food from any recontamination, either from the environment or the package itself. Often, due to the inherent properties of the polymers, they do not support growth of microbes; however, that may not always be true. Therefore, it is essential to achieve the surface sterilization to avoid any health risks to consumers, in addition to the potential economic losses (Misra et al., 2011). Conventional sterilization methods such as dry heat, hot steam, UV light, and chemicals like ethylene oxide and hydrogen peroxide, have traditionally been used in the sterilization of medical instruments and implants, as well as packaging materials in the food industry; certain limitations, however, have motivated the search for new approaches (Pankaj et al., 2014b). Cold plasma treatment has shown great potential as an alternative to the conventional processes for surface sterilization. Cold plasma can be used for sterilization of a wide range of packaging materials without any health risk and in an environment-friendly way. Researchers have reported the use of cold plasma treatment for sterilization of different packaging materials such as PET foils, polystyrene, and multilayer packaging based on PET/PVDC/PE-LD, resulting in a minimum of a 2 log_{10} reduction with minimal changes in the packaging functionality (Muranyi et al., 2008, 2010). A summary of some notable studies addressing plasma-aided sterilization of packaging materials is presented in Table 3.

3.4 In-Package Decontamination

This is one of the novel applications of cold plasma in the food industry. The idea is to generate plasma species in prepackaged food products; the reactive species will carry out the package sterilization and food decontamination and later will revert back to the original gas composition during the storage period, leaving no chance for postprocessing recontamination (discussed in preceding chapters). The concept has been studied by various researchers and promising results for microbiological decontamination of various food products has been encountered (Misra et al., 2014a,b,c; Patil et al., 2014; Ziuzina et al., 2014).

Since in this type of processing the packaging material will also act as the dielectric barrier, it is essential to access the compatibility of different polymeric films to such applications. Traditional polymers like polyethylene, polypropylene,

TABLE 3 The Sterilization Effects of Various Plasma Processes on a Variety of Microorganisms

Polymer	Micro-Organism	Plasma Process	Salient Result	Reference
Polypropylene	*E. coli* K12	One atmosphere uniform glow discharge plasma (OAUGDP)	$\geq$5 log reduction after 24 s exposure	Gadri et al. (2000)
Polypropylene	*P. aeruginosa*	OAUGDP	$\geq$6 log reduction after 30 s exposure	Gadri et al. (2000)
Polypropylene	*S. aureus*	OAUGDP	$\geq$6 log reduction after 22 s exposure	Gadri et al. (2000)
Nitrocellulose	*Bacillus stearothermophilus* endospores	OAUGDP	$\geq$5 log reduction after 5.5 min exposure	Gadri et al. (2000)
Polyethylene terephthalate foil	*Bacillus subtilis*, and *Aspergillus niger* spores	Atmospheric pressure DBD; electric power density at the electrodes = 7 W/cm^2; 20–50 kHz	More than four orders of magnitude reduction within few seconds	Heise et al. (2004)
Polyethylene terephthalate bottles and foils	*Bacillus atrophaeus* spores	Low-pressure microwave plasma reactor; Gas mixture, H_2:N_2:O_2 = 1.2:1:0.8; Power = 2000 W, pressure = 30 Pa; on-off cycle = 1 ms	About 6 $\log_{10}$ reduction after only 5 s exposure	Deilmann et al. (2009)
Polyethylene terephthalate foils	*Bacillus subtilis* spores	Low-pressure microwave-driven air plasma; Frequency = 2.45 GHz	Up to 4 order of magnitude reduction of the spores within 0.5–4 s of treatment	Schneider et al. (2005)
Polyethylene terephthalate foils	*Salmonella* serotype Mons, *Staphylococcus aureus* and *Escherichia coli*	Cascaded DBD in air; equipped with a 282 nm excimer lamp; power input = 130 W	5–6 log reduction in vegetative cells was observed within 1-3 s	Muranyi et al. (2007)
Polyethylene terephthalate foils	*Bacillus atrophaeus*, *Aspergillus niger*, and *Clostridium botulinum* spores	Cascaded DBD in air; equipped with a 282 nm excimer lamp; power input = 130 W	*Aspergillus niger* was the most resistant test strain with an inactivation rate of about 5 $\log_{10}$ in 5 s	Muranyi et al. (2007)
Glass, polyethylene, polypropylene, nylon, and paper foil	*Escherichia coli* O157:H7, *S. typhimurium*, *Staphylococcus aureus*	Low-pressure air plasma; pressure = 0.5–5.0 Torr; power density = 12.4–54.1 mW/cm^3	Up to 4-log reduction observed within 300 s	Lee et al. (2015)

polystyrene, PET, and bio-based films like PLA, sodium caseinate, zein, and starch have shown promising results for use with in-package plasma treatments (Pankaj et al., 2014a,c,d, 2015).

4 Conclusion

Cold plasma has been used for decades in the polymer industry. It has been used for various applications in the modification of the polymer surface properties. It has also shown immense potential in creating novel functional polymers, coatings, and antimicrobial systems specifically for the food-packaging industry. Many of the problems and limitations of food-packaging polymers itself should be explored further to leverage the surface functionalization, deposition, and sterilization using cold plasma technologies. Cold plasma has shown immense potential as a simple, safe, and environmental friendly alternative to various chemical processes used in the food-packaging industry.

References

Ataeefard, M., Moradian, S., Mirabedini, M., Ebrahimi, M., Asiaban, S., 2009. Investigating the effect of power/time in the wettability of Ar and O_2 gas plasma-treated low-density polyethylene. Prog. Organ. Coat. 64, 482–488.

Connolly, J., Valdramidis, V.P., Byrne, E., Karatzas, K.A., Cullen, P.J., Keener, K.M., Mosnier, J.P., 2013. Characterization and antimicrobial efficacy against *E. coli* of a helium/air plasma at atmospheric pressure created in a plastic package. J. Phys. D Appl. Phys. 46, 035401.

Deilmann, M., Halfmann, H., Steves, S., Bibinov, N., Awakowicz, P., 2009. Silicon oxide permeation barrier coating and plasma sterilization of PET bottles and foils. Plasma Process. Polym. 6, S695–S699.

Del Nobile, M., Cannarsi, M., Altieri, C., Sinigaglia, M., Favia, P., Iacoviello, G., D'agostino, R., 2004. Effect of Ag-containing nano-composite active packaging system on survival of *Alicyclobacillus acidoterrestris*. J. Food Sci. 69, E379–E383.

Denes, F., 2004. Macromolecular plasma-chemistry: an emerging field of polymer science. Prog. Polym. Sci. 29, 815–885.

Dixon, D.J., Meenan, B., 2012. Atmospheric dielectric barrier discharge treatments of polyethylene, polypropylene, polystyrene and poly(ethylene terephthalate) for enhanced adhesion. J. Adhes. Sci. Technol. 26, 2325–2337.

Fernández-Gutierrez, S., Pedrow, P.D., Pitts, M.J., Powers, J., 2010. Cold atmospheric-pressure plasmas applied to active packaging of apples. IEEE Trans. Plasma Sci. 38, 957–965.

Gadri, R.B., Roth, J.R., Montie, T.C., Kelly-Wintenberg, K., Tsai, P.P.Y., Helfritch, D.J., Feldman, P., Sherman, D.M., Karakaya, F., Chen, Z., 2000. Sterilization and plasma processing of room temperature surfaces with a one atmosphere uniform glow discharge plasma (OAUGDP). Surf. Coat. Technol. 131, 528–541.

Heise, M., Neff, W., Franken, O., Muranyi, P., Wunderlich, J., 2004. Sterilization of polymer foils with dielectric barrier discharges at atmospheric pressure. Plasmas Polym. 9, 23–33.

Joerger, R.D., Sabesan, S., Visioli, D., Urian, D., Joerger, M.C., 2009. Antimicrobial activity of chitosan attached to ethylene copolymer films. Packag. Technol. Sci. 22, 125–138.

Lee, T., Puligundla, P., Mok, C., 2015. Inactivation of foodborne pathogens on the surfaces of different packaging materials using low-pressure air plasma. Food Control 51, 149–155.

Lei, J., Yang, L., Zhan, Y., Wang, Y., Ye, T., Li, Y., Deng, H., Li, B., 2014. Plasma treated polyethylene terephthalate/polypropylene films assembled with chitosan and various preservatives for antimicrobial food packaging. Colloids Surf. B: Biointerf. 114, 60–66.

Marsh, K., Bugusu, B., 2007. Food packaging—roles, materials, and environmental issues. J. Food Sci. 72, R39–R55.

Misra, N., Patil, S., Moiseev, T., Bourke, P., Mosnier, J., Keener, K., Cullen, P., 2014a. In-package atmospheric pressure cold plasma treatment of strawberries. J. Food Eng. 125, 131–138.

Misra, N.N., Keener, K.M., Bourke, P., Mosnier, J.P., Cullen, P.J., 2014b. In-package atmospheric pressure cold plasma treatment of cherry tomatoes. J. Biosci. Bioeng. 118, 177–182.

Misra, N.N., Moiseev, T., Patil, S., Pankaj, S.K., Bourke, P., Mosnier, J.P., Keener, K.M., Cullen, P.J., 2014c. Cold plasma in modified atmospheres for post-harvest treatment of strawberries. Food Bioprocess Technol. 7, 3045–3054.

Misra, N.N., Pankaj, S.K., Walsh, T., O'Regan, F., Bourke, P., Cullen, P.J., 2014d. In-package nonthermal plasma degradation of pesticides on fresh produce. J. Hazard. Mater. 271, 33–40.

Misra, N.N., Tiwari, B.K., Raghavarao, K.S.M.S., Cullen, P.J., 2011. Nonthermal plasma inactivation of food-borne pathogens. Food Eng. Rev. 3, 159–170.

Muranyi, P., Wunderlich, J., Heise, M., 2007. Sterilization efficiency of a cascaded dielectric barrier discharge. J. Appl. Microbiol. 103, 1535–1544.

Muranyi, P., Wunderlich, J., Heise, M., 2008. Influence of relative gas humidity on the inactivation efficiency of a low temperature gas plasma. J. Appl. Microbiol. 104, 1659–1666.

Muranyi, P., Wunderlich, J., Langowski, H.C., 2010. Modification of bacterial structures by a low-temperature gas plasma and influence on packaging material. J. Appl. Microbiol. 109, 1875–1885.

Pankaj, S.K., Bueno-Ferrer, C., Misra, N.N., Bourke, P., Cullen, P.J., 2014a. Zein film: effects of dielectric barrier discharge atmospheric cold plasma. J. Appl. Polym. Sci 131, 40803.

Pankaj, S.K., Bueno-Ferrer, C., Misra, N.N., Milosavljević, V., O'Donnell, C.P., Bourke, P., Keener, K.M., Cullen, P.J., 2014b. Applications of cold plasma technology in food packaging. Trends Food Sci. Technol. 35, 5–17.

Pankaj, S.K., Bueno-Ferrer, C., Misra, N.N., O'Neill, L., Jiménez, A., Bourke, P., Cullen, P.J., 2014c. Characterization of polylactic acid films for food packaging as affected by dielectric barrier discharge atmospheric plasma. Innov. Food Sci. Emerg. Technol. 21, 107–113.

Pankaj, S.K., Bueno-Ferrer, C., Misra, N.N., O'Neill, L., Tiwari, B.K., Bourke, P., Cullen, P.J., 2014d. Physicochemical characterization of plasma-treated sodium caseinate film. Food Res. Int. 66, 438–444.

Pankaj, S.K., Bueno-Ferrer, C., Misra, N.N., O'Neill, L., Tiwari, B.K., Bourke, P., Cullen, P.J., 2015. Dielectric barrier discharge atmospheric air plasma treatment of high amylose corn starch films. LWT—Food Sci. Technol. 63, 1076–1082.

Patil, S., Moiseev, T., Misra, N.N., Cullen, P.J., Mosnier, J.P., Keener, K.M., Bourke, P., 2014. Influence of high voltage atmospheric cold plasma process parameters and role of relative humidity on inactivation of *Bacillus atrophaeus* spores inside a sealed package. J. Hosp. Infect. 88, 162–169.

Schneider, J., Baumgärtner, K.M., Feichtinger, J., Krüger, J., Muranyi, P., Schulz, A., Walker, M., Wunderlich, J., Schumacher, U., 2005. Investigation of the practicability of low-pressure microwave plasmas in the sterilisation of food packaging materials at industrial level. Surf. Coat. Technol. 200, 962–966.

Starostin, S.A., Creatore, M., Bouwstra, J.B., van de Sanden, M.C.M., de Vries, H.W., 2015. Towards roll-to-roll deposition of high quality moisture barrier films on polymers by atmospheric pressure plasma assisted process. Plasma Processes Polym. 12, 545–554.

Tenn, N., Follain, N., Fatyeyeva, K., Valleton, J.-M., Poncin-Epaillard, F., Delpouve, N., Marais, S., 2012. Improvement of water barrier properties of poly(ethylene-co-vinyl alcohol) films by hydrophobic plasma surface treatments. J. Phys. Chem. C 116, 12599–12612.

Van Deynse, A., De Geyter, N., Leys, C., Morent, R., 2014. Influence of water vapor addition on the surface modification of polyethylene in an argon dielectric barrier discharge. Plasma Processes Polym. 11, 117–125.

Vartiainen, J., Rättö, M., Paulussen, S., 2005. Antimicrobial activity of glucose oxidase-immobilized plasma-activated polypropylene films. Packag. Technol. Sci. 18, 243–251.

Walton, S.G., Lock, E.H., Ni, A., Baraket, M., Fernsler, R.F., Pappas, D.D., Strawhecker, K.E., Bujanda, A.A., 2010. Study of plasma-polyethylene interactions using electron beam-generated plasmas produced in Ar/SF_6 mixtures. J. Appl. Polym. Sci. 117, 3515–3523.

Whittle, J.D., Short, R.D., Steele, D.A., Bradley, J.W., Bryant, P.M., Jan, F., Biederman, H., Serov, A.A., Choukurov, A., Hook, A.L., Ciridon, W.A., Ceccone, G., Hegemann, D., Körner, E., Michelmore, A., 2013. Variability in plasma polymerization processes—an international round-Robin study. Plasma Processes Polym. 10, 767–778.

Ziuzina, D., Patil, S., Cullen, P.J., Keener, K.M., Bourke, P., 2014. Atmospheric cold plasma inactivation of *Escherichia coli*, *Salmonella enterica serovar Typhimurium* and *Listeria monocytogenes* inoculated on fresh produce. Food Microbiol. 42, 109–116.

Chapter 13

Nonthermal Plasma for Effluent and Waste Treatment

B. Jiang[*,†], J. Zheng[†], M. Wu[†]
[*]*Qingdao University Technology, Qingdao, P.R. China,* [†]*China University of Petroleum, Qingdao, P.R. China*

1 Introduction

The food-processing industries are often associated with discharges of volatile emissions, wastewaters, and byproducts that can potentially impact the natural environment and human health (Maxime et al., 2006). Water is substantially utilized in processing raw materials, generating steam, cleaning packing materials, and washing equipment and plants. Thus, a distinguishing feature of the discharged water from the food-processing industry is one of high organic loads, which cannot be efficiently removed by traditional physiochemical or biological processes, eg, coagulation/flocculation, anaerobic, and aerobic methods (Barrera-Díaz et al., 2006). In addition, such pollutants may cause interference to the microbiological processes in wastewater treatment plants, leading to a reduced or even complete failure of treatment efficacy. Consequently, appropriate treatment of the discharged waste from the food industries is essential, which requires developing more viable water purification technologies.

Cold Plasma in Food and Agriculture. http://dx.doi.org/10.1016/B978-0-12-801365-6.00013-5

Nonthermal plasma can be generated by introducing electrical energy into a reaction zone in a selective manner, which leads to various physical and chemical effects, such as primary formation of oxidizing species including: radicals ($H^{\cdot}$, $O^{\cdot}$, $OH^{\cdot}$, etc.) and molecules (H_2O_2, O_3, etc.), shockwaves, ultraviolet light, and electrohydraulic cavitation (Jiang et al., 2014). The formed reactive species can subsequently attack pollutant molecules and decompose them into more environmentally friendly products (Bai et al., 2009). Thus, electrical plasma technology is widely considered as a promising process for aqueous and gaseous pollutants removal.

The performance of nonthermal plasma for pollutant removal is mainly dependent on the production of highly active species. The patterns of electrical discharge, such as pulsed corona discharge, dielectric barrier discharge, and gliding arc discharge, govern the energy efficiency of active species generation for the various devices (Jiang et al., 2014; Locke et al., 2006). Meanwhile, the yield of highly active species is also closely related with the reactor configuration, eg, electrode properties, reactor geometry, and gas input. These issues have been covered by some previous reviews involving plasma technology and its environmental applications (Jiang et al., 2014; Locke et al., 2006; Malik, 2010; Shi et al., 2009; Tijani et al., 2014). It is well known that, in plasma-chemical degradation processes, $OH^{\cdot}$ ($E^0 = 2.8\ V_{NHE}$) is the strongest oxidative species for pollutant destruction; however, other slower reaction pathways such as those induced by O_3 ($E^0 = 2.07\ V_{NHE}$) or H_2O_2 ($E^0 = 1.77\ V_{NHE}$) are shunted or even bypassed (Jiang et al., 2014). Thus, the incorporation of a catalyst into the electrical plasma process can effectively activate mild oxidative products (eg, H_2O_2 and O_3) into more active species, ie, $OH^{\cdot}$, thereby enhancing the destruction of various pollutants. For example, adding iron salts into electrical plasma systems is an attractive strategy, considering the fact that ferrous or ferric ions can catalytically improve the formation of $OH^{\cdot}$ from H_2O_2 via the Fenton reaction. The performance of the combined plasma/Fenton processes are being evaluated for the remediation of organic pollutant-bearing acidic wastewater. In addition to such added catalysts, several other substances, eg, TiO_2, powdered oyster shells, carbon materials, and pyrite, have also been introduced into plasma reactors to promote the formation of $OH^{\cdot}$ or increase the probability of active species attacking the target pollutants (de Brito Benetoli et al., 2012; Grymonpré et al., 2004; Hao et al., 2006; Kušić et al., 2005; Lukes and Locke, 2005; Sano et al., 2011; Wang et al., 2009).

This chapter explores the degree of success achieved by electrical plasma processes in eliminating water and gas pollutants, to illustrate its feasibility in waste treatment for the food-processing industry. In addition, to determine economical feasibility, the applications of various alternative plasma processes for pollutant removal is also reviewed. Furthermore, the synergistic mechanisms of the utilized catalysts to enhance the economic viability of electrical plasma techniques are discussed.

2 Degradation of Pesticides

Modern agriculture is heavily dependent on the use of pesticides, with current annual pesticide use estimated at approximately 2.5 million tons (Gavrilescu, 2005). These substances help farmers to increase production, either by diminishing external problems (ie, preventing the occurrence of insects, fungi, viruses, etc.), acting as plant growth regulators, or by preventing the growth of weeds. Due to the nonbiodegradable nature of many pesticides, these compounds may be introduced to the environment and move to surface and groundwater, giving rise to serious environmental problems and health concerns. The World Health Organization (WHO) estimates that pesticides poison 2–3 million people and cause between 20,000 and 200,000 accidental deaths each year, primarily in developing countries (Rodrigo et al., 2014).

2.1 Phenolic Compounds

Phenol and phenolic compounds are extensively used as raw materials in the manufacture of pesticides; thereby, such contaminants are commonly present in the related industrial effluents (Feng et al., 2014; Grymonpre et al., 1999; Hoeben et al., 2000; Krugly et al., 2015; Lesage et al., 2014; Li et al., 2007a; Lukes and Locke, 2005; Marotta et al., 2012; Qu et al., 2013; Sano et al., 2002; Tang et al., 2009; Wang et al., 2012). In this regard, their effective removal from wastewater streams is a serious environmental concern. Notably, phenolic compounds are usually selected as models to estimate the oxidative mechanism of $OH^{\cdot}$ produced by nonthermal plasma in liquid or in a gas/liquid hybrid phase. The preferable performance of electrical plasma technologies can be obtained for the degradation of phenolic compounds in various reactors. An example of the fast removal of phenolic compounds by this technology was reported by Zhang et al. (2007b) for 4-chlorophenol, a common phenolic derivative. Almost 100% removal of the compound (500 mL, 60 mg/L) was found after only 7 min of continuous plasma treatment with an O_2 flow rate of 0.4 m^3/h, input voltage of 16 kV, and a frequency of 100 Hz. The superior removal efficiency was also confirmed in the case of phenol degradation behavior using a liquid discharge reactor (Wang and Chen, 2011). The constant rate and energy efficiency of the process was 3.1×10^{-2} min^{-1} and 14.6×10^{-9} mol/J, respectively. Furthermore, these two indices of phenol removal were enhanced by 68% and 59%, respectively, when used in combination with TiO_2-mediated photocatalysis.

In the plasma-chemical phenol degradation process, due to the electron-donating character of the phenolic OH group and the electrophilicity of the active radicals, some hydroxylation products, ie, resorcinol, 1,4-benzoquinone, hydroxyhydroquinone, *o*-dihydroxybenzene and *p*-dihydroxybenzene are commonly detected and can be quantitatively determined using HPLC (Grymonpre et al., 1999; Tezuka and Iwasaki, 1997). Also, some small byproducts, such as acetic acid, formic acid, and oxalic acid, are also identified and can remain stable until the organic load becomes negligible (Hoeben et al., 2000). As depicted in Fig. 1, a phenol degradation mechanism was proposed by Shen et al. (2008) with analyses of the intermediates generated in a liquid discharge reactor without any gas bubbling and chemical catalyst addition. In the series of oxidative reactions, $OH^{\cdot}$ electrophillically attacks phenol molecules, initially producing dihydroxycyclohexadienyl radicals, which can ultimately transform to dihydroxy benzoic acids (DHBs). Formation of hydroquinone and catechol follow the reactions of DHBs through elimination of a H_2O molecule. With further oxidations under the action of the active radicals, triphenols are formed as the intermediate products. The opening of the aromatic rings results in the generation of low molecular weight organic acids, carbon dioxide, and water. However, when the reaction is carried out with addition of Fe^{2+}, the hydroxyl groups of aromatic rings are the identified attack sites, which can be oxidized to yield benzoquinones. It has been noted that under the condition of the bubbling of O_3 gas, phenol degradation occurs through not only hydroxylation, but also direct/indirect ozonation.

In contrast, when high-voltage pulsed corona discharge takes place in an argon atmosphere over an aqueous-phenol solution, the active argon ions and metastable species produced by corona discharge can lead to the formation of $OH^{\cdot}$ via the dissociation of water molecules, but no O_3 can be formed (Hoeben et al., 2000). The hydroxylation reaction is the main pathway of phenol degradation, and hydroxybenzenes will be

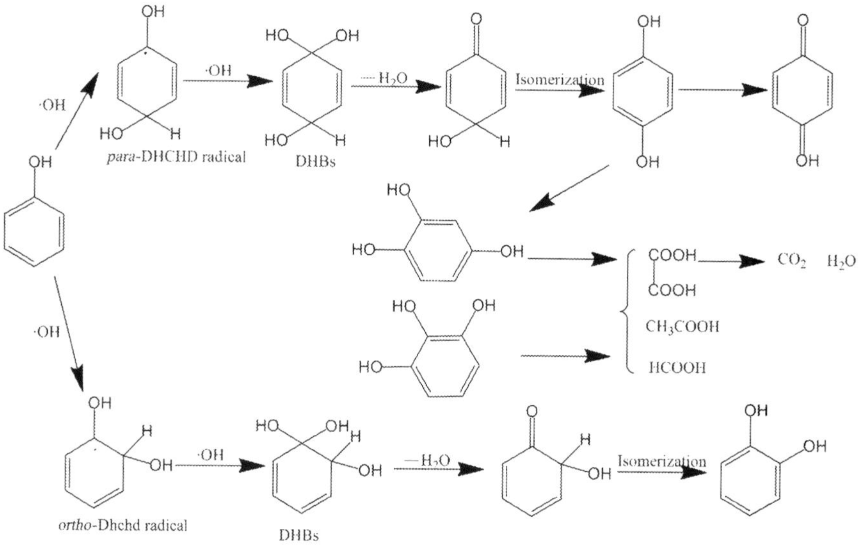

Fig. 1 The degradation pathway of phenol in a liquid discharge reactor without any gas bubbling and chemical catalyst addition (Shen et al., 2008).

detected in much higher concentration than in the case of an air atmosphere. Based on the results reported in the literature (Hoeben et al., 2000; Kušić et al., 2005), it can be concluded that degradation pathways differ greatly based on conditions employed, such as the discharge pattern, gas atmosphere, chemical additions, and solution properties.

Another typical phenolic compound, 4-chlorophenol, is also widely selected as a degradation target by electrical discharge plasma, with $OH^{\cdot}$ principally responsible for its oxidation; formations of 4-chlorocatechol, hydroquinone, and 4-chlororesorcinol are the products in its first degradation step (Bian et al., 2011). Bian et al. (2011) demonstrated that the production of 4-chlorocatechol was about twice as much as that of hydroquinone. The *ortho* position of the hydroxyl group was more active in comparison with that of the Cl group for the 4-chlorophenol molecule, leading to a lower production of 4-chloresorcinol. The ring cleavage reaction could be caused by the direct cleavage of the C_1–C_2 or C_5–C_6 bond of 4-chlorophenol, producing 4-chloro-2,4-hexadien-1-ol, which is probably derived from other reaction routes induced by energetic electrons, local high temperature, or intense shock waves.

The decomposition of phenol can be enhanced in an electrical corona discharge device, in which a cylindrical stainless steel electrode, acting as the anode, is covered with multiwalled carbon nanotubes (Sano et al., 2011). The surface area of the multiwalled carbon nanotubes covering the anode was approximately 332 times larger than that of the bare stainless steel anode. This increase in the anode surface area can greatly enhance the oxidation of phenol. The increased effectiveness can be explicated by considering that the oxidation of phenol by O_3 was improved due to the concentrating effects of O_3 and phenol on the surfaces of the multiwalled carbon nanotubes.

When TiO_2 was introduced into a pulsed discharge, a synergistic effect of plasma and TiO_2 photocatalysis was observed for phenol degradation, through increased generation of oxidative species ($OH^{\cdot}$, $O^{\cdot}$, and H_2O_2) (Wang et al., 2008). The combined process also exhibited improved performance for parachlorophenol removal in a pulsed electrical liquid discharge reactor (Hao et al., 2006). The rate constant of parachlorophenol degradation was significantly improved with the presence of TiO_2, from 1.56×10^{-3} to $2.81 \times 10^{-3}\ s^{-1}$, with approximately double the energy efficiency for parachlorophenol removal. This enhancement was attributed to an increased production of chemically active oxygen species (eg, O_3, H_2O_2, and especially $OH^{\cdot}$).

2.2 PESTICIDES

Recently, many pesticides have been investigated as targets, using electrical discharge plasmas to minimize the associated potential health risks (Li et al., 2013; Wohlers et al., 2009). Diuron, a nonselective herbicide, is widely applied for weed eradication and its degradation behavior has been investigated using a gas-liquid hybrid electrical discharge (Feng et al., 2009). Diuron degradation behavior induced by plasma followed first-order kinetics and its removal efficiency increased with increasing energy input and Fe^{2+} concentration. The presence of Cu^{2+} retarded the oxidation efficiency of diuron during the first 4 min of treatment, while increased oxidation efficacy was observed during the last 10 min of treatment. In this process, the generated products, such as oxalic acid, acetic acid, formic acid, chloride ion, ammonium ion, and nitrate ion were qualitatively detected and based on which, the degradation mechanism of diuron was proposed as depicted in Fig. 2. The methyl group and the aromatic ring of diuron are the pregnable sites that are usually attacked by the active species. $OH^{\cdot}$ reacts with one aliphatic methyl group of diuron by abstracting a hydrogen atom, leading to the formation of a radical ($R_1^{\cdot}$). In aqueous solution, $R_1^{\cdot}$ reacts with oxygen preferentially, yielding a peroxyl radical ($R_1OO^{\cdot}$). The disproportionation of $R_1OO^{\cdot}$, probably passing through a tetraoxide transient, leads to the formations of two oxidation products compounds 1 and 2. Compound 5 is identified as the product formed via $OH^{\cdot}$ attacking the methyl group of compound 2. In addition, compounds 1 and 4 are detected as the degradation products of compound 5. Once the first methyl group is eliminated, $OH^{\cdot}$ will attack the second methyl group of diuron and that results in the production of compound 3. The formation mechanism of compound 6 is similar to that of compound 2. Cleavage of the aromatic ring is suggested by the generation of organic acids. Finally, mineralization is realized by producing H_2O and CO_2 as the final products.

In a pulsed electrical liquid discharge system, the oxidation of atrazine was initiated by $OH^{\cdot}$ and at pH 3, with 90% of atrazine (2×10^{-5} M) removed within 60 min at a voltage of 45 kV and pulse repetition frequency of 60 Hz (Mededovic and Locke, 2007). When the Fe^{2+} ion was utilized as a catalytic electrolyte, atrazine was rapidly oxidized within 10 min, owing to the significant additional $OH^{\cdot}$ formation via a Fenton reaction at the cost of H_2O_2. When a platinum electrode was selected as an anode, in contrast to NiCr, the removal of atrazine was accelerated as sodium persulfate was used as an electrolyte. However, total organic carbon (TOC) measurements indicated no mineralization for any electrode or electrolyte within 60 min of treatment. All byproducts were qualitatively identified as ammelines by using HPLC and GC/MS with electron spray ionization, and the mechanism is suggested in Fig. 3. A similar degradation pathway to atrazine was also demonstrated by Zhu et al. (2014) using a wire-cylinder reactor dielectric barrier discharge. In addition, considering the pesticide concentration in wastewater and natural water typically ranging up to μg/L only, Vanraes et al. (2014) designed a combined nanofiber membrane/dielectric barrier discharge process with liquid flowing as a thin layer on one of the electrodes, and estimated its treatment efficiency for micropollutant atrazine removal (30 μg/L). In this process, the addition of a nanofiber membrane close to the plasma-active region led to a higher local atrazine concentration near the plasma-liquid interface. Thus,

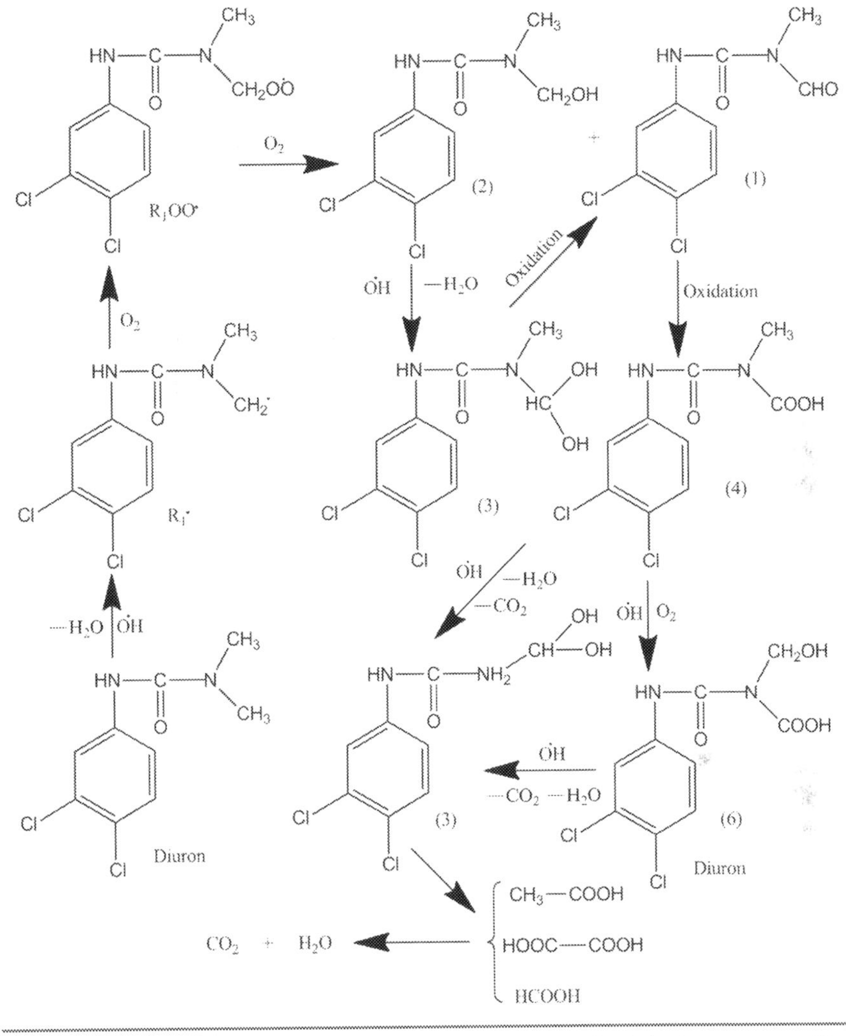

CH_3
HN—C—N
O
$CH_2O\dot{O}$
Cl
Cl
$R_1OO^{\cdot}$
O_2
CH_3
HN—C—N
O
CH_2OH
Cl
Cl
(2)
+
CH_3
HN—C—N
O
CHO
Cl
Cl
(1)
O_2
CH_3
HN—C—N
O
$CH_2^{\cdot}$
Cl
Cl
$R_1^{\cdot}$
$\dot{O}H$ —H_2O
Oxidation
Oxidation
CH_3
HN—C—N OH
HC
OH
O
Cl
Cl
(3)
CH_3
HN—C—N
O
COOH
Cl
Cl
(4)
—H_2O $\dot{O}H$
CH_3
HN—C—N
O
CH_3
Cl
Cl
Diuron
$\dot{O}H$ —H_2O
—CO_2
OH
CH—OH
HN—C—NH_2
O
Cl
Cl
(3)
$\dot{O}H$ O_2
CH_2OH
HN—C—N
O
COOH
Cl
Cl
(6)
Diuron
$\dot{O}H$
—CO_2 —H_2O
CH_3—COOH
CO_2 + H_2O
HOOC—COOH
HCOOH

Fig. 2 The degradation pathway of diuron in a gas-liquid hybrid discharge reactor (Feng et al., 2009).

atrazine decomposition can be significantly accelerated with the introduction of a membrane. With the membrane, 85% of the atrazine was removed, in comparison to only 61% removal without the membrane under the same operating parameters. The byproducts of atrazine decomposition identified by HPLC-MS are deethylatrazine and ammelide. Formation of these byproducts is more pronounced when the membrane is added. These results indicate the synergetic effects of the plasma discharge process and pollutant adsorption behavior, which is attractive for future applications of water treatment.

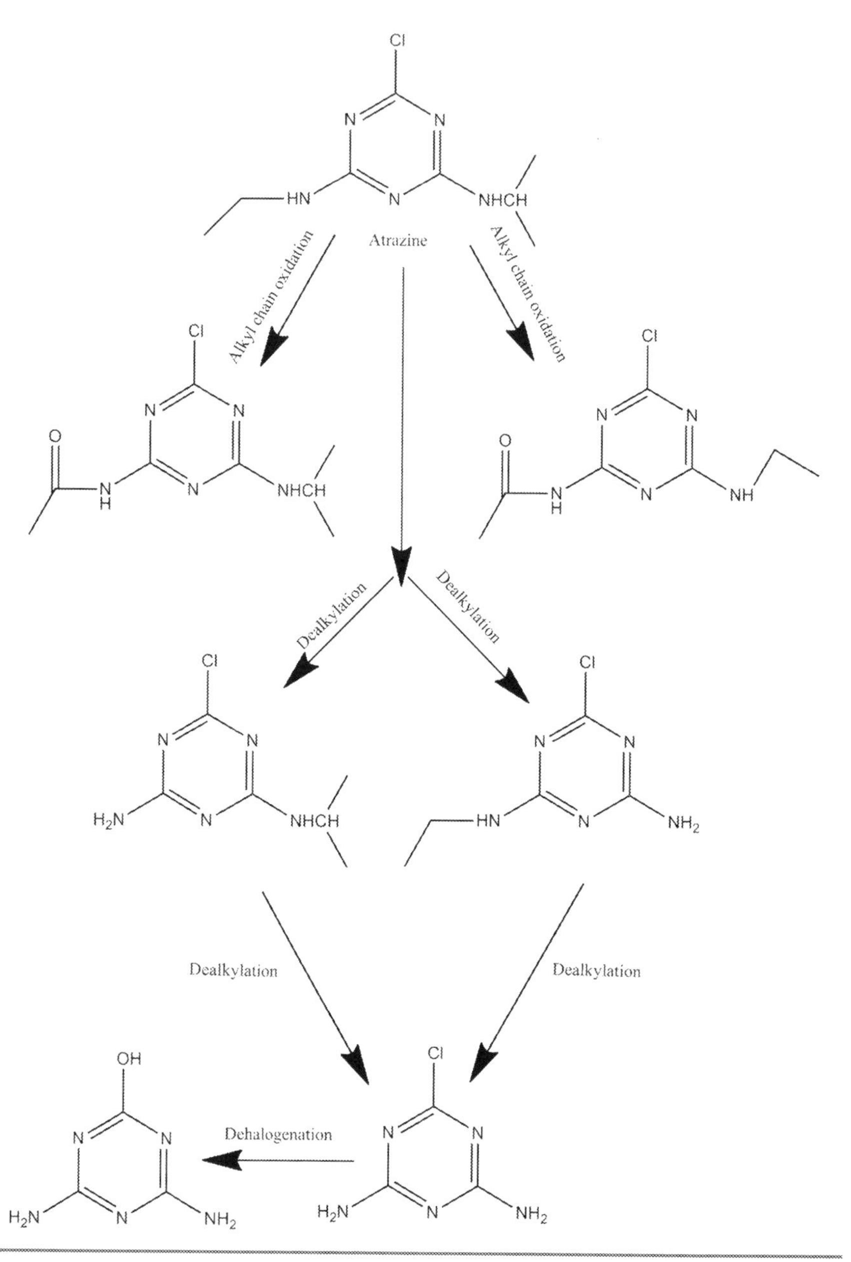

Fig. 3 The degradation pathway of atrazine in a pulsed electrical discharge reactor (Percy and Locke, 2007).

Acetamiprid, a neonicotinoid insecticide, is an anitromethylene heterocyclic compound which has been widely used in agriculture because of its insecticidal efficacy. However, it is associated with serious environment hazards (Li et al., 2010). It was found that acetamiprid could be effectively removed from the wastewater solution (via a first-order kinetics) using a dielectric barrier discharge plasma process with a removal efficiency of 83% at 200 min treatment time for an energy input of 170 W and an initial concentration of 50 mg/L (Li et al., 2014). The addition of $Na_2B_4O_7$ as a radical scavenger greatly inhibited the degradation process, validating that the acetamiprid degradation was partially dominated by the attack of $OH^{\cdot}$ on acetamiprid molecules. The production of O_3 cannot only oxidize acetamiprid directly, but it also transforms to $OH^{\cdot}$ to degrade acetamiprid indirectly, which was proved as a more significant role. In this reaction system, acetamiprid could be decomposed into inorganic carbon and inorganic nitrogen, if the discharge time is sufficiently long.

The recalcitrant organophosphate compounds, eg, the molecules present in some pesticides or chemical warfare regents, threaten environmental safety due to their high toxicity or their toxic degradation products. Pascal et al. (2010) used a gliding arc discharge reactor to degrade triethylphosphate solute at atmospheric pressure and reported that with applied energy of 2.4 kW, a 59% TOC removal and a 34% triethylphosphate degradation, represented by the concentration of the formed inorganic phosphate, can be achieved after 6 h of treatment. The kinetics of triethylphosphate degradation behavior obeyed zero order laws with a 3.4×10^{-4} mol/h degradation rate. In the gliding arc discharge reactor, solution recycling and the nozzle design led to the substantial enhancement of the triethylphosphate removal rate. The suggested reaction pathway in this reactor is dependent on the diffusion of $OH^{\cdot}$, as the rate-determining species, in the plasma-generating zone. Peroxynitrite may also account for the degradation of triethylphosphate, the diffusion of which in the liquid phase should control the degradation rate.

Jović et al. (2013) applied a dielectric barrier discharge process to investigate the degradation of mesotrione and they compared the mechanisms with three other advanced oxidation processes, namely ozonization, photocatalysis, and the Fenton processes. It was proposed that the degradation mechanisms in the aforementioned four technologies were not based only on the production of $OH^{\cdot}$, but similarities in the degradation mechanisms were observed between ozonation and dielectric barrier discharge, which suggested that O_3 plays a significant role for pollutant removal in a dielectric barrier discharge process. Mesotrione is very harmful and toxic to aquatic organisms. However, after treatment with the dielectric barrier discharge, the treated pesticide's toxicity levels were found to be "low" to "nonexistent." The effect of the water matrix on the pollutant removal was also investigated for both distilled water and in real water systems. The degradation of mesotrione in river water was retarded due to the competition of organic/inorganic matter for reactive species. It was found that compared with the real river water matrix, mesotrione degradation was faster in distilled water (approximately 20% faster) when subjected to dielectric barrier discharge treatment.

Dimethoate, one of a broad spectrum dithiophosphate pesticide, has also been studied as a target pollutant and its degradation behavior investigated using dielectric barrier discharge process. The optimal efficiency of greater than 96% was obtained for the following treatment parameters: initial concentration = 20 mg/L, applied power = 85 W, air-gap distance = 5 mm, and treatment time = 7 min (Hu et al., 2013). The degradation rate can be accelerated by adding radical promoters (eg, hydrogen peroxide and Fe(II) ion). In this process, $OH^{\cdot}$ plays an important role in the degradation of dimethoate, and the suggested pathway is shown in Fig. 4. Dimethoate degradation induced by this plasma process begins with the attack of $OH^{\cdot}$ at the bond of P=S, leading to the formation of P=O of omethoate. Subsequently, omethoate can be further rapidly oxidized to C_2 and C_3 by $OH^{\cdot}$ through the scission of P—S bonds and the release of N-methyl acetamide groups, respectively. Meanwhile, the scission of S—C bonds by $OH^{\cdot}$ probably results in the production of compounds C_1 and C_4, through route 3 and the successive attack of $^{\cdot}CH_3$ and $^{\cdot}OCH_3$. And these intermediates can be finally converted into small nontoxic products, eg, PO_4^{3-}, H_2O, and CO_2.

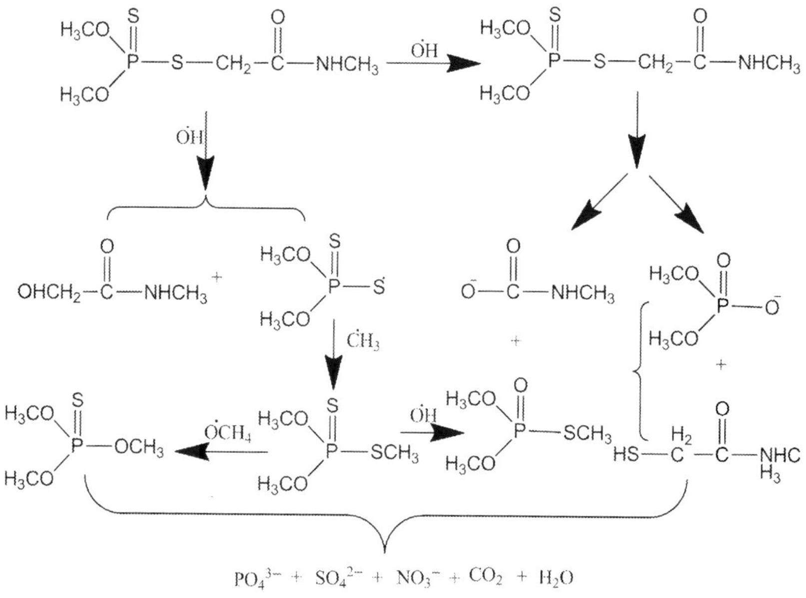

Fig. 4 The degradation pathway of dimethoate in a dielectric barrier discharge reactor (Hu et al., 2013).

The removal of organic pesticides, such as atrazine, chlorfenvinfos, and lindane, from aqueous solutions (1–5 mg/L) were also assessed in two atmospheric pressure dielectric barrier discharge reactors, ie, a conventional batch reactor and a coaxial thin falling water film reactor, at laboratory scale (Hijosa-Valsero et al., 2013). A first-order degradation kinetics was fitted for both experiments. The kinetic constants were slightly faster in the conventional batch reactor (0.534 min^{-1} for atrazine; 0.567 min^{-1} for chlorfenvinfos; and 0.389 min^{-1} for lindane) than in the coaxial thin falling water film reactor (0.104 min^{-1} for atrazine; 0.523 min^{-1} for chlorfenvinfos; and 0.294 min^{-1} for lindane). However, energy efficiencies were about one order of magnitude higher in the coaxial thin falling water film reactor [89 mg/(kWh) for atrazine; 447 mg/(kWh) for chlorfenvinfos; and 50 mg/(kWh) for lindane], than those in the conventional batch reactor. Some degradation intermediate products were identified by GC-MS in distilled water. They were deethylatrazine, deisopropylatrazine, atrazine amide, and 4-chloro-6-(ethenylamino)-1,3,5-triazin-2-ylacetate for atrazine; 2-(2,4-dichlorophenyl)-2-oxoacetyl chloride for chlorfenvinfos; bromophenolisomer, phenol, and ethyl 4-ethoxybenzoate for 2,4-dibromophenol; and 1,3,5-trimethylbenzene and 1,2,4-trimethylbenzene for lindane.

2.3 Pesticides in Soils

Pesticides are widely used to protect plants from disease, weeds, and insect damage, however, many are found to be persistent soil contaminants (Wang et al., 2011a). An interesting contribution of Li's group (Lou et al., 2012; Wang et al., 2010, 2011a,b, 2014) has been evaluating the potential application of electrical plasma technologies (eg, pulsed corona discharge process and dielectric barrier discharge) for remedying pesticide-contaminated soils.

In the case of pentachlorophenol-contaminated soil, its remediation efficacy with pulsed corona discharge plasma was enhanced with increasing peak pulse voltage or pulse frequency (Wang et al., 2010). The removal efficiencies of pentachlorophenol in oxygen and air atmospheres were 92% and 77%, respectively, after a 45-min plasma treatment at a voltage of 14.0 kV, while for argon and nitrogen atmospheres only 19% and 8% respectively, were found. Under an oxygen atmosphere, the input energy was mainly consumed to produce reactive oxygen species, whose concentrations were larger than those under air atmosphere, and thus led to larger degradation efficiencies for pentachlorophenol. Moreover, O radicals and O_3 have much stronger oxidative potentials than those of atomic N, N_2^+, and N^+; consequently, degradation efficiency of pentachlorophenol was higher under air atmosphere than that under the N_2 atmosphere. Based on the analysis of HPLC/MS and ion chromatography, the main intermediates formed during the treatment process were determined to be tetrachlorohydroquinone, tetrachlorocatechol, formic acid, acetic acid, and oxalic acid.

A plasma-TiO_2 technique, where the catalytic effect of TiO_2 was activated by the pulsed electrical discharge plasma, was applied to study the remediation of soil contaminated by organic pollutants (Wang et al., 2011a). In this process, reactive species produced in electrical plasma system probably acted directly on the active sites of TiO_2, and then favored TiO_2 catalytic triggering oxidation reactions. Furthermore, the strong electric field in electrical plasma generation systems can relieve the recombination of holes and electrons on the surface of TiO_2. It was inferred that the roles of O_3, H_2O_2, and $OH^{\cdot}$ were primarily responsible for pollutant degradation. Taking *p*-nitrophenol as an example (Wang et al., 2011a), its hydroxylated intermediates could be produced via electrophilic attack of $OH^{\cdot}$. The aromatic ring of *p*-nitrophenol contains two functional groups, ie, hydroxyl and nitro. The hydroxyl group is an electron donor and an *ortho-/para*-director, while —NO_2 exhibits as an electron-withdrawing group and meta-director. Thus, $OH^{\cdot}$ preferentially attacks the *ortho-* or *para*-position with respect to the OH group due to its electrophilic nature. In addition, $OH^{\cdot}$ probably eliminate nitrous acid from *p*-nitrophenol to yield 1,4-benzosemiquinone as an intermediate, which subsequently disproportionates into hydroquinone and benzoquinone. It is well known that oxidation of organic pollutants by O_3 mainly proceeds through nucleophilic, electrophilic, and cycloaddition reactions (Jiang et al., 2014; Locke et al., 2006). The *ortho-* and *para*-positions with respect to the OH group on *p*-nitrophenol are vulnerable to the nucleophilic and electrophilic attack of O_3, generating the same hydroxylated intermediates as those in the case of $OH^{\cdot}$. Furthermore, cycloaddition intermediates can also be formed via the cycloaddition reactions of O_3 to the *p*-nitrophenol molecule structure.

3 Degradation of Coloring Matter

Due to their low price, high effectiveness, and excellent stability, food dyes are regarded as the most prevalent group of food additives, and the attractiveness of a product for consumers is significantly influenced by its color (Kucharska and Grabka, 2010). However, due to their large-scale production and extensive application, substantial quantities of effluents are generated, which are likely to cause environmental problems if discharged without prior treatment. The direct discharge of colored textile effluents into freshwater bodies adversely affects the aesthetic merit, water transparency, and its dissolved oxygen content. These dyes exhibit highly complex structures and low biodegradability, which account for the toxic effects on flora and fauna present in the water bodies (Dasgupta et al., 2015). Therefore, a wide variety of coloring matters, including azo derivatives, anthraquinone, indigoid, reactive, etc., have been used as substrates for plasma treatment and some typical cases are discussed below.

3.1 Glow Discharge Electrolysis

A specific feature of glow discharge electrolysis is its significantly non-Faradaic chemical effect, which has two reaction zones: the plasma in the surrounding space of the electrode, and the solution near the interface between the plasma and the aqueous electrolyte. Within the primary reaction zone around the electrode, H_2O vapor molecules are electrolytically dissociated at a high temperature with the generation of active species, such as $OH^{\cdot}$, $H^{\cdot}$, and $HO_2^{\cdot}$. Moreover, H_2O+ gas ions generated in the gas phase can bombard the gas-liquid interface via the acceleration of the large electric field, and then react with liquid H_2O molecules to form additional $OH^{\cdot}$ and $H^{\cdot}$. The dissociation of H_2O molecules in glow discharge electrolysis has been described using Eq. (i), in which the value of n was deduced as 7.9–10.6 mol or 12 mol $OH^{\cdot}$ for the passage of each mol electron of electricity by using Fe^{2+} or Fe^{3+}, respectively, as $OH^{\cdot}$ scavengers. In this process, the primary production of $OH^{\cdot}$ in the liquid was validated by an ESR technique using 5,5-dimethylpyrroline-1-oxide as the spin trap, which is regarded as the predominant oxidative species of pollutant removal (Jiang et al., 2014; Wang et al., 2012).

$$H_2O^+ + nH_2O \rightarrow H_3O^+ + (n-1)H^{\cdot} + nOH^{\cdot} \quad \text{(i)}$$

Thus, as for the decolorization of various organic dyes, the nonselective nature of $OH^{\cdot}$ accounts for the cleavage of chromophore groups in most of the plasma treatment processes, which can be indicated by the kinetic behavior:

$$OH^{\cdot} + \text{Chromophore} \rightarrow \text{Products} \quad \text{(ii)}$$

$$OH^{\cdot} + OH^{\cdot} \rightarrow H_2O_2 \quad \text{(iii)}$$

Dye decolorization can be given as:

$$\frac{d[\text{Chromophore}]}{dt} = -k_1[\text{Chromophore}][OH^{\cdot}] \quad \text{(iv)}$$

where k_1 is the rate constant of a $OH^{\cdot}$ reaction with dye chromophore.

As for $OH^{\cdot}$, a steady state will be maintained:

$$\frac{d[OH^{\cdot}]}{dt} = a - 2k_2[OH^{\cdot}] - k_1[\text{Chromophore}][OH^{\cdot}] = 0 \quad \text{(v)}$$

$$[OH^{\cdot}] = \frac{\sqrt{k_1{}^2[\text{Chromophore}]^2 + 8ak_2} - k_1[\text{Chromophore}]}{4k_2} \quad \text{(vi)}$$

It can be seen from Eq. (v) that the concentration of $OH^{\cdot}$ reduces as the dye concentration increases. Therefore, the decolorization of dye does not obey first-order reaction kinetics.

And when [Chromophore]$\ll\sqrt{8ak_2}/k_1$, the equation becomes

$$[\mathrm{OH}^{\cdot}] = \frac{a}{2k_2} \quad \text{(vii)}$$

Therefore, when the initial dye concentration is very low, the dye decolorization fits first-order kinetic as follows:

$$\frac{d[\mathrm{Chromophore}]}{dt} = -\frac{ak_1}{2k_2}[\mathrm{Chromophore}] = -k[\mathrm{Chromophore}] \quad \text{(viii)}$$

For example, with an initial polar brilliant B concentration of 50 mg/L, 56% of the color was removed within 30 min of treatment using glow discharge electrolysis (Wang, 2009). The decolorization does not comply with first-order reaction kinetic in inert solutions. In comparison with other electrical plasma systems, glow discharge electrolysis processes provide a much more preferable condition for additional $OH^{\cdot}$ production via a typical Fenton reaction owing to its higher yield of H_2O_2, which significantly accelerates the abatement of the target pollutant (Gao et al., 2003, 2008). Gao et al. (2003) compared the catalytic effect of various metals, including; Mn^{2+}, Cu^{2+}, Co^{2+}, Fe^{3+}, and Fe^{2+} on the transformation of H_2O_2 to $OH^{\cdot}$, and found that among them, $FeSO_4$ exhibited the most evident catalytic effect for the decolorization of flavine G and brilliant red B. When the initial concentrations of the above dyes were 32 mg/L, the decolorization rate could be completely decolorized in a few minutes with the presence of Fe^{2+}. In another case, the decolorization rate of methyl orange increased almost eight times in the presence of Fe^{2+} in a glow discharge system (Gong et al., 2008). Based on the HPLC-MS results, the main reaction products in the aqueous solution with the presence of Fe^{2+} using a glow discharge system were determined. The proposed mechanistic scheme is presented in Fig. 5, which illustrates the reaction intermediate products and pathways of methyl orange in aqueous solution.

Salts, such as NaCl, are extensively utilized to improve the exhaustion and fixation capacity of dye onto textile materials. Hence, chloride ions are abundantly present in textile wastewater. Ramjaun et al. (2011) reported that Cl^- ions, present at low concentration levels in the solution, exhibited an inhibitory effect on dye decolorization, eg, Reactive Black 5, Reactive Red 239, and Reactive Yellow 176, by $OH^{\cdot}$. However, a higher load of Cl^- ions (>0.02 M) greatly improved the decolorization efficiency of the aforementioned selected dyes during the anodic contact glow discharge process, but did not increase dye mineralization. The defect of this treatment process is the formation of adsorbable organic halogens (AOX), which displays some toxicological properties and is strictly regulated in discharges. Thus, it should be better to pretreat the textile wastewater by a desalination process, or to prolong its treatment time, to ensure the total abatement of the toxic intermediates.

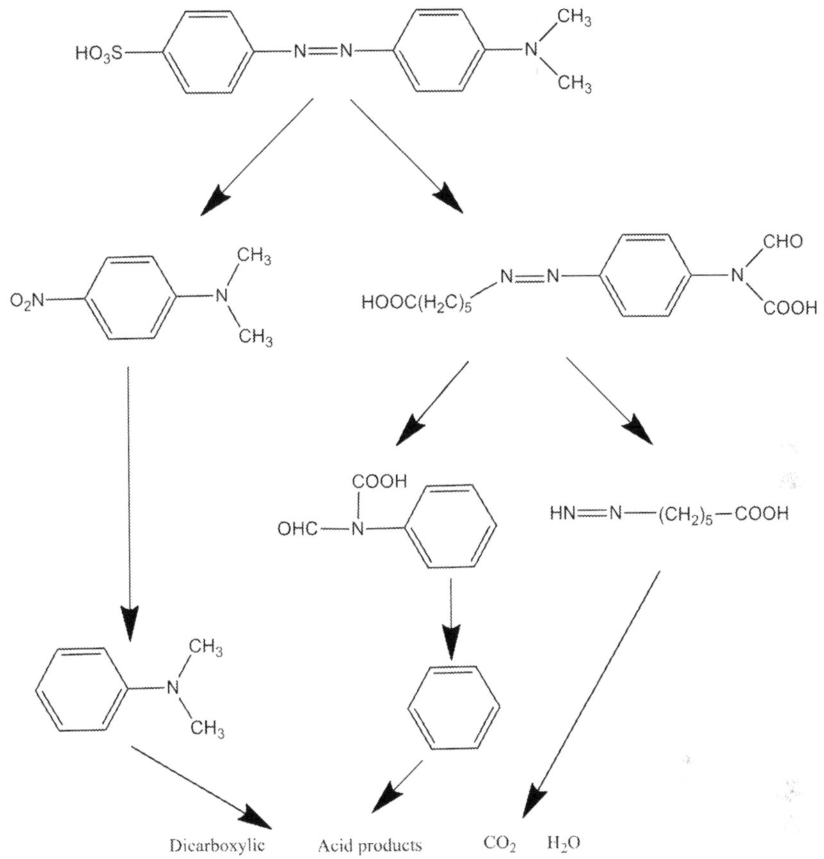

Fig. 5 The degradation pathway of methyl orange in a glow discharge system (Gong et al., 2008).

3.2 Gliding Arc Discharge

The application of a gliding arc discharge for decolorization and mineralization of Crystal Violet was reported by Abdelmalek et al. (2006). In this process, the bleaching of the Crystal Violet solution and its COD (chemical oxygen demand) abatement was reported as 89% and 45%, respectively, after a 60 min. The degradation of the target dye solutions proceeds via two main parallel steps: (a) the bleaching of the dye pollutant with the productions of achromatistous intermediates and (b) the further oxidation of the organic intermediates to CO_2 and H_2O. In their other study (Doubla et al., 2008), bromothymol blue was found to be rapidly decolorized within 10 min. For treatment times above 10 min, alkaline bromothymol blue solutions changed from blue to yellow, probably induced by both acidic effects and fading reactions. In addition, adding H_2O_2

markedly improved decolorization from 60.6% to 94.0% after 5 min treatment. Furthermore, postdischarge oxidation reactions initiated by the reactive species with a relatively long life span increased the decolorization efficiency by 10%. In this reaction system, two stages of dye bleaching, namely a slow H_2O_2 oxidation process and a fast step, attributable to $OH^{\cdot}$, were observed. In this process, when dye solutions are exposed to electrical discharge in an air atmosphere, nitrite ions can be formed and then transformed to an active intermediate, namely peroxonitrite (Eqs. ix, x), which was also propitious to the degradation of the dye molecule. This can explain the synergistic effects of the addition of H_2O_2 on the decolorization reaction rate.

$$NO_2^- + H_2O_2 \rightarrow ONO_2^- \quad \text{(ix)}$$

$$HNO_2 + H_2O_2 \rightarrow HOONO + H_2O \ (\text{in moderately acidic medium}) \quad \text{(x)}$$

A typical gas-liquid gliding arc discharge reactor was applied for acid orange II removal, in which the dye degradation rate decreased, but the absolute degradation quantity was enhanced with increasing initial concentration (Yan et al., 2008). In the degradation process, most organic nitrogen converted to molecular nitrogen via $OH^{\cdot}$ attacking the azo linkage-bearing carbon of a hydroxy-substituted ring, and then the formed nitrogen molecule volatilized or was further transferred to NO_2^- and NO_3^-. The initial acid orange II solution has poor biodegradability because of its very low BOD_5/COD ratio (approximately 0.02). However, after 15 min of plasma oxidation, the BOD_5/COD increased from 0.02 to 0.32, which further increased to 0.43 with extension of the treatment time to 20 min. This result was also validated by Liu and Si (2012) for acid orange II, methyl orange, direct fast black and reactive red K-2BP which were all promoted to values above 0.4. These results suggest it is more useful to combine gliding arc plasma with traditional biological process for the remediation of wastewaters contaminated by generally nonbiodegradable dye.

3.3 Dielectric Barrier Discharge

Better insights into the degradation mechanisms of dyes have been also elaborated (Huang et al., 2010, 2012; Xue et al., 2008). Studies on methylene blue, alizarin red, and methyl orange using a typical gas-liquid planar dielectric barrier discharge reactor revealed that the degradation process of these dyes mainly contains three reaction stages, ie, bond-breaking oxidation, ring-opening oxidation, and complete oxidation. In the case of alizarin red as a target, the primary steady intermediates, including glyoxylic acid and vinyl formate, etc., in solutions were identified using GC-MS.

The decolorization of some reactive dyes (ie, Reactive Black 5, Reactive Blue 52, Reactive Yellow 125, and Reactive Green 15) were reported by Dojčinović et al. (2011) using a novel dielectric barrier discharge with a coaxial water falling film. The $OH^{\cdot}$ generated in this plasma reactor induced the formation of longer-lived oxiditive species, resulting in almost complete decolorization of the tested dyes with the energy input of 90 kJ/L after 24 h residence time. Specifically, in the most effective decolorization, approximately 97% was achieved by adding 10 mM H_2O_2 in the case of 80 mg/L Reactive Black 5 with 45 kJ/L, after a residence time of 24 h. Toxicity tests, conducted by using A. salina, indicated that after a plasma treatment with 50% and 90% decolorization, respectively, the toxicity of Reactive Blue 52 and Reactive Yellow 125 were greatly reduced to approximately 0% mortality. However, for 100 mg/L Reactive Green 15, the solution at 90% decolorization exhibited a slight increase of toxicity of 4% mortality, in comparison with its virgin solution. For Reactive Black 5, there was no evident variation in toxicity after the plasma treatment, except for its initial concentration of 50 mg/L and 50% bleaching, as toxicity decreased to 0% mortality.

Tichonovas et al. (2013) developed a novel gas dielectric barrier discharge reactor for plasma generation that can be cooled by wastewater and air (feed gas). In this reactor, the plasma was formed in a quartz tube with a central gas-filled glass electrode immersed in the liquid; ambient air was used as a feed gas for the plasma generation. UV light produced in the quartz tube provides another possibility of intensifying this treatment process due to its synergistic action with the oxidizing species produced in situ or with adscititious photocatalysts, such as TiO_2. In this reactor, ozone was produced with a concentration ranging from 0.19 to 0.46 mg/s, with the corresponding discharge power ranging from 3 to 33 W. For four groups (Astrazon, Realan, Lanaset, and Optilan) of industrial textile dyes (of 13 tested overall), the demand of energy for separate dye decolorization ranged from 18.7 to 866 kJ/g. A majority of the 13 dyes (up to 50 mg/L) were bleached by up to 95% after 300 s of plasma treatment. The FTIR analysis suggested that the degradation byproducts of these dyes mainly consisted of carboxylic acids, nitrates, amides and amines. With ecotoxicity tests, it was found that the degradation byproducts were less toxic than the stable dyes in the effluent immediately after dying and easily they were bio-degradable, and they can be further oxidized via the subsequent biological treatment.

A prototype dielectric barrier discharge plasma source coupled with TiO_2 was designed by Ghezzar et al. (2013) to treat flowing liquid under a gravity falling film. The optimized concentration of TiO_2 (4 g/L) to the dye solution enhanced the discoloration rate from 11% to 96% for an energy density of 220 kJ/L. In the presence of the photocatalyst, the concentration of TOC decreased by 80%, while no mineralization was observed for the case free of the photocatalyst. The proposed synergistic effect of plasma-catalytic mechanism is: (i) dye molecule adsorption by the photocatalyst, (ii) degradation on the catalyst surface, and finally (iii) desorption of intermediate products into the bulk solution. Decolorization is induced by cleavage of the chromophore azo —N=N— group indicated by the maximum absorption wavelength of 427 nm. Mineralization means that all carbon bonds (C—C, C=C,

C—N, C—S, C=N) are broken, leading to the production of CO_2, SO_4^{2-}, NO_3^-. In a plasma/photocatalysis combined process, whatever the amount of added TiO_2 in the suspension, the mineralization of aqueous dye can be improved. For example, approximately 80% of organic carbon was eliminated with 4 g/L TiO_2 at an energy density of 220 kJ/L.

The above combined plasma/photocatalysis process potentially permits the removal of solution recirculation and thus reduces the required energy input. Consequently, the DBD/TiO_2 process can be regarded as an economical, efficient and promising industrial technology.

3.4 Pulsed Electrical Discharge

In pulsed electrical liquid discharge system, the spark-streamer hybrid mode is known to be the most effective process for aqueous dye decolorization in comparison to separate streamers or a spark discharge mode (Sugiarto et al., 2003). The spark-streamer discharge regime is characterized by a great amount of electrical discharge channels distributed in a bush like fashion with a large volume in the gap between the discharge electrodes. Under this condition, $OH^{\cdot}$ is favorably produced via electron impact dissociation in the electrical discharge channels, which can account for the highest efficiency in the decolorization of dyes (Jiang et al., 2014).

Electrical discharges can also occur in the gas phase above the liquid surface, which reduces the discharge inception voltage and enhances the production of active oxygen species, eg, O_3, under an oxygen-containing gas atmosphere. Thereby, the removal of aqueous contaminants can be intensified by these abundant active species and the treatment efficiency is governed by the diffusion velocity at the interface of the gas and liquid. Consequently, a gas-liquid hybrid discharge reactor typically displays a higher efficacy for pollutant destruction over separate liquid or gas electrical discharge reactors. For example, the apparent rate constant for acid orange II removal in a hybrid gas-liquid reactor (9.14×10^{-2} min^{-1}) was significantly higher than the sum of that obtained in a liquid discharge reactor (1.50×10^{-2} min^{-1}) and a gas discharge reactor (3.06×10^{-2} min^{-1}). To quantitatively evaluate the enhancement in the degradation rate with the hybrid gas-liquid reactor, a promoted factor was calculated to be about 302% (Yang et al., 2005).

The decolorization of dyes, eg, methyl orange, basic orange, methylene blue and eosin yellowish, were examined by Liu et al. (2014b), Xiao et al. (2012), and Zhang et al. (2008, 2010b) using a hybrid gas-liquid reactor. In this reactor, a modified activated carbon fiber (ACF) significantly enhanced the degradation of dyes with high energy efficiency (Jiang et al., 2013a; Zhang et al., 2007a, 2010a). Specifically, a TiO_2-modified ACF did not only adsorb the pollutants forming concentrated center on the surface of the ACF, but also photocatalytically produce highly active species, ie, $OH^{\cdot}$, with a high-voltage discharge zone acting as the lamphouse (Li et al., 2007b;

Zhang et al., 2010a). Thus, for methyl orange, the COD removal increased to 93% in the combined treatment, which was much higher than that of 57% for the pulsed electrical discharge treatment alone. The surface morphology of TiO_2-modified ACF did not evidently change, even after several cycles for the methyl orange decolorization process in the pulsed discharge reactor.

The decolorization of methyl orange has also been investigated by Jiang et al. (2012) using a novel circulatory airtight reactor. After 20 min treatment, 92% of methyl orange was bleached with the energy efficiency of 11.68 g/(kWh), at the gas velocity of 0.08 m^3/h, and input energy of 5.67 W. In comparison to the other three reactor systems, namely liquid discharge, gas discharge, and hybrid gas-liquid discharge, this circulatory airtight reactor system displays a prominent energy efficiency and organic pollutants removal rate. In this process, the energetic electrons emitted from the high-voltage needle electrode directly attacked the aqueous dye molecules and simultaneously dissociated H_2O molecules to $OH^{\cdot}$ in the gas phase or at the gas-liquid interface, contributing greatly to the dye bleaching effect. In addition, due to the circulation of the liquid above the earthed electrode, there was continual fresh solution exposure to the highly reactive plasma zone, which effectively relieved mass transfer resistance to some extent. Furthermore, unused long-lived reactive species, mainly O_3, can be pressured into the reservoir, initiating further oxidation of organics.

4 Destruction of Off-Odors

Estimates from the food-manufacturing sectors, including baking, vegetable oil extraction, solid fat processing, etc., have indicated that approximately 260,000 metric tons of volatile organic compounds (VOCs) per year are discharged in Europe alone (Preis et al., 2013). Although the relatively low concentrations of foul-smelling compounds may not constitute a toxic hazard, the deterioration of air quality inevitably poses undeniable effects on people's health.

In 1990–2010, to effectively eliminate olfactory and odor pollution, nonthermal plasma technology has drawn the attention of scientists. The electric energy delivered to the device can be effectively consumed for electron acceleration. They gain a typical temperature of 10,000–250,000 K, while the background gas remains at ambient temperature. In this condition, the primary energetic electrons collide with background molecules generating secondary electrons, photons, ions, radicals, etc. Specifically, the pulsed electrical streamer discharge accelerates the charged electrons, owing the energy of approximately 5–20 eV, which can trigger the vibrational/rotational excitation of water (threshold energy < 1 eV), dissociation of water (threshold energy ≈ 7.0 eV), or even ionization of water (threshold energy ≈ 13 eV) as shown in Eqs. (xi)–(xvii), when H_2O molecule is present in the background gas (Jiang et al., 2014). When the

O_2 molecule is exposed to these energetic electrons, an O atom can be formed via dissociation of O_2 and thus enhance the production of $OH^{\cdot}$ and O_3 (Jiang et al., 2014). This active species is highly nonselective, creating a chemically reactive environment, in which the harmful substances in odorous and toxic gases can be attacked and finally converted into nontoxic molecules.

Dissociation:

$$H_2O + e \rightarrow OH^{\cdot} + H^{\cdot} + e \tag{xi}$$

Ionization:

$$H_2O + e \rightarrow 2e + H_2O^+ \tag{xii}$$

$$H_2O^+ + H_2O \rightarrow OH^{\cdot} + H_3O^+ \tag{xiii}$$

Vibrational/rotational excitation:

$$H_2O + e \rightarrow H_2O^* + e \tag{xiv}$$

$$H_2O^* + H_2O \rightarrow H_2O + H^{\cdot} + OH^{\cdot} \tag{xv}$$

$$H_2O^* + H_2O \rightarrow H_2 + O^{\cdot} + H_2O \tag{xvi}$$

$$H_2O^* + H_2O \rightarrow 2H^{\cdot} + O^{\cdot} + H_2O \tag{xvii}$$

Several types of cold plasma processes have been applied for off-odor remediation (Mizuno, 2007; Vandenbroucke et al., 2011), including: (a) pulsed electrical plasma (ie, corona, streamer, and spark), which can be generated using pulse high voltages with nanosecond rising rates introduced to an electrode with nonuniform geometry; (b) volume dielectric barrier discharge, which utilizes at least one dielectric material between the discharge gap with the plasma, generally produced between planar electrodes or in cylindrical configurations; (c) surface discharge, which can occur adjacent to a strip metal electrode attached to the surface of an insulating dielectric. The strip electrode is driven by an AC high voltage with respect to a counter electrode with an insulating dielectric in the middle; and (d) when a sufficiently large DC positive/negative voltage is applied to the sharp curvature of an electrode, corona discharge occurs due to the local increase of the electric field. In this case, the mechanism of the electron avalanche formation differs physically, which is dependent on the polarity of the electrode with the low curvature radius (Chang et al., 1991).

The incorporation of heterogeneous photocatalysts into a plasma reactor can increase the retention of the adsorbate molecules in the plasma-generating zone, and therefore result in the enhanced selectivity toward the oxidation of total organics (Van Durme et al., 2007). Furthermore, the additional short-lived oxidizing species formed on the surface of the catalyst leads to synergetic effects for adsorbate pollutant oxidation. For example, atmospheric electrical discharge emits UV radiation with wavelengths between 290 and 400 nm, which falls in the absorption range of TiO_2 (Duten et al., 2002; Sano et al., 2006). This implies that placing photocatalysts, such as $SrTiO_3$, TiO_2, ZnO, ZnS, and CdS, into the plasma radiation zone can accelerate the decomposition of VOCs. In addition, O_3 is a commonly produced species by electrical discharge in oxygen or air atmospheres. The oxidation capacity of O_3 can be further exploited by adding catalysts, exclusively applied for O_3 activation, inside or in close vicinity to the electrical discharge zone. It does not only improve the energy utilization for pollutant removal, but also removes harmful O_3 from the outlet gas stream. It has been reported that the oxides of some metal, such as Mn, Co, Ni, and Ag, have O_3 catalytic decomposing properties (Futamura et al., 2002; Li et al., 2002; Magureanu et al., 2005; Park et al., 2008; Song et al., 2002). These oxides are often supported on some inert materials, eg, γ-Al_2O_3, TiO_2, SiO_2, zeolites, and activated carbon, or on their combinations, which enhances the physical stability and chemical activity of the catalysts. For example, MnO_2 catalytically decomposes O_3 for the VOCs abatement through the following mechanism (Jarrige and Vervisch, 2009; Preis et al., 2013):

(a) the adsorbed atomic oxygen O* can be formed via the catalytic decomposition of O_3 on the surface of MnO_2;
(b) the adsorbed O* heterogeneously reacts with the adsorbate pollutants on the MnO_2 surface; and
(c) O* is desorbed, and the released O* atoms can be further utilized for destroying VOCs in the gas phase.

In the aforementioned combined processes, the electrical plasma interaction with the surfaces of the catalyst provides the opportunity for in situ regeneration of VOC-saturated solid catalysts. This alternate process of adsorption and desorption can effectively change the exhaust gas with a large flow rate and low VOC concentration into one with a low flow rate and high concentration on the catalyst surface, leading to considerable utilization of formed oxidative species. Thus, this combined process can be applied as an economical VOC abatement strategy for small- and medium-scale plants that discharge diluted VOC wastes.

4.1 Pulsed Electrical Discharge

As for off-odors removal, the decomposition of the gases probably influences the treatment efficiency of nonthermal plasma technology. The decomposition of carbon tetrachloride in a N_2 atmosphere with addition of H_2 and O_2 was estimated in a tubular pulsed corona reactor (Huang et al., 2001). The results suggested that the oxidation of carbon tetrachloride is superior with 2% H_2 addition and lower with 2% O_2 addition in comparison with that in the pure N_2 atmosphere. This difference is attributed to the difference in the plasma properties of H_2 and O_2 and their activities in the process of CCl_4 conversion to stable products. The predominant products of CCl_4 decomposition are Cl_2 and ClCN in an N_2 atmosphere; HCl and hydrocarbons, eg, CH_4 in the mixture H_2 (2%)/N_2 gas; as for the mixture of O_2(2%)/N_2, CO_2, CO, $COCl_2$, and probably CCl_3O_2 were produced from CCl_4 decomposition. To remove the toxic byproducts (ClCN and $COCl_2$ derived from the plasma decomposition of CCl_4) in situ absorption by $Ca(OH)_2$ coated on the surface of the grounding electrode was employed with the plasma process. Results showed that $Ca(OH)_2$ in the plasma zone was an effective scavenger in the reaction of CCl_4 decomposition, via the capture of the unwanted byproducts.

The influence of H_2O vapor and the added fly ash on the conversion efficiencies of NO and SO_2 was evaluated in a wire-plate positive pulsed discharge reactor under atmospheric conditions (Yu et al., 2009). The conversion efficiencies of NO and SO_2 are primarily induced by the plasma-produced active oxygen species, such as $OH^{\cdot}$, O atoms, O_3, $HO_2^{\cdot}$, and H_2O_2. With the presence of H_2O vapor, the conversion of SO_2 is accelerated, but NO conversion is inhibited. As for the addition of fly ash, low fly ash concentration enhances the oxidation of NO and SO_2; however, their conversion efficiencies are retarded by high fly ash concentrations. The synergistic effects of H_2O vapor and fly ash strengthen the chemical adsorption capacity of the fly ash, and lead to a considerable enhancement in the conversion of NO and SO_2.

The destruction effectiveness of different olfactory pollutants, such as hydrogen sulfide (H_2S), dimethyl sulfide (DMS), and ethanethiol (C_2H_5SH) in air, were investigated by applying a pulsed corona discharge (Jarrige and Vervisch, 2007). It was found that the aforementioned three selected sulfide compounds, at diluted concentrations, can be completely decomposed when a sufficient energy density was introduced into the plasma reactor. DMS showed the energy input was around 30 eV/molecules, which was less than both for C_2H_5SH (45 eV/molecules) and H_2S (115 eV/molecules). SO_2 was the only identified byproduct in the process of H_2S decomposition, but in spite of not being detected, the formation of SO_3 can be inferred based on the mass balance analysis of sulfur. The identification of byproducts during the degradation of DMS and C_2H_5SH enabled a reaction pathway to be proposed, starting with the attacking of active radicals and breaking of C—S bonds. The degradation of DMS through reaction with atomic oxygen proceeds via the formation of a $CH_3S(O)CH_3$ adduct, which is then quickly decomposed to $CH_3^{\cdot}$ and $CH_3SO^{\cdot}$ as depicted in Eq. (xviii). Methyl radical ($CH_3^{\cdot}$) was considered as an important intermediate in the formation of HCHO in the presence of oxygen. Due to its far lower reactivity than DMS toward atomic oxygen, the removal of HCHO was slower and occurred only when DMS was completely decomposed. The oxidation of HCHO proceeded via the formation of formyl radicals ($HCO^{\cdot}$), which finally turned into CO and to a lesser extent CO_2.

$$CH_3SCH_3 + O \rightarrow CH_3SO^{\cdot} + CH_3^{\cdot} \quad \text{(xviii)}$$

During the removal process of DMS, the primarily observed byproducts were HCHO, SO_2, CO, and CO_2. A second organic product, namely acetaldehyde (CH_3CHO), has also been detected in the case of C_2H_5SH. Unlike DMS, the first reaction step initiated by atomic oxygen was followed by three different pathways as shown in Eqs. (xix)–(xxi). The formed ethyl radical ($C_2H_5^{\cdot}$) primarily converted to acetaldehyde after reaction with oxygen molecules and self-reaction of the $C_2H_5O_2$ radical in Eqs. (xxii)–(xxv).

$$C_2H_5SH + O \rightarrow C_2H_5^{\cdot} + HSO^{\cdot} \quad \text{(xix)}$$

$$C_2H_5SH + O \rightarrow C_2H_5SO^{\cdot} + H^{\cdot} \quad \text{(xx)}$$

$$C_2H_5SH + O \rightarrow C_2H_5S^{\cdot} + OH^{\cdot} \quad \text{(xxi)}$$

$$C_2H_5^{\cdot} + O_2 \rightarrow C_2H_5O_2 \quad \text{(xxii)}$$

$$C_2H_5O_2 + C_2H_5O_2 \rightarrow C_2H_5OH + CH_3CHO \quad \text{(xxiii)}$$

$$C_2H_5O_2 + C_2H_5O_2 \rightarrow 2C_2H_5O^{\cdot} + O_2 \quad \text{(xxiv)}$$

$$C_2H_5O^{\cdot} + O_2 \rightarrow CH_3CHO + HO_2^{\cdot} \quad \text{(xxv)}$$

Although "end-of-pipe" application of nonthermal plasma for the treatment of VOCs has been frequently proposed, the coemission of potentially toxic byproducts and unutilized O_3 are serious roadblocks toward industrial implementation. To overcome these issues, a compact ACF-based filter was used downstream from the discharge reactor. For the removal of H_2S, methyl mercaptan (MM), and DMS in pulsed corona discharge reactor, ACF filter can effectively adsorb H_2S, MM, and DMS when the reactor was not in operation, but this process gradually lost its efficacy for pollutant removal due to the accumulation of the adsorbed compounds (Yan et al., 2006). For the sole plasma oxidation process, the removal efficiencies of H_2S (120 mg/m^3), MM (120 mg/m^3), and DMS (110 mg/m^3) were approximately 90%, 69%, and 52%, respectively, when the input power was 5.6 W and gas residence time was 5.1 s. As for the adsorption process of the ACF filter, the removal efficiency of H_2S, MM, and DMS all decreased to 15% after 3 h. However, when the plasma reactor was used operationally, the removal efficiency of the pollutants was almost 100% by combining the electrical plasma and the downstream ACF filter, and remained at more than 98% for at least 5 h. The results indicated that the ACF filter was beneficial for the enhancement of pollutant removal efficiency, the retrenchment of energy cost, and the elimination of ozone or other byproducts. Specifically, the energy cost was approximately 3 W h/m^3 for a H_2S removal efficiency of 90% for the sole plasma oxidation process, while it was decreased below 1.2 W h/m^3 with incorporating an ACF filter. At the same time, O_3, SO_2, and dimethyl disulfide can be effectively removed simultaneously, and about 90% of O_3 was removed in the effluent gas with the residual O_3 found to be less than 5 mg/m^3. Though the direct reaction between H_2S (or MM, DMS) and O_3 in gas phase was slow, the adsorbed states of these substances probably reacted with each other more easily. In general, the ACF filter enhanced pollutant removal and can be ascribed to multiple effects, not merely adsorption with the chemical reaction shown to occur on the surface of the ACF.

4.2 Volume Dielectric Barrier Discharge

The majority of research involving the application of dielectric barrier discharge for VOCs removal are concerned with coaxial reactors, with a central rod inside a tube, or with planar configuration reactors, a rod between parallel plates, which is driven by an AC electrical supply (Baojuan et al., 2008; Fan et al., 2015; Fang et al., 2007; Liu et al., 2014a; Preis et al., 2013; Subrahmanyam et al., 2006; Vandenbroucke et al., 2011). Removal of various dilute VOCs or their mixtures has been carried out with dielectric barrier discharge reactors and can be further promoted with integrating various transition metal oxide catalysts into the plasma treatment system. For example, for the removal of toluene, benzene, and chlorobenzene, the maximal removal efficiencies were reported in the presence of MnO_x catalysts for a coaxial reactor with a sintered-metal fiber as the inner catalytic electrode, whose activity can be further improved with AgO_x deposition (Karuppiah et al., 2012). With a specific input energy of 60 J/L, MnO_x/SMF, the oxidation efficiency was reported as 30%, 50%, and 60% for chlorobenzene, benzene, and toluene, respectively. Attractively, the AgO_x/MnO_x/sintered-metal fiber catalyst showed superior performance for the above targets over a MnO_x/sintered-metal fiber. At 60 J/L, efficacies of 45%, 60%, and 75% for chlorobenzene, benzene, and toluene can be achieved using a AgO_x/MnO_x/sintered-metal fiber catalyst. In addition, it was demonstrated that H_2O vapor improves the treatment performance of the DBD plasma reactor, probably owing to the production of $OH^\cdot$, whereas O_3 probably converts to nascent oxygen via an in situ decomposition reaction on the catalyst surface.

Martin et al. (2007) built two volume dielectric barrier discharge reactors and evaluated their capacity for the treatment of 40–50 m^3/h naphtenic and parafinic bitumen fumes. In a first step, polycyclic aromatic hydrocarbons and polyaromatic sulfured hydrocarbons in the bitumen fumes were mildly oxidized by the active species of the discharge. The deodorization of the foul smell proceeds by the destruction of the C—S bonds of the VOC molecules via the generated active species. In a following step, target molecules were adsorbed with high efficiency using cheap solid mineral catalyst filters. Furthermore, it was demonstrated that the affinity of pollutants for the mineral filter can be improved by the plasma treatment. The concurrence of adsorption and desorption of the products ensures its durable trapping and catalytic capacity. For example, after 400 h of working, no obvious variation of the trapping capacity was observed.

A dielectric barrier discharge consisting of a wire-cylindrical tube reactor with a nano-TiO_2/sintered-metal fiber electrode was developed for xylene abatement (Ye et al., 2012). The performance of the reactor using nano-TiO_2/sintered-metal fibers as an electrode was superior in abating xylene over reactors using resistance wire or sintered-metal fiber electrodes. The removal efficiency of xylene (431.8 ± 5 mg/m^3) achieved 92.7% in this reactor using a photocatalytic electrode at a relatively high voltage of 23.6 kV, an input frequency of 300 pps and gas velocity of 800 mL/min, which was much higher than 64.7% in the traditional reactors employing resistance wire or sintered-metal fiber electrodes. The selectivity of CO_2 of the reactor using the nano-TiO_2/sintered-metal fiber electrode was observed to be 86.6%, which was about twice as large as that of the reactor using a resistance wire electrode. In this process, the target or its intermediate products are adsorbed on the surface of TiO_2 and may then be further oxidized by $OH^{\cdot}$. This was validated in an another study with trichloroethylene being effectively destroyed, giving CO_x and H_2O, in a dielectric barrier discharge reactor with a catalytic sintered-metal fiber electrode modified by both MnO_2 and TiO_2 catalysts (Subrahmanyam et al., 2007a). The sintered-metal fiber electrodes modified with other oxides, eg, Mn and Co, were also proven efficient during the destruction of toluene, isopropanol, and trichloroethylene (Subrahmanyam et al., 2007b).

By investigating the plasma oxidation of various organic/inorganic pollutants, either in gaseous emission or immobilized on carriers, the advantages of packed-bed materials in the plasma-generating zone have been demonstrated. It was reported that catalytically active ferroelectric can modify a homogeneous gas plasma and cause a remarkable influence on the pollutants' conversion pathways. Within the ferroelectric layer, the external electric field intensity is considerably augmented in the vicinity of the anomalous edges and corners of the packed-bed grains. Consequently, the electrons can be increasingly accelerated and thus gain more energy, leading to an enhanced pattern of plasma reactivity. In comparison with a catalytic posttreatment, in which direct or indirect reactions of O_3 mainly account for the oxidation of targets, plasma in situ activating solid catalyst processes can occur when the catalyst is placed in the plasma zone, which involves the enhanced formation of highly reactive species in the interior of porous catalysts. When $LaCoO_3$ is used as a solid catalyst, a reservoir of reactive species capable of oxidizing CO is produced during plasma pretreatment (Roland et al., 2002). Based on these results, in situ plasma catalysis can be regarded as a viable method to enhance the energy efficiency of the plasma oxidation process, especially by improving the plasma selectivity to CO_2.

4.3 Surface-Plasma Discharge

A surface-plasma discharge was applied for NO_x removal and to improve the oxiditive conversion efficiency a coil was inserted into the electric circuit (Jolibois et al., 2012). The coil, inserted in series between the power supply and plasma reactor, improved the efficiency of NO_x removal. This can be ascribed to two effects. On the one hand, the presence of coil in the electrical circuit can slightly increase the voltage across the terminals of the plasma device at a given energy density. Thus, the relatively large slew rate (dV/dt) of the voltage waveform, due to the presence of the coil, results in the enhancement of radical species production. On the other hand, the effectiveness of the cleaning device can be improved by the coil via the reduction in power consumption.

The performance of a surface-plasma reactor for nitric oxide (NO) oxidation has been compared with that of a volume-plasma reactor (Malik et al., 2011). For both reactors, significantly more energy was consumed for NO removal at higher concentrations of oxygen and initial NO. The results showed that the lowest energy cost for NO removal was obtained in mixtures of N_2 and NO with relatively high initial NO concentrations. It was reported that the energy cost for 50% NO removal (EC_{50}) from an air stream was estimated to be 120 eV/molecule for the volume-plasma reactor, whereas it was only 70 eV/molecule in the case of the surface-plasma reactor. In an N_2 atmosphere, in spite of a smaller difference in energy cost, a higher efficiency for NO removal was obtained, in which extra NO derived from the oxidation of N_2 is inhibited due to the lack of oxygen. In addition, the energy efficiency of the surface-plasma reactor was found to be almost independent of the electrical energy applied in the plasma device, whereas the efficiency for the volume-plasma reactor reduced considerably with increasing energy input. This indicates the potential benefits of surface-plasma discharges operating in more compact reactors and at higher energy densities over conventional volume-plasma discharges.

Recently, Li et al. (2013) developed a new hybrid surface/packed-bed discharge (HSPBD) plasma reactor for benzene degradation. The HSPBD reactor was demonstrated to have remarkably superior performance for benzene removal over separate surface or packed-bed discharge reactors. In the HSPBD reactor, the energy yield increased by 3.9 and 5.5 g/(kWh) respectively, in comparison with that of a surface discharge reactor and a packed-bed discharge reactor driven at an energy density of 280 J/L. It should be noted that the application of a HSPBD reactor for benzene degradation displays a distinct synergistic enhancement rather than a simple additive of the effectiveness of the surface discharge and packed-bed discharge reactors. Furthermore, in a HSPBD reactor, the production of byproducts, eg, NO_2, was restrained, while O_3 was promoted.

4.4 DC Corona Discharge

Various patterns of corona electrical discharges, produced by DC for either polarity or positive pulsed high voltages, were utilized for toluene removal through oxidation reactions under ambient conditions (Schiorlin et al., 2009). Although treatment efficiency increases in the order of $+DC < -DC < +$pulsed, higher product selectivity is achieved with $-$DC, with CO_2 and CO responsible for approximately 90% of all reacted carbon. All experimental results are in agreement with the proposal that in the case of +DC, corona plasma oxidation of toluene is triggered by the reactions with ions, such as $O_2^{+\cdot}$, H_3O^+, and NO^+, both in dry as well as humid air. Conversely, for −DC, no result was obtained for confirming any significant oxidation of toluene induced by the negative ions. In addition, it can be concluded that in humid air, $OH^{\cdot}$ participates in the initial oxidation stage of toluene caused by −DC.

A corona reactor consisting of pin-to-mesh electrodes incorporating a heterogeneous catalyst was developed as a two-phase treatment system, namely in-plasma (IPC) or post-plasma catalysis (PPC), for the remediation of indoor air purification (Van Durme et al., 2007). It was noted that introduction of a TiO_2 catalyst into the plasma system was ineffective for O_3 reduction. However, adding only 10 g of $CuOMnO_2/TiO_2$ in the PPC resulted in the abatement of outlet O_3 concentration by a factor of 7. Humidity is known to have a limiting influence on O_3 abatement rates. In moisture-free air, 0.5 ppmv toluene was removed three times more efficiently by introducing TiO_2 (IPC). In humid conditions (RH=27%), the performance of this IPC was reduced: toluene removal was only 1.5 times more efficient than that of the sole plasma oxidation. In the case of dry air, placing 10 g $CuOMnO_2/TiO_2$ downstream of the plasma reactor modules (PPC) increases toluene removal efficiencies by 40 times. In humid gas streams, toluene removal efficiency in PPC decreased by competitive adsorption between H_2O vapor and the target pollutant molecules. NO_x production by the corona discharge was monitored. In dry air, the NO_2 outlet concentration was 1500 ppb_v for an energy density of 10 J/L, while this was three times lower at 50% relative humidity. Both heterogeneous catalysts (ie, TiO_2 and $CuOMnO_2/TiO_2$) were capable of removing NO_x by up to 90% in the gas stream. Inactivation of the applied catalysts may be attributed to the affinity of the formed HNO_3 for the surface of the catalyst in the electrical plasma zone. It was determined that surface nitrate ion concentrations were approximately 0.184 mg/m^2 and 0.143 mg/m^2 for TiO_2 and MnO_2-CuO/TiO_2, respectively.

5 Concluding Remarks

Although a large number of experiment results are reported in the studies regarding utilization of electrical plasmas for pollutant treatment, the application of such technology for industrial pollutant removal still requires overcoming significant technical problems. As for large volumes of wastes in various industries, designing the most suitable plasma process requires technical solutions in chemical engineering, material science, and electricity. In addition, the industrial application of such technology demands an effective scale-up to develop the innovative devices with the right reactor dimensions and configuration of the electrodes to obtain high energy efficiencies for pollutant removal.

The pulsed electrical discharge usually consists of streamers, for which the ionization zone fills the entire electrode gap. Although this is propitious for reducing the pressure drop and scale-up of the technology, the requirement of large power, reliable, and inexpensive pulsed voltage sources hampers the industrial application of this technology (Jiang et al., 2014). Unlike pulsed discharges, gliding arc discharge allows for the introduction of high electrical power and consequently results in the large yield of short-lived active species in the plasma zone. Thus, for high concentrations of pollutants, gliding arc plasma may be the leading choice for the electrical plasma. Hybrid gas-liquid plasma reactors are especially suitable for simultaneous removals of gas- and liquid-phase pollutants. DC glow discharge maybe the first choice option for high conductivity water remediation. But this technology requires high energy consumption and may not be compatible with gaseous pollutants removal due to its peculiar property of nonfaradic electrolysis, which may restrict its up-scaling.

In general, the advantage of dielectric barrier discharge over other discharges lies in its comparatively straightforward scale-up to large dimensions for pollutants elimination. For example, in the remediation of wastewater, the interaction between plasma and water is fundamentally similar to the major pathways in other advanced oxidation processes that utilize O_3, H_2O_2, and/or UV. As an illustration, a plasma reactor with the falling water film can be comparable in energy efficiency to conventional advanced oxidation processes, such as O_3 and the Fenton process. However, the main limitation of this technology is how to enhance the generation of oxidative species, economically utilize these species for pollutant removal and sensibly combine the process with other processes to retrench the energy consumption. When applying plasma for off-odor removal, incomplete oxidation of VOCs leads to the formation of various intermediate and unwanted byproducts. Thus, further investigations are required to obtain the derived information about the distribution of byproducts, which would dictate the appropriate catalyst to increase the efficiency of the hybrid system. In addition, in the case of a plasma-catalytic system, a better understanding of the synergistic mechanisms of action would facilitate the development of well-designed combined instruments. Therefore, further progress on this subject is expected in the future.

References

Abdelmalek, F., Ghezzar, M.R., Belhadj, M., Addou, A., Brisset, J.L., 2006. Bleaching and degradation of textile dyes by nonthermal plasma process at atmospheric pressure. Ind. Eng. Chem. Res. 45, 23–29.

Bai, Y.H., Chen, J.R., Li, X.Y., Zhang, C.H., 2009. Non-thermal plasmas chemistry as a tool for environmental pollutants abatement. Reviews of Environmental Contamination and Toxicology, vol. 201. Springer, New York, pp. 117–136.

Baojuan, D., Jian, L., Wenjun, L., Tao, Z., Yili, L., Yuquan, J., Lijuan, H., 2008. Volatile organic compounds (VOCs) removal by using dielectric barrier discharge, bioinformatics and biomedical engineering. In: The 2nd International Conference on ICBBE 2008, pp. 3945–3948.

Barrera-Díaz, C., Roa-Morales, G., Ávila-Córdoba, L., Pavón-Silva, T., Bilyeu, B., 2006. Electrochemical treatment applied to food-processing industrial wastewater. Ind. Eng. Chem. Res. 45, 34–38.

Bian, W., Song, X., Liu, D., Zhang, J., Chen, X., 2011. The intermediate products in the degradation of 4-chlorophenol by pulsed high voltage discharge in water. J. Hazard. Mater. 192, 1330–1339.

Chang, J.S., Lawless, P.A., Yamamoto, T., 1991. Corona discharge processes. IEEE Trans. Plasma Sci. 19, 1152–1166.

Dasgupta, J., Sikder, J., Chakraborty, S., Curcio, S., Drioli, E., 2015. Remediation of textile effluents by membrane based treatment techniques: a state of the art review. J. Environ. Manag. 147, 55–72.

de Brito Benetoli, L.O., Cadorin, B.M., Baldissarelli, V.Z., Geremias, R., de Souza, I.G., Debacher, N.A., 2012. Pyrite-enhanced methylene blue degradation in non-thermal plasma water treatment reactor. J. Hazard. Mater. 237, 55–62.

Dojčinović, B.P., Roglić, G.M., Obradović, B.M., Kuraica, M.M., Kostić, M.M., Nešić, J., Manojlović, D.D., 2011. Decolorization of reactive textile dyes using water falling film dielectric barrier discharge. J. Hazard. Mater. 192, 763–771.

Doubla, A., Bello, L.B., Fotso, M., Brisset, J.L., 2008. Plasmachemical decolourisation of bromothymol blue by gliding electric discharge at atmospheric pressure. Dyes Pigments 77, 118–124.

Duten, X., Packan, D., Yu, L., Laux, C., Kruger, C., 2002. DC and pulsed glow discharges in atmospheric pressure air and nitrogen. IEEE Trans. Plasma Sci. 30, 178–179.

Fan, Y., Cai, Y., Li, X., Yin, H., Chen, L., Liu, S., 2015. Regeneration of the HZSM-5 zeolite deactivated in the upgrading of bio-oil via non-thermal plasma injection (NTPI) technology. J. Anal. Appl. Pyrol. 111, 209–215.

Fang, H.J., Hou, H.Q., Xia, L.Y., Shu, X.H., Zhang, R.X., 2007. A combined plasma photolysis (CPP) method for removal of CS_2 from gas streams at atmospheric pressure. Chemosphere 69, 1734–1739.

Feng, J., Liu, R., Chen, P., Yuan, S., Zhao, D., Zhang, J., Zheng, Z., 2014. Degradation of aqueous 3,4-dichloroaniline by a novel dielectric barrier discharge plasma reactor. Environ. Sci. Pollut. Res. 1–13.

Feng, J., Zheng, Z., Luan, J., Li, K., Wang, L., Feng, J., 2009. Gas-liquid hybrid discharge-induced degradation of diuron in aqueous solution. J. Hazard. Mater. 164, 838–846.

Futamura, S., Zhang, A., Einaga, H., Kabashima, H., 2002. Involvement of catalyst materials in nonthermal plasma chemical processing of hazardous air pollutants. Catal. Today 72, 259–265.

Gao, J., Wang, X., Hu, Z., Deng, H., Hou, J., Lu, X., Kang, J., 2003. Plasma degradation of dyes in water with contact glow discharge electrolysis. Water Res. 37, 267–272.

Gao, J., Yu, J., Lu, Q., He, X., Yang, W., Li, Y., Pu, L., Yang, Z., 2008. Decoloration of alizarin red S in aqueous solution by glow discharge electrolysis. Dyes Pigments 76, 47–52.

Gavrilescu, M., 2005. Fate of pesticides in the environment and its bioremediation. Eng. Life Sci. 5, 497–526.

Ghezzar, M.R., Ognier, S., Cavadias, S., Abdelmalek, F., Addou, A., 2013. DBD plate-TiO_2 treatment of Yellow Tartrazine azo dye solution in falling film. Sep. Purif. Technol. 104, 250–255.

Gong, J., Wang, J., Xie, W., Cai, W., 2008. Enhanced degradation of aqueous methyl orange by contact glow discharge electrolysis using Fe^{2+} as catalyst. J. Appl. Electrochem. 38, 1749–1755.

Grymonpré, D.R., Finney, W.C., Clark, R.J., Locke, B.R., 2004. Hybrid gas-liquid electrical discharge reactors for organic compound degradation. Ind. Eng. Chem. Res. 43, 1975–1989.

Grymonpre, D.R., Finney, W.C., Locke, B.R., 1999. Aqueous-phase pulsed streamer corona reactor using suspended activated carbon particles for phenol oxidation: model-data comparison. Chem. Eng. Sci. 54, 3095–3105.

Hao, X.L., Zhou, M.H., Zhang, Y., Lei, L.C., 2006. Enhanced degradation of organic pollutant 4-chlorophenol in water by non-thermal plasma process with TiO_2. Plasma Chem. Plasma Process. 26, 455–468.

Hijosa-Valsero, M., Molina, R., Schikora, H., Müller, M., Bayona, J.M., 2013. Removal of priority pollutants from water by means of dielectric barrier discharge atmospheric plasma. J. Hazard. Mater. 262, 664–673.

Hoeben, W., Van Veldhuizen, E., Rutgers, W., Cramers, C., Kroesen, G., 2000. The degradation of aqueous phenol solutions by pulsed positive corona discharges. Plasma Sources Sci. Technol. 9, 361.

Hu, Y., Bai, Y., Li, X., Chen, J., 2013. Application of dielectric barrier discharge plasma for degradation and pathways of dimethoate in aqueous solution. Sep. Purif. Technol. 120, 191–197.

Huang, F., Chen, L., Wang, H., Feng, T., Yan, Z., 2012. Degradation of methyl orange by atmospheric DBD plasma: analysis of the degradation effects and degradation path. J. Electrostat. 70, 43–47.

Huang, F., Chen, L., Wang, H., Yan, Z., 2010. Analysis of the degradation mechanism of methylene blue by atmospheric pressure dielectric barrier discharge plasma. Chem. Eng. J. 162, 250–256.

Huang, L., Nakajyo, K., Hari, T., Ozawa, S., Matsuda, H., 2001. Decomposition of carbon tetrachloride by a pulsed corona reactor incorporated with in situ absorption. Ind. Eng. Chem. Res. 40, 5481–5486.

Jarrige, J., Vervisch, P., 2007. Decomposition of gaseous sulfide compounds in air by pulsed corona discharge. Plasma Chem. Plasma Process. 27, 241–255.

Jarrige, J., Vervisch, P., 2009. Plasma-enhanced catalysis of propane and isopropyl alcohol at ambient temperature on a MnO_2-based catalyst. Appl. Catal. B Environ. 90, 74–82.

Jiang, B., Zheng, J., Liu, Q., Wu, M., 2012. Degradation of azo dye using non-thermal plasma advanced oxidation process in a circulatory airtight reactor system. Chem. Eng. J. 204, 32–39.

Jiang, B., Zheng, J., Lu, X., Liu, Q., Wu, M., Yan, Z., Qiu, S., Xue, Q., Wei, Z., Xiao, H., 2013a. Degradation of organic dye by pulsed discharge non-thermal plasma technology assisted with modified activated carbon fibers. Chem. Eng. J. 215, 969–978.

Jiang, B., Zheng, J., Qiu, S., Wu, M., Zhang, Q., Yan, Z., Xue, Q., 2014. Review on electrical discharge plasma technology for wastewater remediation. Chem. Eng. J. 236, 348–368.

Jiang, N., Lu, N., Shang, K., Li, J., Wu, Y., 2013b. Innovative approach for benzene degradation using hybrid surface/packed-bed discharge plasmas. Environ. Sci. Technol. 47, 9898–9903.

Jolibois, J., Takashima, K., Mizuno, A., 2012. Application of a non-thermal surface plasma discharge in wet condition for gas exhaust treatment: NO_x removal. J. Electrostat. 70, 300–308.

Jović, M., Manojlović, D., Stanković, D., Dojčinović, B., Obradović, B., Gašić, U., Roglić, G., 2013. Degradation of triketone herbicides, mesotrione and sulcotrione, using advanced oxidation processes. J. Hazard. Mater. 260, 1092–1099.

Karuppiah, J., Reddy, E.L., Reddy, P.M.K., Ramaraju, B., Karvembu, R., Subrahmanyam, C., 2012. Abatement of mixture of volatile organic compounds (VOCs) in a catalytic non-thermal plasma reactor. J. Hazard. Mater. 237, 283–289.

Krugly, E., Martuzevicius, D., Tichonovas, M., Jankunaite, D., Rumskaite, I., Sedlina, J., Racys, V., Baltrusaitis, J., 2015. Decomposition of 2-naphthol in water using a non-thermal plasma reactor. Chem. Eng. J. 260, 188–198.

Kucharska, M., Grabka, J., 2010. A review of chromatographic methods for determination of synthetic food dyes. Talanta 80, 1045–1051.

Kušić, H., Koprivanac, N., Locke, B.R., 2005. Decomposition of phenol by hybrid gas/liquid electrical discharge reactors with zeolite catalysts. J. Hazard. Mater. 125, 190–200.

Lesage, O., Roques-Carmes, T., Commenge, J.-M., Duten, X., Tatoulian, M., Cavadias, S., Mantovani, D., Ognier, S., 2014. Degradation of 4-chlorobenzoïc acid in a thin falling film dielectric barrier discharge reactor. Ind. Eng. Chem. Res. 53, 10387–10396.

Li, D., Yakushiji, D., Kanazawa, S., Ohkubo, T., Nomoto, Y., 2002. Decomposition of toluene by streamer corona discharge with catalyst. J. Electrostat. 55, 311–319.

Li, J., Sato, M., Ohshima, T., 2007a. Degradation of phenol in water using a gas-liquid phase pulsed discharge plasma reactor. Thin Solid Films 515, 4283–4288.

Li, J., Zhou, Z., Wang, H., Li, G., Wu, Y., 2007b. Research on decoloration of dye wastewater by combination of pulsed discharge plasma and TiO_2 nanoparticles. Desalination 212, 123–128.

Li, L., Chen, X., Zhang, D., Pan, X., 2010. Effects of insecticide acetamiprid on photosystem II (PSII) activity of *Synechocystis* sp. (FACHB-898). Pestic. Biochem. Physiol. 98, 300–304.

Li, S., Jiang, Y., Cao, X., Dong, Y., Dong, M., Xu, J., 2013. Degradation of nitenpyram pesticide in aqueous solution by low-temperature plasma. Environ. Technol. 34, 1609–1616.

Li, S., Ma, X., Jiang, Y., Cao, X., 2014. Acetamiprid removal in wastewater by the low-temperature plasma using dielectric barrier discharge. Ecotoxicol. Environ. Saf. 106, 146–153.

Liu, S.Y., Mei, D.H., Shen, Z., Tu, X., 2014a. Nonoxidative conversion of methane in a dielectric barrier discharge reactor: prediction of reaction performance based on neural network model. J. Phys. Chem. C 118, 10686–10693.

Liu, Y.-N., Si, A.H., 2012. Gliding arc discharge for decolorization and biodegradability of azo dyes and printing and dyeing wastewater. Plasma Chem. Plasma Process. 32, 597–607.

Liu, Y., Fu, J., Deng, S., Zhang, X., Shen, F., Yang, G., Peng, H., Zhang, Y., 2014b. Degradation of basic and acid dyes in high-voltage pulsed discharge. J. Taiwan Inst. Chem. Eng. 45, 2480–2487.

Locke, B., Sato, M., Sunka, P., Hoffmann, M., Chang, J.-S., 2006. Electrohydraulic discharge and nonthermal plasma for water treatment. Ind. Eng. Chem. Res. 45, 882–905.

Lou, J., Lu, N., Li, J., Wang, T., Wu, Y., 2012. Remediation of chloramphenicol-contaminated soil by atmospheric pressure dielectric barrier discharge. Chem. Eng. J. 180, 99–105.

Lukes, P., Locke, B.R., 2005. Degradation of substituted phenols in a hybrid gas-liquid electrical discharge reactor. Ind. Eng. Chem. Res. 44, 2921–2930.

Magureanu, M., Mandache, N.B., Eloy, P., Gaigneaux, E.M., Parvulescu, V.I., 2005. Plasma-assisted catalysis for volatile organic compounds abatement. Appl. Catal. B Environ. 61, 12–20.

Malik, M.A., 2010. Water purification by plasmas: which reactors are most energy efficient? Plasma Chem. Plasma Process. 30, 21–31.

Malik, M.A., Kolb, J.F., Sun, Y., Schoenbach, K.H., 2011. Comparative study of NO removal in surface-plasma and volume-plasma reactors based on pulsed corona discharges. J. Hazard. Mater. 197, 220–228.

Marotta, E., Ceriani, E., Schiorlin, M., Ceretta, C., Paradisi, C., 2012. Comparison of the rates of phenol advanced oxidation in deionized and tap water within a dielectric barrier discharge reactor. Water Res. 46, 6239–6246.

Martin, L., Ognier, S., Gasthauer, E., Cavadias, S., Dresvin, S., Amouroux, J., 2007. Destruction of highly diluted volatile organic components (VOCs) in air by dielectric barrier discharge and mineral bed adsorption. Energy Fuel 22, 576–582.

Maxime, D., Marcotte, M., Arcand, Y., 2006. Development of eco-efficiency indicators for the Canadian food and beverage industry. J. Clean. Prod. 14, 636–648.

Mededovic, S., Locke, B.R., 2007. Side-chain degradation of atrazine by pulsed electrical discharge in water. Ind. Eng. Chem. Res. 46, 2702–2709.

Mizuno, A., 2007. Industrial applications of atmospheric non-thermal plasma in environmental remediation. Plasma Phys. Control. Fus. 49, A1.

Park, S.Y., Deshwal, B.R., Moon, S.H., 2008. NO_x removal from the flue gas of oil-fired boiler using a multistage plasma-catalyst hybrid system. Fuel Process. Technol. 89, 540–548.

Pascal, S., Moussa, D., Hnatiuc, E., Brisset, J.L., 2010. Plasma chemical degradation of phosphorous-containing warfare agents simulants. J. Hazard. Mater. 175, 1037–1041.

Preis, S., Klauson, D., Gregor, A., 2013. Potential of electric discharge plasma methods in abatement of volatile organic compounds originating from the food industry. J. Environ. Manag. 114, 125–138.

Qu, G., Liang, D., Qu, D., Huang, Y., Liu, T., Mao, H., Ji, P., Huang, D., 2013. Simultaneous removal of cadmium ions and phenol from water solution by pulsed corona discharge plasma combined with activated carbon. Chem. Eng. J. 228, 28–35.

Ramjaun, S.N., Yuan, R., Wang, Z., Liu, J., 2011. Degradation of reactive dyes by contact glow discharge electrolysis in the presence of Cl^- ions: kinetics and AOX formation. Electrochim. Acta 58, 364–371.

Rodrigo, M.A., Oturan, N., Oturan, M.A., 2014. Electrochemically assisted remediation of pesticides in soils and water: a review. Chem. Rev. 114, 8720–8745.

Roland, U., Holzer, F., Kopinke, F.D., 2002. Improved oxidation of air pollutants in a non-thermal plasma. Catal. Today 73, 315–323.

Sano, N., Kawashima, T., Fujikawa, J., Fujimoto, T., Kitai, T., Kanki, T., Toyoda, A., 2002. Decomposition of organic compounds in water by direct contact of gas corona discharge: influence of discharge conditions. Ind. Eng. Chem. Res. 41, 5906–5911.

Sano, N., Yamane, Y., Hori, Y., Akatsuka, T., Tamon, H., 2011. Application of multiwalled carbon nanotubes in a wetted-wall corona-discharge reactor to enhance phenol decomposition in water. Ind. Eng. Chem. Res. 50, 9901–9909.

Sano, T., Negishi, N., Sakai, E., Matsuzawa, S., 2006. Contributions of photocatalytic/catalytic activities of TiO_2 and γ-Al_2O_3 in nonthermal plasma on oxidation of acetaldehyde and CO. J. Mol. Catal. A Chem. 245, 235–241.

Schiorlin, M., Marotta, E., Rea, M., Paradisi, C., 2009. Comparison of toluene removal in air at atmospheric conditions by different corona discharges. Environ. Sci. Technol. 43, 9386–9392.

Shen, Y., Lei, L., Zhang, X., Zhou, M., Zhang, Y., 2008. Effect of various gases and chemical catalysts on phenol degradation pathways by pulsed electrical discharges. J. Hazard. Mater. 150, 713–722.

Shi, J., Bian, W., Yin, X., 2009. Organic contaminants removal by the technique of pulsed high-voltage discharge in water. J. Hazard. Mater. 171, 924–931.

Song, Y.H., Kim, S.J., Choi, K.I., Yamamoto, T., 2002. Effects of adsorption and temperature on a nonthermal plasma process for removing VOCs. J. Electrostat. 55, 189–201.

Subrahmanyam, C., Magureanu, M., Laub, D., Renken, A., Kiwi-Minsker, L., 2007a. Nonthermal plasma abatement of trichloroethylene enhanced by photocatalysis. J. Phys. Chem. C 111, 4315–4318.

Subrahmanyam, C., Magureanu, M., Renken, A., Kiwi-Minsker, L., 2006. Catalytic abatement of volatile organic compounds assisted by non-thermal plasma: Part 1. A novel dielectric barrier discharge reactor containing catalytic electrode. Appl. Catal. B Environ. 65, 150–156.

Subrahmanyam, C., Renken, A., Kiwi-Minsker, L., 2007b. Novel catalytic non-thermal plasma reactor for the abatement of VOCs. Chem. Eng. J. 134, 78–83.

Sugiarto, A.T., Ito, S., Ohshima, T., Sato, M., Skalny, J.D., 2003. Oxidative decoloration of dyes by pulsed discharge plasma in water. J. Electrostat. 58, 135–145.

Tang, Q., Lin, S., Jiang, W., Lim, T., 2009. Gas phase dielectric barrier discharge induced reactive species degradation of 2,4-dinitrophenol. Chem. Eng. J. 153, 94–100.

Tezuka, M., Iwasaki, M., 1997. Oxidative degradation of phenols by contact glow discharge electrolysis. Denki Kagaku Oyobi Kogyo Butsuri Kagaku 65, 1057–1060.

Tichonovas, M., Krugly, E., Racys, V., Hippler, R., Kauneliene, V., Stasiulaitiene, I., Martuzevicius, D., 2013. Degradation of various textile dyes as wastewater pollutants under dielectric barrier discharge plasma treatment. Chem. Eng. J. 229, 9–19.

Tijani, J.O., Fatoba, O.O., Madzivire, G., Petrik, L.F., 2014. A review of combined advanced oxidation technologies for the removal of organic pollutants from water. Water Air Soil Pollut. 225, 1–30.

Van Durme, J., Dewulf, J., Sysmans, W., Leys, C., Van Langenhove, H., 2007. Efficient toluene abatement in indoor air by a plasma catalytic hybrid system. Appl. Catal. B Environ. 74, 161–169.

Vandenbroucke, A.M., Morent, R., De Geyter, N., Leys, C., 2011. Non-thermal plasmas for non-catalytic and catalytic VOC abatement. J. Hazard. Mater. 195, 30–54.

Vanraes, P., Willems, G., Daels, N., Van Hulle, S.W., De Clerck, K., Surmont, P., Lynen, F., Vandamme, J., Van Durme, J., Nikiforov, A., 2014. Decomposition of atrazine traces in water by combination of non-thermal electrical discharge and adsorption on nanofiber membrane. Water Res. 72, 361–371.

Wang, H.J., Chen, X.Y., 2011. Kinetic analysis and energy efficiency of phenol degradation in a plasma-photocatalysis system. J. Hazard. Mater. 186, 1888–1892.

Wang, H., Chu, J., Ou, H., Zhao, R., Han, J., 2009. Analysis of TiO_2 photocatalysis in a pulsed discharge system for phenol degradation. J. Electrostat. 67, 886–889.

Wang, H., Li, J., Quan, X., Wu, Y., 2008. Enhanced generation of oxidative species and phenol degradation in a discharge plasma system coupled with TiO_2 photocatalysis. Appl. Catal. B Environ. 83, 72–77.

Wang, L., 2009. Aqueous organic dye discoloration induced by contact glow discharge electrolysis. J. Hazard. Mater. 171, 577–581.

Wang, T., Lu, N., Li, J., Wu, Y., Su, Y., 2011a. Enhanced degradation of *p*-nitrophenol in soil in a pulsed discharge plasma-catalytic system. J. Hazard. Mater. 195, 276–280.

Wang, T.C., Lu, N., Li, J., Wu, Y., 2010. Degradation of pentachlorophenol in soil by pulsed corona discharge plasma. J. Hazard. Mater. 180, 436–441.

Wang, T.C., Lu, N., Li, J., Wu, Y., 2011b. Plasma-TiO_2 catalytic method for high-efficiency remediation of *p*-nitrophenol contaminated soil in pulsed discharge. Environ. Sci. Technol. 45, 9301–9307.

Wang, T.C., Qu, G., Li, J., Liang, D., 2014. Evaluation of the potential of soil remediation by direct multi-channel pulsed corona discharge in soil. J. Hazard. Mater. 264, 169–175.

Wang, X., Zhou, M., Jin, X., 2012. Application of glow discharge plasma for wastewater treatment. Electrochim. Acta 83, 501–512.

Wohlers, J., Koh, I.-O., Thiemann, W., Rotard, W., 2009. Application of an air ionization device using an atmospheric pressure corona discharge process for water purification. Water Air Soil Pollut. 196, 101–113.

Xiao, H., He, J., Zhang, Y., Li, Y., Li, Y., Shen, F., Yang, G., Yang, X., Deng, S., Wang, Y., Li, L., 2012. Study of a novel high voltage pulsed discharge reactor with porous titanium electrodes. J. Taiwan Inst. Chem. Eng. 43, 597–603.

Xue, J., Chen, L., Wang, H., 2008. Degradation mechanism of Alizarin Red in hybrid gas-liquid phase dielectric barrier discharge plasmas: experimental and theoretical examination. Chem. Eng. J. 138, 120–127.

Yan, J., Liu, Y., Bo, Z., Li, X., Cen, K., 2008. Degradation of gas-liquid gliding arc discharge on Acid Orange II. J. Hazard. Mater. 157, 441–447.

Yan, N.Q., Qu, Z., Jia, J.P., Wang, X.P., Wu, D., 2006. Removal characteristics of gaseous sulfur-containing compounds by pulsed corona plasma. Ind. Eng. Chem. Res. 45, 6420–6427.

Yang, B., Zhou, M., Lei, L., 2005. Synergistic effects of liquid and gas phase discharges using pulsed high voltage for dyes degradation in the presence of oxygen. Chemosphere 60, 405–411.

Ye, Z., Wang, C., Shao, Z., Ye, Q., He, Y., Shi, Y., 2012. A novel dielectric barrier discharge reactor with photocatalytic electrode based on sintered metal fibers for abatement of xylene. J. Hazard. Mater. 241, 216–223.

Yu, C.J., Xu, F., Luo, Z.Y., Cao, W., Wei, B., Gao, X., Fang, M.X., Cen, K.F., 2009. Influences of water vapor and fly ash addition on NO and SO_2 gas conversion efficiencies enhanced by pulsed corona discharge. J. Electrostat. 67, 829–834.

Zhang, Y., Deng, S., Sun, B., Xiao, H., Li, L., Yang, G., Hui, Q., Wu, J., Zheng, J., 2010a. Preparation of TiO_2-loaded activated carbon fiber hybrids and application in a pulsed discharge reactor for decomposition of methyl orange. J. Colloid Interface Sci. 347, 260–266.

Zhang, Y., Sun, B., Deng, S., Wang, Y., Peng, H., Li, Y., Zhang, X., 2010b. Methyl orange degradation by pulsed discharge in the presence of activated carbon fibers. Chem. Eng. J. 159, 47–52.

Zhang, Y., Zheng, J., Qu, X., Chen, H., 2007a. Effect of granular activated carbon on degradation of methyl orange when applied in combination with high-voltage pulse discharge. J. Colloid Interface Sci. 316, 523–530.

Zhang, Y., Zheng, J., Qu, X., Chen, H., 2008. Design of a novel non-equilibrium plasma-based water treatment reactor. Chemosphere 70, 1518–1524.

Zhang, Y., Zhou, M., Hao, X., Lei, L., 2007b. Degradation mechanisms of 4-chlorophenol in a novel gas-liquid hybrid discharge reactor by pulsed high voltage system with oxygen or nitrogen bubbling. Chemosphere 67, 702–711.

Zhu, D., Jiang, L., Liu, R.L., Chen, P., Lang, L., Feng, J.W., Yuan, S.J., Zhao, D.Y., 2014. Wire-cylinder dielectric barrier discharge induced degradation of aqueous atrazine. Chemosphere 117, 506–514.

Chapter 14

Future of Cold Plasma in Food Processing

K.M. Keener*, N.N. Misra†
*Iowa State University, Ames, IA, USA, †GTECH, Research & Development, General Mills India Pvt Ltd, Mumbai, India

1 Introduction

The term "disruptive technology" refers to a new technology having a lower cost and better performance (Utterback and Acee, 2005). Cold plasma is a disruptive technology to many current food-manufacturing processes, including thermal processing, chlorine wash, chemical fumigation, etc. The challenge in the introduction of any novel technology into the food industry arises from the perception that it is disruptive, risky, and difficult to implement when faced with day-to-day competitive pressure in a mature and low-margin industry. Success of a technology in the food industry requires both disrupting the use of an incumbent and introducing an alternative. Therefore, for a novel technology like atmospheric cold plasma to obtain successful adoption in the food industry, several nontechnological aspects need to be considered. First, the consumer is the primary driving force. The consumer demand(s) are constantly changing. Currently, consumers desire high-quality, nutritious, "fresh-like" foods with no artificial preservatives or chemicals for themselves,

Cold Plasma in Food and Agriculture. http://dx.doi.org/10.1016/B978-0-12-801365-6.00014-7

their families, and their pets. There is also a subset of consumers that consider price as a lesser concern and are willing to pay more for their wants. A vast majority of consumers assume that the foods available in the marketplace are safe to eat, assuming they are handled and cooked properly. A food business cannot sell a product processed using a novel technology by declaring it safer than what it sold earlier. Food safety and retention of consumer confidence for a specific food category is built over time and is the responsibility of the growers, processors, distributors, retailers, and regulatory agencies. Thus, the food industry's interest in exploring a novel technology such as atmospheric cold plasma is not driven solely by the food-safety factor, but also the consumer factor. How will the consumer perceive the technology? How will it be marketed to the consumer? Other factors which are driving the interests of food producers, processors, distributors, and retailers in exploring and adopting cold plasma technologies include: (1) potential extension of product shelf life and lower consumer food waste; (2) maximum retention of food quality and lower food processing and storage losses; (3) low energy requirement, which is more "green" than current technology; (4) low operational and maintenance costs; there is a need for simple systems with minimal maintenance and sanitation requirements; (5) enhanced chemical safety of foods, including plasma inactivation and the removal of pesticide and chemical residues; and (6) green technology and environmental sustainability, as the technology only needs air and electricity to create an effective plasma.

In order to explain this scenario, we take up the example of strawberries as a fresh produce. Strawberries are a high-value product, which are very perishable due to their susceptibility to fungal contamination and soft texture. Being soft and having a complex surface structure, strawberries cannot be washed during industrial processing as with other fruits or vegetables. It also makes strawberries prone to the physical damage during storage or transportation, the overall result being a short shelf life. Several novel technologies have been explored to extend the shelf life of strawberries, including high-pressure processing (HPP), pulsed electric field (PEF), ultrasound application, ultraviolet light, and ozone application. However, there are several challenges and difficulties which are clearly visible with all these approaches. For example, HPP is not suitable for high porosity foods, which includes strawberries; PEF and ultrasound are suitable for liquid foods; in UV light processing the "shadowing effect" is a major issue; and ozone processing has not been very successful for strawberries (Ashford, 2002). It is under such a situation that cold plasma offers relative advantages and is attracting researchers and food processors. Atmospheric cold plasma in various gas mixtures has been found to significantly reduce bacterial and fungal counts on strawberries, while retaining quality parameters such as color, texture, and sensory attributes (Misra et al., 2014a,c; Schnabel et al., 2015). Most plasma technologies require less power for their operation, and being in a gaseous state, they leave no harmful chemical residues, as has been reported with other chemical sanitizers. In addition, plasma could decrease levels of most pesticide residues on produce (Heo et al., 2014; Misra et al., 2014b). Specifically, the high-voltage atmospheric pressure cold technology inside a sealed package based on dielectric barrier discharge (Klockow and Keener, 2009) can accommodate a large class of foods due to its ability to operate at centimeter scale gaps, and that too in air.

While the results are promising and the technology proves to be advantageous for industry, it is important to identify the factors that are limiting the advancement of the technology to a full-fledge commercial scale. To date, commercial advances in atmospheric plasma technology have been focused on specific needs within the food industry, such as treatment of bottle caps to reduce surface tension for labeling with food grade inks (without organic solvents) or modification of packaging polymers. The three major challenges for widespread use of atmospheric plasma as a food-manufacturing tool are: (1) regulatory approval, (2) designing the plasma source, and (3) process control. The objective of this chapter is to discuss these challenges. The former chapters in this book discussed the fundamental aspects of the technology and the present state of the art. In this closing chapter of the book, it is intended to provide an overview of the research requirements in order to leverage the potential of cold plasma technology at an industrial scale. Technological developments are the key to the faster adoption of novel processing methods; therefore, the challenges that one is likely to encounter in scaling-up various cold plasma sources is briefly touched upon with an aim of exploiting the knowledge available.

2 Regulatory Approval

To explain the subtleties of regulatory requirements for novel food process and products, the food regulations framework is briefly reviewed in this section. Each country has their own individual process for regulatory review and approval of new technologies. Acceptance of data or conclusions reached from a regulatory review may significantly differ between countries. As an example, an overview of the regulatory approval process for the introduction of new technologies of food process interventions in the United States is provided.

Food regulations in the United States are a patchwork of rules and regulations that have developed over time. For a single food, there are numerous government agencies that have inspection roles. At the federal level, the primary agencies with regulatory responsibilities are the Food and Drug Administration (FDA), an agency within the Department of Health and Human Services, and the Food Safety Inspection Service (FSIS), an agency within the United States Department of Agriculture. The FDA has responsibility to ensure the safety of all foods under the Federal Food Drug and Cosmetic Act of 1938 (FFDCA) and the Food Additives Amendment of the FFDCA in 1958 (FDA, 2015a). Federal Food, Drug, and Cosmetic Act Section 201 (f) defines "food" as articles used for food or drink for man or other animals, chewing gum, and articles used for components of any such articles (FDA, 2015a). Additionally, the FSIS has the primary responsibility for food safety of meat, poultry, and egg products under the Meat Product Inspection Act (1906) (FSIS, 2006a), Poultry Product Inspection Act (1957) (FSIS, 2006b) and Egg Product Inspection Act (1970) (FSIS, 2006c). Other federal and state regulatory agencies have supporting roles in various commodities and provide environmental assessment, product grading, and export inspection services.

Regulatory approval for food technology and food process interventions in the United States is regulated by the FDA under authorization from the Food Additives Amendments of the FFDCA (FDA, 2015a). Section 201(s) of the Federal Food, Drug, and Cosmetic Act (FD&C Act) defines a "food additive" as "any substance the intended use of which results or may reasonably be expected to result, directly or indirectly, in its becoming a component or otherwise affecting the characteristic of any food (including any substance intended for use in producing, manufacturing, packing, processing, preparing, treating, packaging, transporting, or holding food); and including any source of radiation ..." (FDA, 2015b). There are three main pathways for FDA regulatory review: (1) Generally Recognized as Safe (GRAS) Self-Affirmation; (2) Food Contact Notification Petition; and (3) Food Additives Petition.

The FDA's voluntary GRAS notification program was established in 1997. It allows the manufacturer to make a scientific assessment of the food technology or food process intervention. If the scientific evidence demonstrates the technology or the process does not create a potential for introduction of a "food additive," or if the food additive is already listed as GRAS, then the manufacturer can conclude the food technology or food process as GRAS. The manufacturer of a GRAS-designated technology or process intervention has the option of going directly to market or to notify the FDA of its GRAS self-determination and await an FDA review. Notification to the FDA allows the FDA to accept the decision with "no questions," or ask for more information and await further review. The FDA has the authority to raise objection to a self-affirmation. Generally, most rejections are due to insufficient data. The manufacturer can affirm the GRAS status themselves for the substance or process and go directly to market, which generally will spur an FDA inquiry and request for supporting data.

Historically known as an indirect food additive, food contact substances (FCSs) are those, including packaging, that indirectly or incidentally contact a food as a result of the manufacturing process, but are not directly added as ingredients. A food contact notification petition is required to be submitted to FDA when scientific data suggest a "substance" or "substance resulting from a technology or process" is not GRAS. Food contact notifications (FCNs) are only required for new uses of FCSs that are food additives. Although a notification is not required for a food contact substance that is GRAS or prior sanctioned for its intended use in contact with food, some companies do choose to notify the FDA's Office of Food Additive Safety in order to clarify the regulatory status of such substances. Manufacturers may also use FCNs to notify the FDA of new uses of FCSs that are GRAS or prior sanctioned. Unlike food additive regulations and the threshold of regulation exemptions, approvals under the FCN process are proprietary.

There is a 120-day mandatory review requirement for a FCN; however, the process may extend beyond 120 days if additional information is requested during the regulatory review. FCN data is considered confidential and approvals are limited to the identified manufacturer or their appointed suppliers and the specific product-process conditions reviewed. For example, if one obtains an FCN for a novel packaging material for raw meat, it is likely another FCN would need to be requested for the use of this packaging material on raw fruit.

A few additional comments on FCNs are felt appropriate. FDA may issue a letter of nonacceptance for an FCN where a bioassay for the FCS is not negative for cancer-causing effects. For those indirect food additives that result in less than 1.5 μg per person per day exposure, FDA has an abbreviated FCN process called "threshold of regulation." The threshold of the regulation process may result in a shorter FDA review time, but the submitted documents are not considered proprietary (FDA, 2015b).

Food additives are characterized based on their composition and intended use. A food additive is intended to supply and/or enhance nutrients, color, flavor, stability, or generally alter a food's characteristics. They are generally listed on the food ingredient label. Substances that alter the characteristics of a food may be categorized into one or more categories based on their intended effect including emulsifiers, binders, sequestrants, stabilization, anticaking agents, enzymes, etc. The food additive petition (FAP), unlike the FCN, does not have a regulatory limit on time to review. There have been past instances where FAPs have been under FDA review for more than five years. In general, FAPs have longer review times than FCNs; however, issued FAPs are codified in the Federal Register as a final rule and include defined tolerances for use where maximum and/or minimum limits are determined for the requested food additive. The final rules are product specific, but not company or process specific. Any information provided to FDA during an FAP review is considered nonproprietary and available for request under the Freedom of Information Act.

The information required for submission in a FAP review is similar to the FCN and outlined in 21 CFR 571.1(c). FAP requires: (1) Name and all pertinent information concerning the food additive, including chemical identity and composition of the additive or manufacturing methods and controls if the chemical identity and composition are not known; (2) intended use, use level (amount of the food additive proposed for use), and labeling (cautions, warnings, shelf life, directions for use); (3) data establishing the intended effect (physical, nutritional, or other technical effect); (4) a description of analytical methods to determine the amount of the food additive in the food; (5) safety evaluation; (6) proposed tolerances for the food additive; (7) proposed regulation; and (8) environmental assessment. Some of the data requested for this review requires involvement with other government agencies such as the Occupational Health and Safety Administration (OSHA) and the Environmental Protection Agency (EPA). Further details on the GRAS Self-Affirmation, Food Contact Notification, and FAP can be found on the FDA's website (FDA, 2015a,b,c).

From the regulatory perspective, approval for a new plasma process, whether it be for direct or indirect food treatment, food-packaging treatment, or food treatment inside a food package, requires a significant amount of data collection, data analysis, and time. Because of the complexity in plasma chemistry there exist a large number of possible chemical effects which must be examined, prioritized, and assessed. Ideally, the results from the assessment will determine which regulatory approval route to follow. As stated in the beginning of this section, the regulatory approval process varies significantly between countries. For example, an atmospheric plasma device can be designed to be an ozone-generation device. In the United States, the FDA has designated ozone as GRAS, so any device claiming ozone generation does not require FDA regulatory review or approval. If one generates reactive gas species other than ozone, then the question of which path to proceed for a regulatory review (and ultimate approval) depends on the market opportunity and the company's willingness to accept regulatory risk. It is recommended the plasma community take an active role in the regulatory review process to speed up commercialization of plasma technology. The plasma community (manufacturers, distributors, scientists, engineers, consumer groups, etc.) should come together to develop guidance documents that can aid atmospheric plasma technology manufacturers in obtaining regulatory approval. Some potential guidance documents could include: a technology overview stating what atmospheric cold plasma is and describing different methods of plasma generation; recommended practices for plasma measurement, data collection, and data analysis; a regulatory "roadmap" for seeking approval of a food plasma treatment process or device; and consumer consideration in the adoption of atmospheric plasma technology in the food industry. The single biggest obstacle to the adoption of plasma technology in the food industry is the regulatory approval process.

3 Design of Plasma Source

At this time, most research work concerned with atmospheric cold plasma treatment of foods seeks to establish the process conditions for maximum microbiological decontamination and minimum damage to the nutritional and sensorial quality of foods. However, there exist several issues yet to be addressed before industrial-scale plasma sources for treatment of foods could be designed. The first among these includes understanding the chemistry of atmospheric cold plasmas, particularly air plasmas. A simple air (oxygen, nitrogen, water vapor) plasma has been documented to produce more than 75 unique chemical species and involve 500 simultaneous chemical reactions, at four different time scales (nanosecond, microsecond, millisecond, and seconds). Second, as is common with any new technology, there is a need to standardize the plasma analytics (eg, power consumption, power density, voltage gradient, excitation frequency, gas flow, optical emission spectroscopy measurement, absorption spectroscopy measurement, etc.) to harmonize the research efforts across the globe and accelerate developments in plasma applications for the food, bio and medical industries. This said, the development of new plasma sources is also important, both for advancing the science and developing innovative applications. Finally, the design of scaled-up plasma processes and devices should not take precedence over health and safety issues of the operators and the work environment. We elaborate upon some of these points in the following sections.

3.1 Features and Design of the Machine

The first and foremost expectation of the food industry for a cold plasma processing machine would be the ability it has to bring about the desirable effects in foods, that is, through antimicrobial action, extension of shelf life, and operation at room temperatures. The success of cold plasma technology in various applications is linked with the type of plasma chemistry conceived.

From a business viewpoint, the machine should be inexpensive. It is difficult to foresee a great future for plasma applications for decontamination of foods, which use costly noble gases due to the low operating margins. Ideally, plasma sources capable of ionizing air at large gaps will be suitable for the food industry. If plasma sources with such capabilities can be made commercially, plasma technology will find broad acceptance. Plasmas generated in noble gas or gas mixtures other than air or common gases could prove to be limited to processing high-value foods or functionalized ingredients. In some applications, noble gases have benefits over air. An example of this includes the use of argon cold plasma for enhancing the extraction from oleogeneous plant materials (Kodama et al., 2014).

The ability to process continuously at high speed and for several days to months with the least maintenance is also an expectation. Finally, the machine should be capable of operating with a wide range of gases. While it is not impossible to meet all these requirements, there are several other challenges which will be faced by the engineers involved in machine fabrication. First, they have to ensure that there are robust measures to control the electrical input parameters to the machine, including several feedback and active control loops. Second, being that it is intended for the treatment of foods, the machine design should follow the principles of hygienic design. It may be noted that in the food industry, machines are often fabricated using stainless steel. However, for a high-voltage plasma processing machine, this may not be true considering the possibility of arcing, induced/eddy currents in metallic parts. In addition, appropriate shielding, insulation and good grounding is also to be ensured. Apart from the basic functional parts of the machine, other essential components would include a system to monitor the ambient "toxic" gas levels in the vicinity of the machine, and an active system to capture or destroy any residual reactive gas species, eg, using carbon filters. It should be noted that some atmospheric cold plasma devices using air can generate greater than 100,000 times of the Occupational Safety and Health Administration's limit for numerous reactive gas species, including ozone (0.1 ppm 8 h exposure maximum), within a few minutes of operation (Keener et al., 2012).

At this point of the discussion, it is worth mentioning some of the emerging organizations involved in the fabrication of cold plasma systems for applications in the food industry. Plasmatreat USA, Inc. is presently a leading organization which provides plasma technology-based solutions to several manufacturing industries. The firm has developed the Openair® nonequilibrium plasma technology for surface modification. This process is commercially used for increasing the surface tension on bottle caps, which allows printing with solvent-free food-safe inks on high speed lines. Anacail Limited, a spin-out from the University of Glasgow, Scotland, is yet another organization which specializes in plasma technology-based in-package ozone treatment of perishable foods (Anacail, 2014). This organization claims to have a strictly "ozone" generation device, but their patent reveals that the technology is based on a plasma coaxial "surface" generator, which creates a hemispherical plasma field that passes through packages. To date, there has been no published scientific data for the Anacail device.

3.2 Reference Plasma Source

At the 1988 Gaseous Electronic Conference (GEC) meeting, it was agreed that there needed to be an easy-to-model "Reference Cell" for making comparative measurements with other "identical" radio-frequency (RF) plasma systems. Accordingly, a collaborative experimental effort was initiated at the workshop on the Design, Calibration and Modeling of RF Plasma Processing Systems, following which researchers from Sandia National Labs developed the blueprints for the GEC reference cell (Olthoff and Greenberg, 1995). The GEC reference cell is essentially a parallel-plate, capacitively coupled, RF plasma reactor. The development of this standard reference cell strongly boosted the study of plasma physics of the etching chemistries and also helped to develop diagnostics to be used on commercial plasma systems. This collaborative effort is a perfect example of the harmonization of the global plasma research for the advancement of plasma processing in electronics.

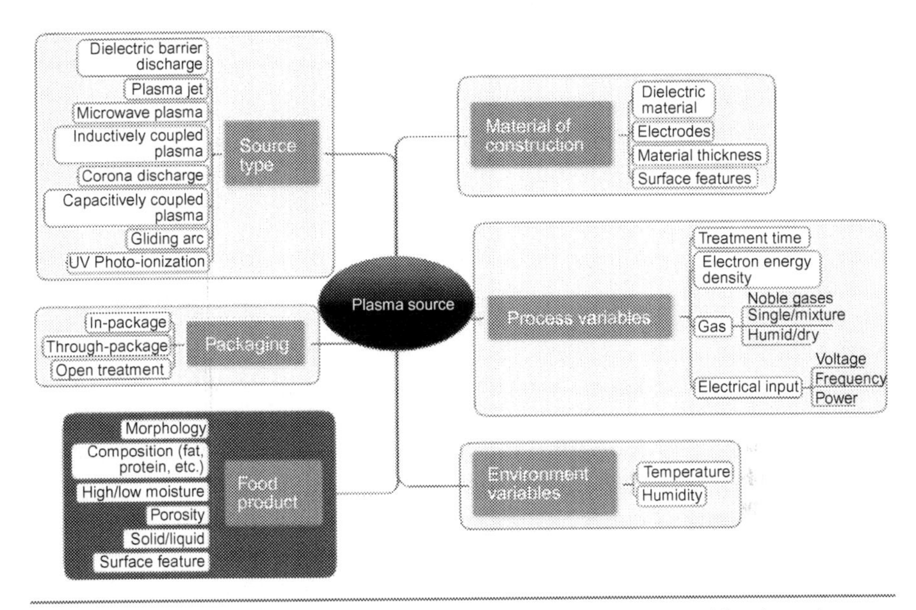

Fig. 1 Some important variables governing the plasma treatment of food articles.

A similar approach should be undertaken for atmospheric pressure nonequilibrium/cold plasmas. A key challenge, however, arises from the range of variables that altogether make up a plasma process for decontamination of foods, as shown in Fig. 1.

It may be argued that there cannot be a single type of nonequilibrium and/or cold plasma source; jets, barrier discharges and microwave plasma sources all have potential merit for the treatment of foods due to wide variety of food products and their unique attributes (eg, enzyme inactivation, microbial reductions, pasteurization, sterilization). Nevertheless, the food research community can still take this motivational example as a lesson for the development of standard cold plasma sources for each major class (DBD, RF plasma jet, and microwave plasma reactor), for validating the microbial inactivation levels and the retention of food quality. In addition, a range for each operational parameter (eg, power density, voltage, current, frequency, etc.) should also be agreed upon. This will significantly reduce (if not eliminate) some of the major factors responsible for variability in the results of various research groups across the globe and make the research data more meaningful. In addition, the referenced cold plasma source could also serve as a successful teaching tool in both graduate and undergraduate labs in novel food-processing technologies. Finally, despite the benefits from standardization, there always remains the natural variability in foods and the environmental factors, such as temperature and humidity.

3.3 New Plasma Sources

At the heart of cold plasma processing are the plasma sources. Several new designs of plasma sources for various applications are under development. A good design engineer must be familiar with both the system and component design aspects. If cold plasma treatment of foods is to become a reality, an efficient means of carrying out continuous treatments on a large scale must be found. Ideally, any processing equipment developed should be able to handle a wide variety and geometries of foods, and do so without inflicting physical damage on the food. Cost is undoubtedly always important, and low running cost is a major advantage of cold plasma processes. A good design engineer should ensure that the cost of an industrial-scale plasma setup should not be drastically high, for example, by developing plasma sources operating at line frequency, rather than radio-frequency power sources, a significant cost management could be realized. This could potentially eliminate the time-to-technology evaluation and lead to widespread adoption. A related example would be that of HPP, where despite the known benefits, high initial cost of the equipment has prevented its extensive use.

In order to highlight the importance of developing new plasma sources, we present two examples:

(1) Takamatsu et al. (2011) have developed an atmospheric pressure, multigas plasma jet source, which prevents any thermal or electrical discharge damage to the target material (see Fig. 2). The main novelty of this plasma jet lies in the number of different plasma gas species it can handle (Takamatsu et al., 2013). It may be noted that most commercially available (laboratory-scale) plasma jet sources are designed for use with only one or two gases (argon, helium, air, or mixtures thereof).

(2) It is well recorded that plasma jets generally produce positive and negative ions, (V)UV radiation, and reactive radical species, which interact with the substrate. With the involvement of a multitude of reactive species, it becomes difficult to isolate the mode of action of each species and the corresponding effects. To address this issue, recently an "X-jet," a modified microscale atmospheric pressure plasma jet, has been developed (Fig. 3) (Lackmann et al., 2013; Schneider et al., 2011, 2012). The X-jet allows to effectively separate the plasma-generated VUV, UV photons, and heavy reactive particles from each other and can be used separately for treatments. This is achieved by introducing a flow of helium into the jet under well-defined flow conditions. The X-jet system permits one to obtain a better understanding of the chemical processes occurring in the plasma effluent.

It is likely that many more additional plasma devices and applications will be invented over time as interest in atmospheric plasma grows exponentially.

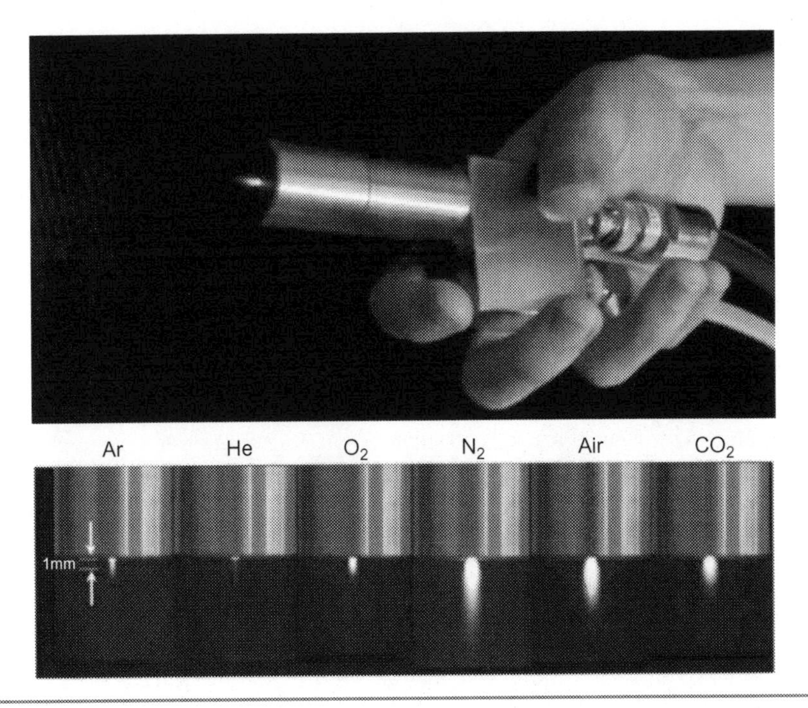

Fig. 2 The multigas plasma jet. *(From Takamatsu, T., Hirai, H., Sasaki, R., Miyahara, H., Okino, A., 2013. Surface hydrophilization of polyimide films using atmospheric damage-free multigas plasma jet source. IEEE Trans. Plasma Sci. 41(1), 119–125.)*

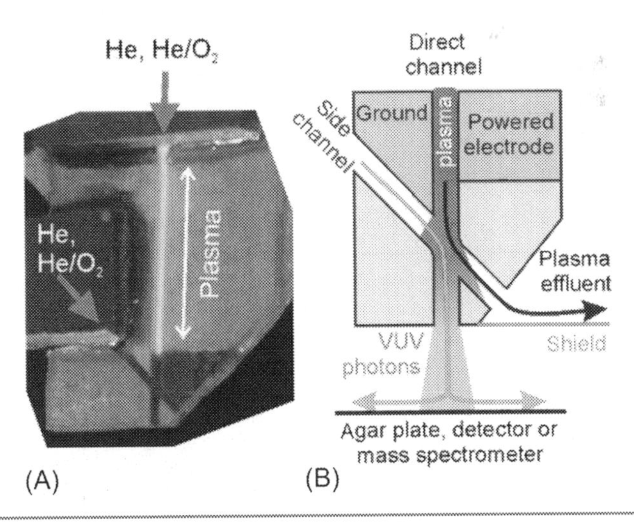

Fig. 3 (A) Photograph of the X-jet with two crossed channels after the nozzle, operated with He/O_2 plasma. The plasma effluent is diverted into a side channel by the additional helium flow. VUV and UV photons propagate in line-of-sight with the plasma through the direct channel. (B) Schematic representation of the gas flows and photon flux in the channel structure. *(From Schneider, S., Lackmann, J.-W., Ellerweg, D., Denis, B., Narberhaus, F., Bandow, J.E., Benedikt, J., 2012. The role of VUV radiation in the inactivation of bacteria with an atmospheric pressure plasma jet. Plasma Processes Polym. 9(6), 561–568 by courtesy of John Wiley & Sons.)*

3.4 Safety Aspects

When prospecting for large-scale applications, besides engineering design, the aspects of health and safety of the operators in an industrial manufacturing scenario cannot be overlooked. Accordingly, health and safety measures for cold plasma devices and processes should be considered by design engineers. When considering the health and safety aspects of cold plasma for food processing, a number of factors come into play. The word "health" calls to mind the well-being of the workers involved in processing foods with plasma. There are two primary worker safety concerns with plasma devices and processes: the high concentrations of reactive gas species and high voltage. Generally, plasma will contain high concentrations of reactive gas species such as ozone, nitrogen oxides, peroxides, and many others depending on the generation conditions. Just as any hazardous gas or strong oxidizing agents, excited and reactive species of plasma are potentially injurious if humans are exposed to hazardous concentrations for considerable time durations. Ample measures for chemical safety should be installed in-place by ensuring effective destruction of reactive species, including any ozone produced, prior to its exhaust into the atmosphere. A common method to achieve this is by the use of carbon filters. Good ventilation standards could also serve as proactive safety measures. Furthermore, appropriate sensing devices for detection of gas leakage and raising safety alarm should be incorporated into the machine and nearby zone. For in-package plasma treatment technologies, the complete absence of any reactive species prior to the food reaching the supermarket shelves or any consumer is an absolute necessity.

The design of the plasma source should consider safety aspects related to all hazards that could arise from high voltages in a wet environment that operates around the clock. The high-voltage regions and electric fields should be shielded from the operators and exposure to manufacturing environment. In addition, due to the possible existence of strong electric and magnetic fields, operators or visitors with pace makers should be restricted from entering into the processing areas. As rule of thumb, the equipment designed for plasma treatment in production lines should be fully automated, be robust, and be reliable for ensuring minimal interference by workers.

4 Process Control

For a thorough analysis of the interaction of cold plasma with foods and microorganisms, a detailed quantitative analysis of the discharge dynamics, as well as the reactive species production, is required. In the literature, knowledge gaps are present in a number of areas, including a rapid, noninvasive means to measure and control the reaction chemistry of cold gas plasmas and their treatment effect(s) on articles. R&D efforts should focus on developing the underpinning science and technology for understanding plasma kinetics and controlling the plasma chemistry. Based on the authors experiences so far, the problem that could limit, if not inhibit the application of in-package cold plasma for treatment of foods is the lack of appropriate in-line diagnostic tools for controlling the plasma chemistry. With the presence of appropriate process monitoring tools in-place, a consistency in the performance of the process can be ensured.

In principle, diagnostic tools commonly applied by plasma physicists should be applicable for monitoring continuous plasma processes. The first example of such methods includes the optical emission spectroscopy widely used in research. Here the intensity of a specific chemical species can be monitored and the power input to the discharge can be controlled accordingly, via a feedback loop, to maintain the concentration at a preset threshold throughout the process. Alternatively, the electrical parameters, such as the power input to discharge and/or the electron energy distribution, may be targeted directly. Acoustic diagnostics is also a noninvasive technique that has been explored for real-time plasma diagnostics during normal operation (Law et al., 2011; McNally et al., 2011). Good success has been demonstrated under laboratory conditions by decoding the acoustic signals using Fourier transformation and multivariate analysis methods.

It appears to be tacitly assumed that the entire surface of the food should be exposed to the cold plasma jet when treating with plasma jets, and while this is true for narrow plasma jets, in sealed package plasma treatments (based on dielectric barrier discharges), the natural convection and gas diffusion per se ensure the uniform surface exposure to active species. Therefore, from a plasma chemical diagnostics point of view, not only is the time-evolution of active species important, but so is the spatial distribution to ensure uniformity of treatments.

Indeed, there exist a number of explicit and implicit diagnostic and analytical tools for monitoring the plasma chemistry (see Chapter 5). The challenge, however, arises from several variables when treating foods, such as humidity, the interference from several gaseous species, and the noisy factory environments, which all complicate the use of acoustic diagnostics, etc. Photometric measurement is also difficult to implement due to the high speed of food manufacturing and the requirement of a precise setup with a fixed path length to identify the reactive species. Indeed, some important questions remain unanswered. For example, which plasma parameters should be measured for process control? What are their specificity for the product/process? Which diagnostic tools are to be used?

5 Future Innovations

In general, there is a lack of knowledge about the impact of cold plasma technology among various sectors within the food industry. So far, the most practical application and the principal driving force behind the development of gas plasma technologies for food applications is focused on food safety. Within this section we intend to provide a broader perspective of the potential future developments in cold plasma treatment of foods, as evidenced from recent studies.

5.1 Food Packaging

For a complete assessment of the in-package cold plasma technologies, it is essential to quantify all possible changes to the packaging induced by the cold plasma. For example, the migration limits of additives, monomers, oligomers, and low molecular weight volatile compounds from the packaging material into the food (following in-package plasma) should be evaluated for food safety reasons (Pankaj et al., 2014b). Fortunately, the results so far appear promising with negligible changes in packaging materials and migrations below those prescribed by regulatory bodies (Pankaj et al., 2014c). Another interesting feature of DBD based in-package cold plasma is that it has been found to be compatible with biopolymer-based packaging materials (Pankaj et al., 2014a,d, 2015).

5.2 Food Structure-Property Modification

In addition to fundamental studies, it is worth recognizing the very real possibility of modifying food structures and imparting specific functionality using cold plasma treatments. The electronics and polymer-processing sectors are already enjoying the benefits of cold plasma applications for structure modification. The idea of cold plasma-aided modification of food structure and therefore, its function and behavior, is very new to food science, and only some recent studies demonstrate these.

Surface property modification: The modulation of hydrophobicity or hydrophilicity of food surfaces for less seepage of oil was recently demonstrated by Misra et al. (2014d). The authors increased the hydrophobicity of the biscuit surface, which resulted in increased spreading of any oil sprayed and therefore there was less seepage. With plasma processing, for a given volume of oil, up to 50% more spreading of oil can be achieved within a few seconds of the process.

Functional property modification: There is a surge of interest in evaluating and understanding the interactions between cold plasma and proteins. The inactivation of enzymes using cold plasma is widely recognized (Pankaj et al., 2013; Surowsky et al., 2013; Tappi et al., 2014). However, another interesting application yet to be explored is in the area of the modification of food proteins and in turn their functionality. In a recent study, Misra et al. (2015) demonstrated that the secondary structure of gluten becomes more stable in atmospheric pressure cold plasma (in air) treated weak wheat flour and corresponding changes in dough rheology are also observed. The effects were found to be a function of applied voltage and treatment time, implying the possibility to control the extent of change. Similar to this study, Bußler et al. (2015) reported an increase in water and fat binding capacities in protein-rich pea flour following cold plasma treatments in air. This research group also attributed these changes to the modification of protein structure evidenced from protein fluorescence spectroscopy.

5.3 Mass-Transfer Enhancement

It is well known that cold plasma treatment leads to surface etching in materials. A similar phenomenon has been observed when treating lemon peels with a plasma discharge, where the microdischarges from a DBD damaged the oil glands, resulting in increased diffusion; the overall effect was enhanced oil extraction (Kodama et al., 2014). With treatments at 30 kV(pp, 50 Hz) for 1 min maximum enhancement in the yield was noticed. While these results open new opportunities; they must be evaluated with the caveat that an oxidation of the oil components can occur, as was observed for lemon oil with O_2, N_2, or air plasma.

While mass-transfer enhancement and food property modification using cold plasmas are two exemplary areas demonstrating the possible innovations, there are certainly many research studies necessary for the future.

6 Consumer Confidence

The food industry is dominated by a highly commoditized marketplace; development of "new" or "different" products may lead to a competitive advantage. New food technologies enable innovations within the food sector, but not all technologies are equally accepted by consumers (Siegrist, 2008; Sapp and Downing-Matibag, 2009). To investigate consumer attitudes toward a specific technology is a key point, which has to be addressed before the new product is fully developed (da Costa et al., 2000). Consumers may perceive new food technologies as more risky than traditional food technologies. For example, food irradiation is in general perceived as an unknown and unacceptable food hazard by a majority of the public (Siegrist et al., 2006). How the public will react to cold plasma technology is, at present, unclear. Also, consumer opinion data is often expensive and slower to generate. The advertising industry could play a vital role in making the public aware of the benefits of cold plasma technology in food processing. A first hand assessment of the consumer opinion about the cold plasma technology can be obtained through public debates and public participation methods. The benign effects of cold plasma treatments (air and electricity) on food, and the economic as well as environmental benefits should be clearly explained. Scientists and engineers involved in the development of cold plasma technologies have a societal responsibility to share their knowledge with the public and policy makers, in order to expand the awareness of the potential for cold plasma in food-processing sector.

7 Closing Remarks

The future looks very promising for the use of cold plasma technology in the food industry. Atmospheric cold plasma has the potential to provide a treatment process for removing biological agents, toxins, or surface contamination from foods and food contact surfaces, modify packaging materials, enhance functionality of food ingredients, and reduce contaminants in water and wastewater. Cold plasma treatment can be delivered as a "dry process," where water or a wet environment is of concern. Atmospheric cold plasma has a unique ability to generate reactive gas from only air and electricity.

To meet the growing demand for food in the world, it is clear that the food industry needs to advance in manufacturing technologies, food safety practices, and environmental sustainability practices. This requires dedicated research and development in new technology. Cold plasma technology is available in a variety of forms and offers a myriad of opportunities to improve manufacturing performance, food safety, and environmental sustainability. Current science only partially explains the highly complex air plasma chemistry involving hundreds of transient species and reactions. Further scientific understanding is vital for regulatory approval and development of efficient, large-scale plasma sources. It has been observed that cold plasma research on food has grown exponentially with more than 100 papers published annually.

It is envisaged that overcoming the regulatory and consumer issues will come from focusing the research efforts and political strategies on open innovation through industry-academia-government collaboration. To fully capture the potential of cold plasma technology in food manufacturing, a concerted effort from the plasma community (academicians, industrialists, marketers, government authorities, etc.) is needed. Research studies at all levels are needed: in the laboratory, followed by prototype development, followed by pilot scale treatment, and finally commercial scale production. There remains a need for researchers and clinicians in the field to work together to evaluate the safety of cold plasma-treated food products.

The ideas and thoughts shared within this chapter are expected to help alleviate the current challenges in the adoption of plasma technology for foods. The opportunities for cold plasma technology in the food industry are only limited by the creativity of the inventor.

References

Anacail, 2014. New ozone Packaging Solution Extends Shelf Life of Grapes and Tomatoes by Up To 300%. Anacail Limited, UK.

Ashford, N.A., 2002. Government and environmental innovation in Europe and North America. Am. Behav. Sci. 45 (9), 1417–1434.

Bußler, S., Steins, V., Ehlbeck, J., Schlüter, O., 2015. Impact of thermal treatment versus cold atmospheric plasma processing on the techno-functional protein properties from *Pisum sativum* 'Salamanca'. J. Food Eng. 167, 166–174.

da Costa, M.C., Deliza, R., Rosenthal, A., Hedderley, D., Frewer, L., 2000. Non conventional technologies and impact on consumer behavior. Trends Food Sci. Technol. 11 (4–5), 188–193.

FDA, 2015a. U.S. Food and Drug Administration. Federal Food, Drug and Cosmetic Act of 1938, http://www.fda.gov/regulatoryinformation/legislation/federalfooddrugandcosmeticactfdcact/ (accessed 01.07.15).

FDA, 2015b. Regulatory Report: FDA's Food Contact Substance Notification Program, http://www.fda.gov/Food/IngredientsPackagingLabeling/PackagingFCS/ucm064161.htm#authors (accessed 01.07.15).

FDA, 2015c. U.S. Food and Drug Administration. Promoting and Protecting Your Health, http://www.fda.gov/Food/default.htm (accessed 01.07.15).

FSIS, 2006a. Meat Products Inspection Act. United States Department of Agriculture Food Safety Inspection Service, Washington, DC.

FSIS, 2006b. Poultry Products Inspection Act. United States Department of Agriculture Food Safety Inspection Service, Washington, DC.

FSIS, 2006c. Egg Products Inspection Act. United States Department of Agriculture Food Safety Inspection Service, Washington, DC.

Heo, N.S., Lee, M.K., Kim, G.W., Lee, S.J., Park, J.Y., Park, T.J., 2014. Microbial inactivation and pesticide removal by remote exposure of atmospheric air plasma in confined environments. J. Biosci. Bioeng. 117 (1), 81–85.

Keener, K.M., Jensen, J.L., Valdramidis, V.P., Byrne, E., Connelly, J.A., Mosnier, J.P., Cullen, P.J., 2012. Decontamination of *Bacillus subtilis* spores in a sealed package using a non-thermal plasma system. In: NATO-Advanced Research Workshop: Plasma for Bio-Decontamination, Medicine and Food Security. Springer, Heidelberg, Germany, pp. 445–455.

Klockow, P.A., Keener, K.M., 2009. Safety and quality assessment of packaged spinach treated with a novel ozone-generation system. LWT—Food Sci. Technol. 42 (6), 1047–1053.

Kodama, S., Thawatchaipracha, B., Sekiguchi, H., 2014. Enhancement of essential oil extraction for steam distillation by DBD surface treatment. Plasma Processes Polym. 11 (2), 126–132.

Lackmann, J.W., Schneider, S., Edengeiser, E., Jarzina, F., Brinckmann, S., Steinborn, E., Havenith, M., Benedikt, J., Bandow, J.E., 2013. Photons and particles emitted from cold atmospheric-pressure plasma inactivate bacteria and biomolecules independently and synergistically. J. R. Soc. Interface 10 (89), 20130591.

Lagaron, J.M., Lopez-Rubio, A., 2011. Nanotechnology for bioplastics: opportunities, challenges and strategies. Trends Food Sci. Technol. 22 (11), 611–617.

Law, V.J., Neill, F.T.O., Dowling, D.P., 2011. Evaluation of the sensitivity of electro-acoustic measurements for process monitoring and control of an atmospheric pressure plasma jet system. Plasma Sources Sci. Technol. 20 (3), 035024.

McNally, P., Law, V.J., Daniels, S., 2011. System for Analysing Plasma, WO2009135919A1. Dublin City University.

Misra, N.N., Moiseev, T., Patil, S., Pankaj, S.K., Bourke, P., Mosnier, J.P., Keener, K.M., Cullen, P.J., 2014a. Cold plasma in modified atmospheres for post-harvest treatment of strawberries. Food Bioprocess Technol. 7 (10), 3045–3054.

Misra, N.N., Pankaj, S.K., Walsh, T., O'Regan, F., Bourke, P., Cullen, P.J., 2014b. In-package nonthermal plasma degradation of pesticides on fresh produce. J. Hazard. Mater. 271, 33–40.

Misra, N.N., Patil, S., Moiseev, T., Bourke, P., Mosnier, J.P., Keener, K.M., Cullen, P.J., 2014c. In-package atmospheric pressure cold plasma treatment of strawberries. J. Food Eng. 125, 131–138.

Misra, N.N., Sullivan, C., Pankaj, S.K., Alvarez-Jubete, L., Cama, R., Jacoby, F., Cullen, P.J., 2014d. Enhancement of oil spreadability of biscuit surface by nonthermal barrier discharge plasma. Innov. Food Sci. Emerg. Technol. 26, 456–461.

Misra, N.N., Kaur, S., Tiwari, B.K., Kaur, A., Singh, N., Cullen, P.J., 2015. Atmospheric pressure cold plasma (ACP) treatment of wheat flour. Food Hydrocoll. 44, 115–121.

Olthoff, J., Greenberg, K., 1995. The gaseous electronics conference RF reference cell—an introduction. J. Res. Natl. Inst. Stand. Technol. 100 (4), 327.

Pankaj, S.K., Misra, N.N., Cullen, P.J., 2013. Kinetics of tomato peroxidase inactivation by atmospheric pressure cold plasma based on dielectric barrier discharge. Innov. Food Sci. Emerg. Technol. 19, 153–157.

Pankaj, S.K., Bueno-Ferrer, C., Misra, N.N., Bourke, P., Cullen, P.J., 2014a. Zein film: effects of dielectric barrier discharge atmospheric cold plasma. J. Appl. Polym. Sci. 131 (18), 40803.

Pankaj, S.K., Bueno-Ferrer, C., Misra, N.N., Milosavljević, V., O'Donnell, C.P., Bourke, P., Keener, K.M., Cullen, P.J., 2014b. Applications of cold plasma technology in food packaging. Trends Food Sci. Technol. 35 (1), 5–17.

Pankaj, S.K., Bueno-Ferrer, C., Misra, N.N., O'Neill, L., Jiménez, A., Bourke, P., Cullen, P.J., 2014c. Characterization of polylactic acid films for food packaging as affected by dielectric barrier discharge atmospheric plasma. Innov. Food Sci. Emerg. Technol. 21, 107–113.

Pankaj, S.K., Bueno-Ferrer, C., Misra, N.N., O'Neill, L., Tiwari, B.K., Bourke, P., Cullen, P.J., 2014d. Physicochemical characterization of plasma-treated sodium caseinate film. Food Res. Int. 66, 438–444.

Pankaj, S.K., Bueno-Ferrer, C., Misra, N.N., O'Neill, L., Tiwari, B.K., Bourke, P., Cullen, P.J., 2015. Dielectric barrier discharge atmospheric air plasma treatment of high amylose corn starch films. LWT—Food Sci. Technol. 63 (2), 1076–1082.

Sapp, S.G., Downing-Matibag, T., 2009. Consumer acceptance of food irradiation: a test of the recreancy theorem. Int. J. Consumer Stud. 33 (2009), 417–424.

Schnabel, U., Niquet, R., Schlüter, O., Gniffke, H., Ehlbeck, J., 2015. Decontamination and sensory properties of microbiologically contaminated fresh fruits and vegetables by microwave plasma processed air (PPA). J. Food Process. Preserv. 39 (6), 653–662.

Schneider, S., Lackmann, J.W., Narberhaus, F., Bandow, J.E., Denis, B., Benedikt, J., 2011. Separation of VUV/UV photons and reactive particles in the effluent of a He/O_2 atmospheric pressure plasma jet. J. Phys. D: Appl. Phys. 44 (29), 295201.

Schneider, S., Lackmann, J.-W., Ellerweg, D., Denis, B., Narberhaus, F., Bandow, J.E., Benedikt, J., 2012. The role of VUV radiation in the inactivation of bacteria with an atmospheric pressure plasma jet. Plasma Processes Polym. 9 (6), 561–568.

Siegrist, M., 2008. Factors influencing public acceptance of innovative food technologies and products. Trends Food Sci. Technol. 19 (11), 603–608.

Siegrist, M., Keller, C., Kiers, H.A.L., 2006. Lay people's perception of food hazards: comparing aggregated data and individual data. Appetite 47 (3), 324–332.

Surowsky, B., Fischer, A., Schlueter, O., Knorr, D., 2013. Cold plasma effects on enzyme activity in a model food system. Innov. Food Sci. Emerg. Technol. 19, 146–152.

Takamatsu, T., Ichikawa, M., Hirai, H., Sasaki, R., Shibata, M., Miyahara, H., Matsumoto, Y., Okino, A., 2011. Sterilization effect of various gas non-thermal plasma. In: Proceedings of the IEEE International Conference on Plasma Science (ICOPS), Chicago, IL.

Takamatsu, T., Hirai, H., Sasaki, R., Miyahara, H., Okino, A., 2013. Surface hydrophilization of polyimide films using atmospheric damage-free multigas plasma jet source. IEEE Trans. Plasma Sci. 41 (1), 119–125.

Tappi, S., Berardinelli, A., Ragni, L., Dalla Rosa, M., Guarnieri, A., Rocculi, P., 2014. Atmospheric gas plasma treatment of fresh-cut apples. Innov. Food Sci. Emerg. Technol. 21, 114–122.

Utterback, J.M., Acee, H.J., 2005. Disruptive technologies: an expanded view. Int. J. Innov. Manage. 09 (01), 1–17.

Index

Note: Page numbers followed by *f* indicate figures, and *t* indicate tables.

Printed in the United States
By Bookmasters